AF524887

Plant Cell Biotechnology

NATO ASI Series

Advanced Science Institutes Series

A series presenting the results of activities sponsored by the NATO Science Committee, which aims at the dissemination of advanced scientific and technological knowledge, with a view to strengthening links between scientific communities.

The Series is published by an international board of publishers in conjunction with the NATO Scientific Affairs Division

A Life Sciences **B Physics**	Plenum Publishing Corporation London and New York
C Mathematical and Physical Sciences **D Behavioural and Social Sciences** **E Applied Sciences**	Kluwer Academic Publishers Dordrecht, Boston and London
F Computer and Systems Sciences **G Ecological Sciences** **H Cell Biology**	Springer-Verlag Berlin Heidelberg New York London Paris Tokyo

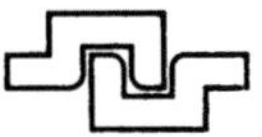

Series H: Cell Biology Vol. 18

Plant Cell Biotechnology

Edited by

M. Salomé S. Pais

Departamento de Biologia Vegetal
Faculdade de Ciências de Lisboa
R. Escola Politecnica 58
1294 Lisboa Codex, Portugal

F. Mavituna

Chemical Engineering Department
UMIST, P.O. Box 88
Manchester M60 1QD, United Kingdom

J.M. Novais

Laboratorio de Engenharia Bioquimica
Instituto Superior Tecnico Av. Rovisco Pais
1000 Lisboa, Portugal

Springer-Verlag
Berlin Heidelberg New York London Paris Tokyo
Published in cooperation with NATO Scientific Affairs Division

Proceedings of the NATO Advanced Study Institute on Plant Cell Biotechnology held in Albufeira, Algarve, Portugal, March 29 – April 10, 1987

ISBN-13: 978-3-642-73159-4 e-ISBN-13: 978-3-642-73157-0
DOI: 10.1007/978-3-642-73157-0

Library of Congress Cataloging-in-Publication Data. NATO Advanced Study Institute on Plant Cell Biotechnology (1987 : Albureira, Portugal) Plant cell biotechnology / edited by M. Salomé S. Pais, F. Mavituna, J. M. Novais. p. cm.—(NATO ASI series. Series H, Cell biology ; vol. 18) "Proceedings of the NATO Advanced Study Institute on Plant Cell Biotechnology held in Albufeira, Algarve, Portugal, March 29 – April 10, 1987"—T.p. verso.

1. Plant biotechnology—Congresses. I. Pais, Maria Salomé S. II. Mavituna, Ferda, 1951-. III. Novais, J. M. IV. Title. V. Series. TP248.27.P55N37 1987 660'.6—dc 19 88-12250

Softcover reprint of the hardcover 1st edition 1988

2131/3140-543210 – Printed on acid-free paper

PREFACE

Plant Cell Biotechnology offers significant potential benefits in the areas of plant improvement and production of fine chemicals. Despite a history going back to the beginning of this century the success stories so far,although significant, are few. The realization of the potential of Plant Biotechnology requires more fundamental research in plant physiology, biochemistry and molecular biology as well as biochemical engineering aspects of production systems. Because of the multi-disciplinary nature of Plant Cell Biotechnology, the field needs the interaction and cooperation of scientists, engineers and professionals of diverse skills and expertise. The field is also short of staff trained in the relevant speciality areas.

In order to review the major topics, report on current research and exchange views by bringing together the experts in Plant Cell Biotechnology with the professionals interested in this field and the young scientists whom the field desperately needs, a NATO Advanced Study Institute was held in Portugal between 29th March and 10th April 1987.

This book is the Proceedings of this course and includes the review and research lectures given, papers based on poster presentation of current research and notes on some of the discussion sessions. It covers many of the diverse topics in this field; applications in agriculture and plant breeding, plant genetic engineering, biochemical and engineering aspects of biotransformations and production of fine chemicals by plant cell and tissue cultures. The sections based on discussion sessions give a critical and realistic review of the failures and success stories and point to the way ahead. The book therefore, should appeal to the researchers, both expert and novice in the field, and professionals from agricultural, pharmaceutical and chemical sectors with a current or potential interest in what Plant Cell Biotechnology can offer.

We would like to acknowledge the NATO Scientific Affairs Division for providing the financial support. The additional financial support was provided by Instituto Nacional de Investigação Científica, Junta Nacional de Investigação Científica e Tecnológica and Fundação Calouste Gulbenkian. Our thanks are also extended to J.M.Novais and J.S. Cabral in the organizing Committee, A.W.Alfermann, P.Brodelius, L.C. Fowke, I. Vasil and Y. Yamada in the Scientific Committee.

We would also like to thank all those who helped with the organisational and secretarial matter in particular M.M.Oliveira, P. Scotti, J. Feijó, H. Velez and L. Aires.

Finally we would like to acknowledge R. Greenbank from the Chemical Engineering Department, UMIST, Manchester for all her help with the indexing.

M. Salomé S. PAIS (Director)

June 1987 Ferda MAVITUNA (Co-Director)

CONTENTS

ORGANIZING COMMITTEE

PAIS, M. S. S.	Departamento de Biologia Vegetal Faculdade de Ciências de Lisboa, R.Escola Politecnica 58, 1294 Lisboa Codex Portugal
MAVITUNA, F.	Chemical Engineering Department UMIST, P. O. Box 88, Manchester M60 1QD, UK
NOVAIS, J. M.	Laboratório de Engenharia Bioquímica, Instituto Superior Técnico Av. Rovisco Pais, 1000 Lisboa Portugal
CABRAL, J. S.	Laboratorio de Engenharia Bioquimica, Instituto Superior Tecnico Av. Rovisco Pais, 100 Lisboa Portugal

SCIENTIFIC COMMITTEE

ALFERMANN, A. W.	Institut für Entwicklung und Molekularbiologie der Pflanzen, Universität Düsseldorf,Universitätstrasse-1 D-4000 Düsseldorf, FRG
BRODELIUS, P.	Institute of Biotechnology, Swiss Federal Institute of Technology, Honggerberg, CH-8093 Zurich , Switzerland
FOWKE, L. C.	Department of Biology, University of Saskatchewan, Saskatoon, Saskatchewan, Canada SJN OWO
VASIL, I. K.	Department of Botany, University of Florida, Gainsville Florida, 32611 USA

CONTRIBUTORS

ALFERMANN, A. W.	(See Scientific Committee)
ANDERSEN, J. B.	Department of Plant Pathology; The Royal Veterinary and Agricultural University, Thorvalosensvej 40 DK-18171 Frederiksberg Denmark
BAETEN, H.	Institut voor Sheikundig Onderzoek, Museumlaan 5, B-1980 Tervuren Belgium
BARZ, W.	Biochemistry of Plants,University of Muenster, Hindenburg Platz 55, 4400 Muenster FRG
BENSON, E. E.	Department of Agriculture and Horticulture, Nottingham University, School of Agriculture, Sutton Bonington, Loughborough, Leics. LER 5RD, UK
BOTTACIN, A.	Dipartimento di Biotechnologie Agrarie, Università di Padova, Via Gradenigo 6, 35131 Padova Italy
BRETELER, H.	NOVAPLANT , Cell Biotechnology Group Research Institute ITAL. P.O. Box 48 - 6700 AA Wageningen The Netherlands
BRINGMANN, G.	Organische-Chemisches Institut, Westefälische Wilhelms-Universitat, Orleansring 23, D-4400 Muenster FRG
BRODELIUS, P.	(See Scientif Committee)
CABRAL, J. M. S.	(See Scientif Committee)
CACCO, G.	Istituto di Chimica Agraria e Forestale, Università di Reggio-Calabria, Italy
CALLEBAUT, A.	Institut voor Sheikundig Onderzoek, Museumlaan 5, B-1980 Tervuren Belgium
CANDELA, M.	Departamiento de Genetica, Fac. Biologia, Universidad Complutense 28048 Madrid Spain

CARUSO,	M. FARMITALIA Carlo Erba, Via dei Gracchi 35 20146 Milano Italy
CASTELLI, S.	Istituto Biosintesi Vegetali - CNR, Via Bassini 15 20133 Milano Italy
CHIN, C.H.	Department of Horticulture and Forestry, Rutgers The State University of New Jersey, New Brunswick, New Jersey 08902 USA
CLOSE, T. J.	Division of Plant Industry,CSIRO , G P.O. Box 1600 Canberra ACT 2601 Australia
COCKING, E. C.	Department of Botany, University of Nottingham, Nottingham NG 7 2RD, UK
CONGER, B. V.	Department of Plant and Soil Science, University of Tennessee, P.O. Box 1071, Knoxville Tennessee 37901, USA
CRESPI - PERELLINO, N.	FARMITALIA Carlo Erba, Via dei Gracchi 35 20146 Milano, Italy
DANIEL, S.	Biochemistry of Plants,University Muenster, Hindenburg Platz , 55, 4400 Muenster FRG
De CAT, W.	Institut voor Sheikundig Onderzoek, Museumlaan 5, B-1980 Tervuren, Belgium
DUARTE, C.	Departamento de Química, Faculdade de Ciências de Lisboa, R. Escola Politécnica 1294 Lisboa Codex Portugal
ESPINO, J.	Departamiento de Genetica, Fac. Biologia, Universidad Complutense 28048 Madrid Spain
ESQUÍVEL, M. G.	Instituto Superior de Agronomia, Tapada da Ajuda 1399 Lisboa Codex Portugal
EUSÉBIO, A.	Departamento de Quimica, Faculdade de Ciências de Lisboa, R. Escola Politécnica, 1294 Lisboa Codex Portugal
FERNANDEZ, B.	Catedra de Fisiologia Vegetal, Universidad Oviedo, Spain

FONSECA, M. M. R. — Laboratório de Engenharia Bioquímica, Instituto Superior Técnico, Av. Rovisco Pais 1000 Lisboa, Portugal

FOWKE, L. C. — (See Scientific Committee)

FOWLER, M. W. — The Wolfson Institute of Biotechnology, University of Sheffield, Sheffield S10 2TN, UK

FUJITA, Y. — Bioscience Research Center, MITSUI Petrochemichal Industries Ltd, Waki-Cho, Kuga-Gun, Yamaguchi 740 Japan

GARCIA - REINA, G. — Departamento de Biologia,Universidad Politecnica de Canarias, Box 550, Las Palmas de Gran Canaria Spain

GAROFANO, L. — FARMITALIA Carlo Erba, Via dei Gracchi 35, 20146 Milano Italy

GRAY, D. J. — AREC,lFAS,University of Florida, P.O.Box 388 Leesburg, Florida 32749, USA

GUICCIARDI, A. — FARMITALIA Carlo Erba, Via dei Gracchi 35, 20146 Milano Italy

HAKMAN, I. — Department of Biology,University of Saskatchewan, Saskatoon, Saskatchewan, Canada 5JN OWO

HAMILTON, R. — Department of Chemical and Biochemical Engineering, Rutgers, The State University of New Jersey, New Brunswick, New Jersey 08902 USA

HAMILL, J. D. — Plant Cell Biotechnology Group, Institute of Food Research Norwich Laboratory, Norwich NR4 7VA, UK

HARKES, P. A. A. — Department of Plant Molecular Biology, University of Leiden, P.O. Box 9502, 2300 RA Leiden, The Netherlands

HELSPER, J. P. F. G. — NOVAPLANT Cell Biotechnology Group Research Institute ITAL, P.O. Box 48 6700 AA Wageningen, The Netherlands

HOENIG, M.	Institut voor Scheikundig Onderzoek, Museumlaan, 5,B- 1980 Tervuren, Belgium
Ten HOOPER, H. J. G.	Department of Biochemical Engineering, Delft University of Technology, The Netherlands
HULST, A. C.	Agricultural University, Department of Food Science, Food and Bioengineering Group, de Drejen 12 6703 BC Wageningen, The Netherlands
HÜSEMANN, W.	Lehrstuhl für Biochemie der Pflanzen, Univerität Muenster, D-4400 Muenster FRG
JANSEN, J. R.	Organisch Chemisches Institut, Westfälische Wilhelms-Universität, Orleansring 23, D-4400 Muenster FRG
JAQUES, V.	Biochemistry of Plants, University of Muenster, Hindenburgplatz 55, 4400 Muenster FRG
KADO, C. I.	Davis Crown Gall Group, Department of Plant Pathology, University of California, Davis USA
KETEL, D. H.	NOVAPLANT Cell Biotechnology Group Research Institute ITAL, P.O. Box 48, 6700 AA Wageningen, The Netherlands
KESSMANN, H.	Biochemistry of Plants,University of Muenster, Hindenburplatz 55, 4400 Muenster FRG
KÖSTER, S.	Biochemistry of Plants,University of Muenster, Hindenburgplatz 55, 4400 Muenster FRG
LANG, J.	Department of Chemical and Biochemical Engineering, Rutgers, The State University of New Jersey, New Brunswick,New Jersey 08902 USA
LIMA - COSTA, E.	Unidade Estrutural de Ciências Exactas, Universidade do Algarve 8000 Faro, Portugal
LÖRZ, H.	Max - Planck - Institut für Züchtungsforschung, D- 5000 Köln 30 FRG

LUQUE, A.	Departamento de Biologia,Universidad Politecnica de Canarias, Box 550, Las Palmas de Gran Canaria, Spain
MAGNIEN, E.	Division of Biotechnology, Directorate of Biology, Comission of the European Communities (DG XII) Brussels, Belgium
MANZOCCHI, L. A.	Istituto Biosintesi Vegetali,CNR Via Bassini, 15 20133 Milano Italy
MAVITUNA, F.	(See Organizing Committee)
MOTTE, J. C.	Institut voor Scheikundig Onderzoek Museumlaan 5,B 1980 Tervuren, Belgium
NOVAIS, J. M.	(See Organizing Committee)
ORDAS, R.	Catedra de Fisiologia Vegetal Universidad Oviedo, Oviedo Spain
OSTHOFF, H.	NATTERMANN Research Laboratories Biotechnology of Plant Cell Cultures, P.O. Box 350120, D-5000 Cologne 30 FRG
OTTO, C.	Biochemistry of Plants,University of Muenster, Hindenburgplatz 55, 4400 Muenster FRG
PAIS, M. S. S.	(See Organizing Committee)
PAREILLEUX, A.	Departement de Genie Biochimique et Alimentaire UA-CNRS 544, Institut National des Sciences Appliquées, Av. de Rangueil 31077 Toulouse, France
PEDERSEN, H.	Department of Chemical and Biochemical Engineering,Rutgers, The State University of New Jersey, New Brunswick, New Jersey 08902 USA
PELAEZ, M. I.	Departamento de Genetica Facultad Biologia, Universidad Leon, Spain
PETERSEN, M.	Institut fur Entwicklungs und Molekularbiologie der Pflanzen, Universitat Düsseldorf, Universitätstrsse 1, D-4000 Dusseldorf FRG

PINTO, L.	Departamento de Quimica, Faculdade de Ciencias de Lisboa, R. Escola Politecnica 1294 Lisboa Codex Portugal
QUAYLE, T. J. A.	Davis Crown Gall Group, Departement of Plant Pathology, University of California, Davis USA
RAUTER, A. P.	Departamento de Duimica, Facul-de de Ciencias de Lisboa, R. Escola Politecnica 1294 Lisboa Codex Portugal
REINHARD, E.	Pharmazeutische Institute der Universität Tübingen, auf der Morgenstelle 8, D 7400 Tübingen FRG
REISCH, B. I.	Department of Horticultural Sciences, New York State Agricultural Experiment Station, Cornell University, Geneva, New York 14456 USA
REUSCHER, H.	Organisch-Chemisches Institut, Westfälische Wilhelms Universität, Orleansring 23, D 4400 Muenster FRG
ROBAINA, R.	Departamento de Biologia, Universidad Politecnica de Canarias Box 550, Las Palmas de Gran Canaria, Spain
RODRIGUEZ, R.	Catedra de Fisiologia Vegetal, Universidad Oviedo,Oviedo, Spain
ROGOWSKY, P.	Davis Crown Gall Group, Department of Plant Pathology,University of California, Davis USA
ROMERO, R. R.	Departamento de Biologia,Universidad Politecnica de Canarias Box 550, Las Palmas de Gran Canaria, Spain
RUEDA, J.	Departamento de Genetica, Facultad Biologia, Universidad Complutense, 28040 Madrid Spain
RUIZ, M. L.	Departamento de Genetica, Facultad Biologia, Universidade Leon, Leon, Spain

SANCHEZ, R.	Catedra de Fisiologia Vegetal, Universidad Oviedo, Spain
SEITZ, H. V.	Institut für Biologie I, Universität Tübingen, auf der Morgenstelle 1, D- 7400 Tübingen FRG
SIMOES, C.	Departamento de Quimica , Faculdade de Ciencias de Lisboa, R. Escola Politecnica 1294 Lisboa Codex Portugal
SCHMITZ, U.	Max - Planck - Institut für Züchtungsforchung D-5000 Köln 30 FRG
SCHUBEL, H.	Lehrstuhl für Pharzeutische Biologie, Universität München, Karlstrasse 29,D- 8000 München 2 FRG
SHULER, M. L.	School of Chemical Engineering, Cornell University, Ithaca, New York 14853 USA
SCRAGG, A. H.	The Wolfson Institute of Biotechnology, University of Sheffield, Sheffield S10 2TN, UK
STABA, E. J.	Department of Medicinal Chemistry and Pharmacognosy,University of Minnesota, Minneapolis, Minnesota 55455 USA
STOCKIGT, J.	Lehrstuhl fur Pharmazeutische Biologie, Universität München, Karlstrasse 29, D-8000 München FRG
TANCHAK, M. A.	Department of Biology,University of Saskatchewan, Saskatoon, Saskatchewann, Canada, 57N OWO
TIEMANN, K.	Biochemistry of Plants, University of Muenster, Hindenburgplatz 55, 4400 Muenster, FRG
TORTI, G.	Istituto Biosintesi Vegetali-CNR Via Bassini,15,20133Milano Italy
TRIGIANO, R. N.	Department of Ornamental Horticulture and Landscape Design, University of Tennessee, P. O. Box 1071, Knoxville, Tennessee 379001 USA

ULBRICH, B. — NATTERMANN Research Laboratories Biotechnology of Plant Cell Cultures, P. O. Box 350120 D-5000 Cologne 30 FRG

UPMEIER, B. — Biochemistry of Plants, University of Muenster, Hindenburgplatz 55, 4400 Muenster FRG

Van IREN, F. — Department of Plant Molecular Biology, University of Leiden, P. O. box 9502, 2300 RA Leiden, The Netherlands

VASIL, I. K. — (See Scientific Committee)

VASQUEZ, A. M. — Departamento Genetica, Faculdad Biologia,Universidade Complutense, 28040 Madrid Spain

VERPOORTE, R. — Department of Plant Molecular Biology, University of Leiden, P.O. Box 9502, 2300 RA Leiden The Netherlands

WEIL, J. H. — Institut de Biologie Cellulaire et Moleculaire, Université Louis Pasteur, 15 R. Descartes 67084 Strasbourg, France

WHEELANS, S. K. — Department of Agriculture and Horticulture, University of Nottingham, School of Agriculture, Sutton Bonington, Loughborough, Leics LE 12 5RD,UK

WIESNER, W. — NATTERMANN Research Laboratories Biotechnology of Plant Cell Cultures, P. O. Box 350120, D-5000 Cologne 30, FRG

WIJNSMA, R. — Center for Bio-Pharmaceutical Sciences, Department of Pharmacognosy, University of Leiden, P.O. Box 9502, 2300 RA Leiden, The Netherlands

WILKINSON, A. R. — Department of Chemical Engineering UMIST,Manchester,M60 1QD, UK

WILLIAMS, P. D. — Department of Chemical Engineering UMIST,Manchester,M60 1QD, UK

WHITHERS, L. A.

Department of Agriculture and Horticulture, University of Nothingham, School of Agriculture, Sutton Bonington, Longhborough Leics. LE 12 5RD, UK

YOSHIOKA, T.

Bioscience Research Center, MITSUI Petrochemical Industries Ltd, Waki-Cho, Kuga-Gun, Yamaguchi 740 Japan

INTRODUCTION TO PLANT BIOTECHNOLOGY

F. Mavituna
Department of Chemical Engineering
University of Manchester Institute of Science and Technology
P.O. Box 88
Sackville Street
Manchester M60 1QD
ENGLAND

INTRODUCTION

The techniques of plant organ, tissue and cell culture have evolved over several decades (Table 1). These techniques combined with recent advances in developmental, cellular and molecular genetics and using conventional plant breeding have turned plant biotechnology into an exciting research field with significant impact on agriculture, horticulture and forestry. There has also been a growing interest in the use of suspension and immobilised plant cell cultures and organ cultures for the production of fine chemicals and some specific biotransformation reactions.

TABLE 1: Brief summary of the progress of plant biotechnology

1902	Haberlandt	Maintenance of single cells isolated from plant tissues in simple nutrient media
1904	Hannig	Embryo culture
1934	White	Establishment of actively growing clone of tomato roots
1937	White	The discovery of the importance of the B vitamins for the growth of cultured roots
1937	Went and Thimann	The role of auxin in the control of plant growth
1939	White, Gautheret	Establishment of callus cultures
1954	Muir, Hildebrandt and Riker	Establishment of suspension cultures
1960	Bergman	Single cell cloning
1960	Cocking	Isolation of protoplasts
1964	Morel	Micropropagation

A broad classification of types of aseptic cultures of plant origin includes plant cultures (the culture of seedlings or larger plants), organ cultures (root tips, stem tips, leaf

NATO ASI Series, Vol. H18
Plant Cell Biotechnology. Edited by M.S.S. Pais et al.

primordia, primordia or immature parts of flowers, immature fruits), embryo cultures (isolated mature or immature embryos), tissue or callus cultures (tissues resulting from the proliferation of pieces of plant organs) and suspension cultures (isolated cells or cell aggregates growing dispersed in liquid media). The inter-relationship and interconversions between various types of cultures, and treatments used are summarized in Figure 1.

PLANT MICROPROPAGATION

Plant micropropagation is now an established commercial application. It involves: 1) the selection of suitable explants, their sterilisation and transfer to nutrient medium, 2) the proliferation of shoots and 3) the transfer of shoots to a rooting medium and planting out (Murashige 1974 and Hussey 1983).

In breeding, micropropagation is useful for the maintenance and multiplication of special genotypes. Other uses include rapid bulking up of new cultivars to the levels of several thousands in months instead of years, production of disease-free plants, rapid cloning of heterozygous plants and F_1 hybrids.

EMBRYO CULTURES

An excellent review of embryo cultures is given by Hu and Wang (1986). The removal and culture of embryos of higher plants was one of the earliest techniques in plant tissue and organ culture. To obtain embryo cultures, the ovules, seeds or fruits are surface sterilised and the embryos are aseptically excised from the surrounding tissues and transferred to culture medium (Reinert and Yeoman, 1982). Embryos excised at a mature stage are autotrophic. They can germinate and grow on simple inorganic medium containing an energy source.

Embryo culture has been used to study nutritional and physical requirements for embryonic development, to bypass seed dormancy in order to shorten the breeding cycle, to test seed viability, to provide microcloning source material and to

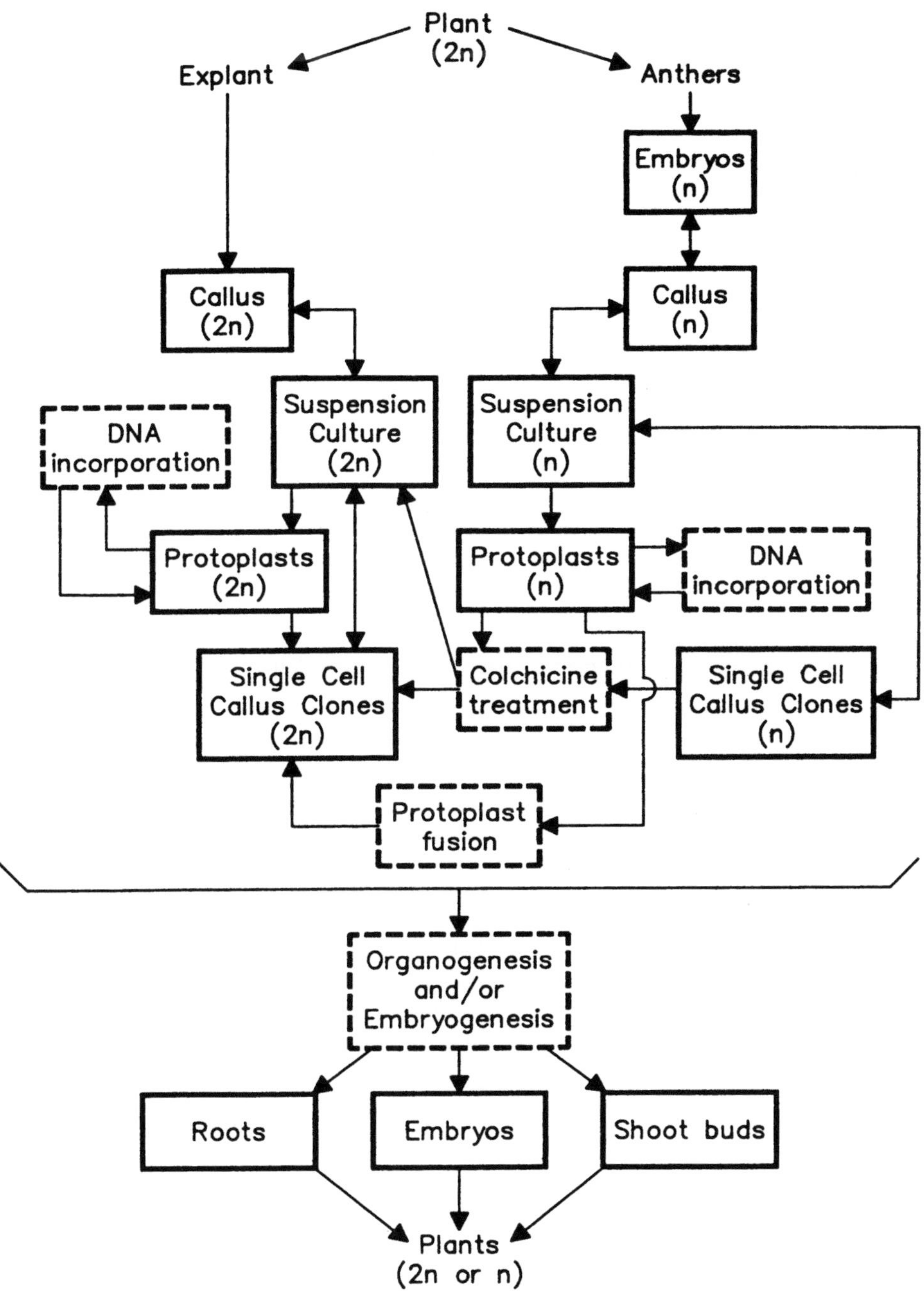

Fig.1. The inter-relationship and inter-conversions between various types of cultures and treatments used. 2n = diploid chromosome number, n = haploid chromosome number.

rescue immature hybrid embryos from incompatible crosses.

PROTOPLAST CULTURES

Plant protoplasts are cells the walls of which have been removed either mechanically or enzymatically (Evans and Cocking, 1977). Leaves have been the most amenable sources for isolation of protoplast. The technique involves surface sterilisation, plasmolysis, removal of cell wall by enzymes, washing and recovery of protoplasts and transfer to culture medium (Reinert and Yeoman, 1982).

The removal of the cell wall leaves the cell surface membrane as the only barrier between the external environment and the interior of the cell hence rendering protoplasts suitable material for genetic manipulations. The protoplasts can regenerate a new cell wall under suitable conditions and revert to normal suspension cultures. The protoplasts can be used for the production of somatic cell hybrids through fusion, as a source of organelles and macromolecules, for genetic transformation via organelle transfer, Ti-plasmids and cocultivation with Agrobacterium, for single cell cloning and to study membrane transport (Evans et al., 1986).

PLANT IMPROVEMENT THROUGH GENETIC ENGINEERING

Recent advances in recombinant DNA and protoplast technology when combined with classical plant breeding techniques may provide useful tools to revolutionise plant improvement. Plant cells contain DNA in nuclei, plastids and mitochondria. Plant cell transformation can be achieved either by introduction of naked DNA directly into plant cell protoplasts or by splicing of DNA to be transferred into a vector before introducing it into protoplasts (Whitaker and Evans, 1986; Wullems et al., 1986). The gene vectors can be Agrobacterium Ti-plasmids, plant DNA viruses, and plant RNA viruses (Grierson and Covey, 1984). Some examples of particular characteristics of plants which might be amenable for genetic manipulation are pest and disease resistance, herbicide resistance, nitrogen fixation, improved photosynthetic efficiency, drought resistance, improved nutritional value and storage life, tolerance to high salinity, flooding and cold.

PRODUCTION OF FINE CHEMICALS BY PLANT CELL CULTURES

Plants are a valuable source of a large range of chemical compounds; such as flavours, fragrances, pigments, enzymes, antimicrobials and industrial feedstocks, most of which are secondary metabolites (Staba, 1980; Fowler, 1983; Whitaker and Hashimoto, 1986). The main reasons for using plant cell cultures as an alternative route to natural production are: independence from environmental factors such as climatic and seasonal constraints, pests and diseases; defined production systems giving close control of market supply; more consistent product quality and yield; reduction in land use for cash crops and freedom from political constraints. In addition to suspension and immobilised plant cell cultures, organ cultures (Hamill et al., 1986) can also be used for the production of valuable secondary metabolites *in vitro*. Biotransformation of some cheap precursors by plant cells into valuable compounds (Furuya, 1978; Reinhard and Alfermann, 1980) is another example of the synthetic potential of plant cell cultures.

IMMOBILISED PLANT CELLS

Immobilisation of plant cells can lead to increased production levels of some valuable secondary metabolites in comparison to suspension cultures (Lindsey and Yeoman, 1984; Brodelius, 1986). There may be several reasons why immobilisation should have such an effect. Cell-to-cell contact appears to play an important role in plant cell metabolism due to the transfer of materials from one cell to another through the plasmodesmata (Hall et al., 1981). This contact may induce cyto-differentiation which can be observed only in callus or the whole plant and seems to be related to secondary metabolism (Yeoman et al., 1982). Since secondary metabolites are produced in slow growing or stationary cultures (Rokem and Goldberg, 1985) growth and production phases can be decoupled easily in an immobilised system and this can be controlled by chemical and physical stress conditions. Finally, since plant cells have low tolerance to shear, immobilisation offers a micro-environment protected from shear.

Figure 2 shows the choices that exist for the use of callus, suspension and immobilised plant cell cultures for the production of intra or extracellular products.

PROBLEMS ASSOCIATED WITH THE PRODUCTION OF FINE CHEMICALS

Main problems with the production of fine chemicals by plant cell cultures stem from the lack of some fundamental knowledge about plant biochemistry, physiology and genetics. The necessity to select high-yielding cell lines, instability of these cell lines and their preservation add to the problems. Most of the products are not released by the cells into the liquid medium. It is therefore desirable to permeabilise the cells without affecting their viability if possible. Problems of process engineering nature include lack of fundamental biochemical engineering data on plant cell cultures, susceptibility to contamination, bioreactor design, scale-up, downstream processing and economics.

DEVELOPMENT OF A PROCESS FOR THE PRODUCTION OF FINE CHEMICALS BY IMMOBILISED PLANT CELLS

A process for the production of valuable metabolites by immobilised plant cells is being developed in our laboratory. Using the production of capsaicin, the hot chilli flavour, which is excreted into the medium by immobilised cells of *Capsicum frutescens* as a model system, we study the following: initiation and maintenance of callus and suspension cultures, inoculum preparation, in situ immobilisation in the bioreactor, bioreactor design and performance, downstream processing, scale-up and economics (Mavituna et al., 1987a).

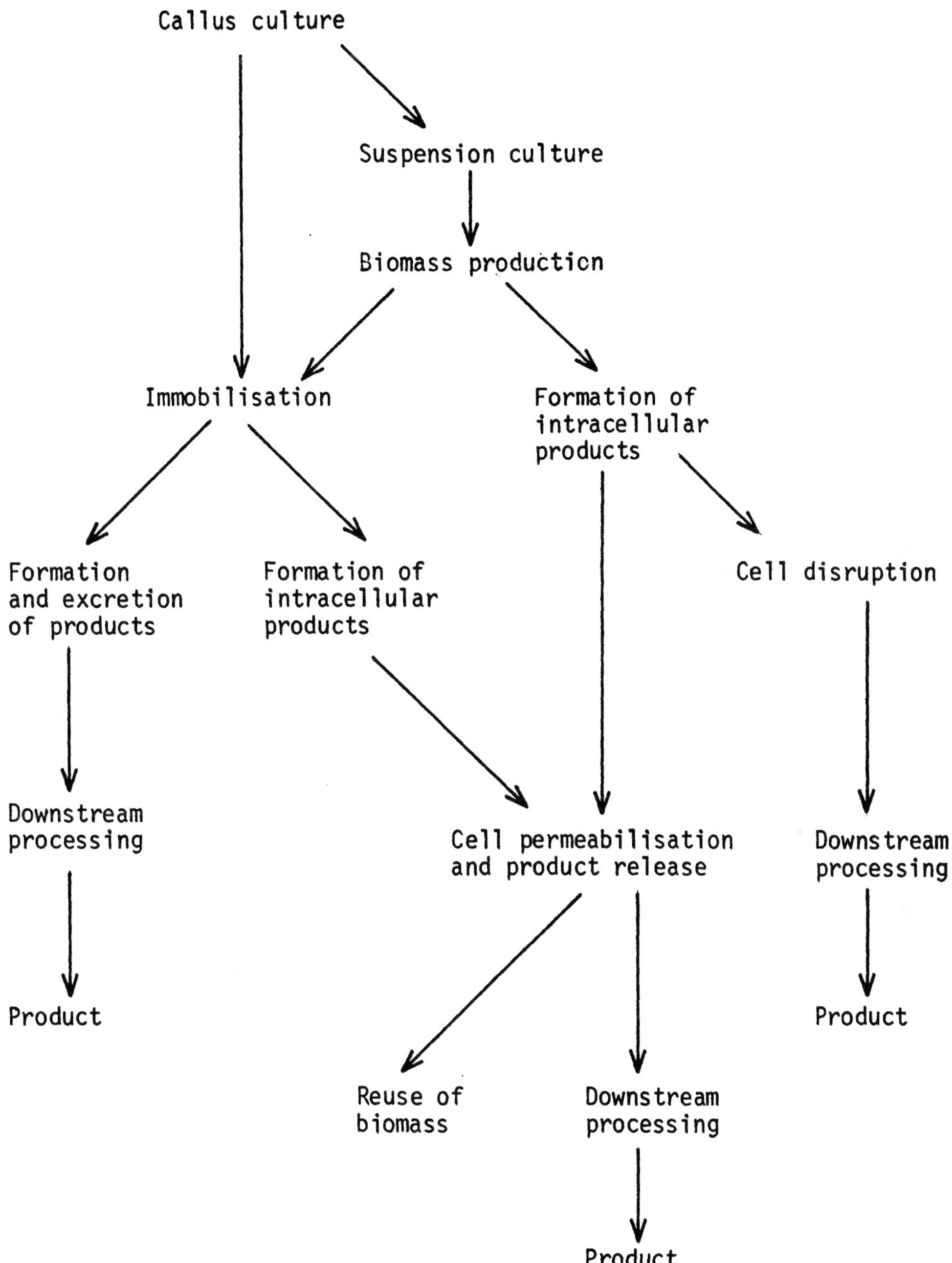

Fig.2. Use of callus, suspension and immobilised cultures for the production of intra- or extra-cellular products.

MATERIALS AND METHODS

Callus, freely suspended and immobilised cultures of Capsicum frutescens were grown in Schenk and Hildebrandt medium at pH 5.8 and 25°C. The reticulated polyurethane foam material used for immobilisation was obtained from Declon (Corby, U.K.). The medium sugars and capsaicin were analysed by HPLC. Capsaicin used in the standard solutions for analysis was supplied by Sigma.

IMMOBILISATION

Microbial contamination is a major problem with plant cell cultures due to the slow growth and rich media requirements. Therefore, for the realisation of a commercial scale process, the immobilisation step and bioreactor operation must be strictly aseptic. This contamination problem is greatly reduced in the bioreactors we have developed due to the fact that immobilisation is carried out in situ within the bioreactor. The inoculum of freely suspended cell aggregates are immobilised in the porous polyurethane foam matrices (Lindsey et al., 1983) due to physical entrapment by filtration from the bulk liquid. Plant cells then grow and/or adhere to each other to fill the porous matrices leaving the bulk liquid clear.

RAPID PRODUCTION OF FINE SUSPENSION CULTURES

Experiments have shown that the main factor affecting the efficiency of immobilisation is the cell aggregate size distribution compared with the average pore size of the foam matrix. We have developed a method which allows the rapid production of fine suspension cultures form coarse and clumpy suspension cultures, such as those of C.frutescens (Mavituna et al., 1987b). Basically, the aggregates are subjected to high shear for a very short period of time. Although there is an initial loss of viability, the resulting fine suspensions recover and continue to grow yielding fine suspensions with normal viability levels.

FUNDAMENTAL BIOCHEMICAL ENGINEERING DATA

Growth and sugar uptake of suspension and immobilised cultures are shown in Figure 3 (Mavituna and Park, 1985). The maximum specific growth rate of the freely suspended cells was found to be slightly greater at 0.245 day^{-1} than that of immobilised cells at 0.144 day^{-1}. The freely suspended batch culture was almost completely dead after 8 weeks from inoculation, whereas the immobilised culture remained over 90% viable in batch cultures for over 12 weeks.

It was found that the typical value of the specific oxygen uptake rate was 1-2 x 10^{-3} g O_2/g dry weight -h and the optimum k_La range was 10-20 h^{-1} for the maximum specific growth rate.

The initial disappearance of sucrose from the medium was concomittant with the appearance of glucose and fructose in the medium. These were then taken up by the cells at the same time, but with a preference towards glucose.

The effective diffusion coefficients of glucose in callus and immobilised cells were experimentally determined to be in the ranges of 0.028 x 10^{-5} to 0.28 x 10^{-5} cm^2/s and 1.39 x 10^{-5} to 13.9 x 10^{-5} cm^2/s respectively, compared with the diffusion coefficient of 0.69 x 10^{-5} cm^2/s for glucose in water (Mavituna et al., 1987c).

BIOREACTOR DESIGN

The bioreactor is an adaptation of a 7 litre LH Fermentation vessel the agitator of which has been removed and an off-centre air sparger fitted. About 1000 x 1 cm^3 foam particles are held between two movable stainless steel grids covering the complete horizontal cross-section of the bioreactor. By pushing these grids in, the foam particles are held as a packed bed within the bioreactor during inoculation by freely suspended cultures and immobilisation. The cells recycle through the bed by the circulating action of the bulk liquid induced by off-centric air sparging. The cell aggregates, while continuing to grow, begin to filter into the foam matrices. After immobilisation, the grids are retracted

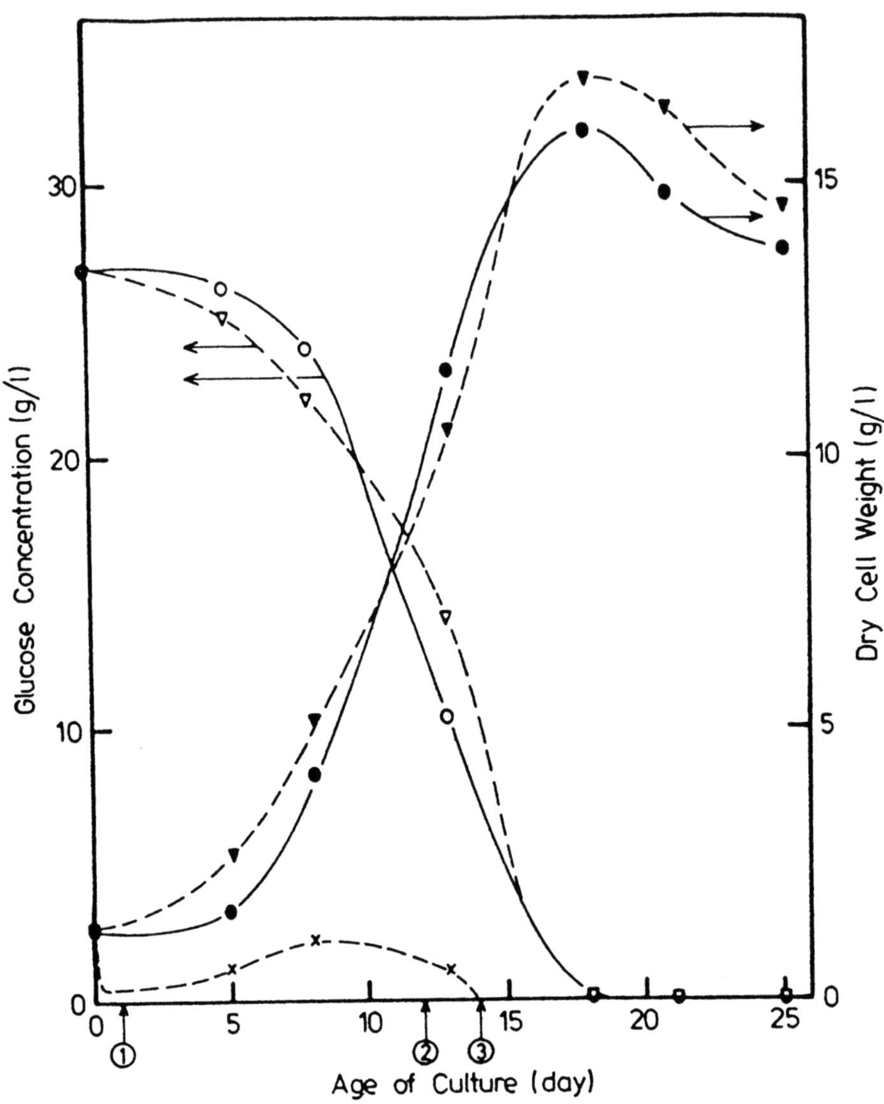

Fig.3. The time course of batch growth and glucose consumption: -0-, -●-, suspension culture; -▽-, -▼-, immobilised culture; -X-, cells outside the foam particles, (1) very large and small aggregates (0.3-0.5 mm) still remained outside the foams, (2) small cell aggregates were all entrapped and only very few large aggregates were still in suspension, (3) even large aggregates were attached to the foams.

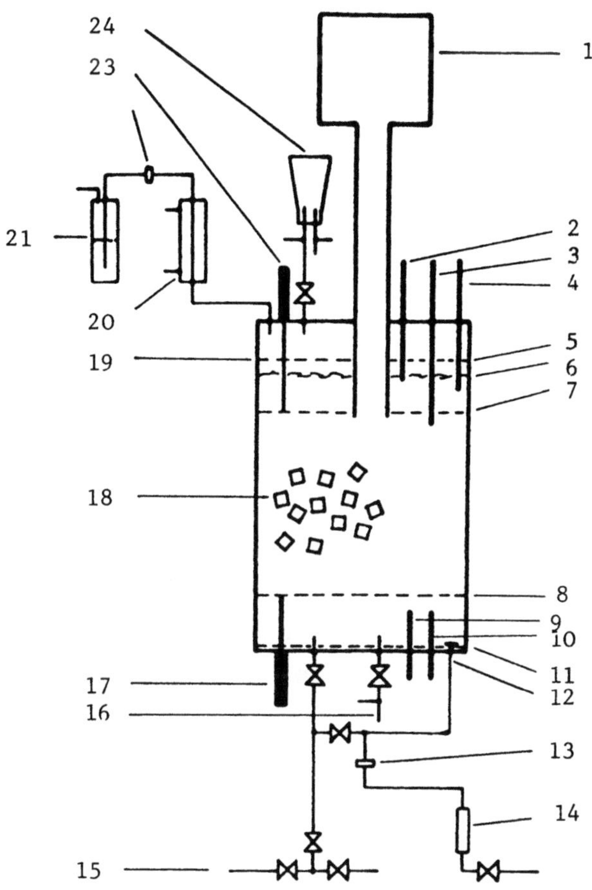

Fig.4. Schematic representation of the bioreactor

KEY:

1 Biomass sampler vessel
2 Antifoam probe
3 pH probe
4 Oxygen prove
5 Grid (raised position)
6 Liquid level
7 Grid (lowered position)
8 Grid (raised position)
9 Heater
10 Temperature sensor
11 Grid (lowered position)
12 Air input through sparge
13 Filter
14 Rotameter
15 Steam point
16 Liquid sample point
17 Handle to move grid
18 Foam particles
19 Grid (raised position)
20 Condensor
21 Air outlet
22 Filter
23 Handle to move grid
24 Inoculation vessel

and the operating is continued as a circulating bed bioreactor (Figure 4).

BIOREACTOR PERFORMANCE

An inoculum of freely suspended cells was immobilised in foam particles in situ in the bioreactor which was operated initially in packed bed then in circulating bed mode for periods of up to 100 days. During these runs, levels of capsaicin up to a maximum of 17.34 mg/1 and yields of capsaicin up to a maximum of 0.45 mg/g dry weight were obtained in depleted Schenk and Hildebrandt growth medium without any precursors. It was established that the production of capsaicin took place only during the period when the aeration was stopped and the D.O. level fell from an average of 65% saturation to almost zero. During the time when the bioreactor was not aerated the productivity of the system was 1.62 capsaicin mg/1-day. A further interesting observation was the rapid loss of the product from the medium when aeration was restored (Wilkinson et al., 1987).

PRODUCT RECOVERY

Preliminary experiments indicate that the production of capsaicin can be further increased and the loss from the medium prevented by continuous extraction of the product into sunflower oil. A maximum production of 1.54 mg capsaicin/g dry weight biomass has been achieved when extraction was integrated with production. The maximum specific production rate of the immobilised cells with simultaneous oil extraction at 0.1 mg capsaicin/g dry weight-day compares well with the rate of 0.5 mg/g dry weight-day during the accumulation of capsaicin in the ripening *Capsicum frutescens* fruit.

CONCLUSION

Micropropagation, improvement of some agricultural and horticultural plants, and shikonin production (Yamada and Fujita, 1983) are already commercial applications of plant biotechnology. With more fundamental research on plant biochemistry, physiology and genetics plant biotechnology will not only remain as an exciting potential solution to some economic and environmental problems but may also introduce some novel processes and products to chemical and pharmaceutical industries.

ACKNOWLEDGEMENTS

The author would like to thank the Biotechnology Directorate of the SERC and Albright & Wilson Ltd. for financial support and Professor Yeoman from the Botany Department of the University of Edinburgh for his invaluable assistance for her research with immobilised plant cell cultures. The research work reported here was carried out by J.M. Park, A.K. Wilkinson and P.D. Williams.

REFERENCES

Brodelius, P (1986) in "Handbook of Plant Cell Culture" DA Evans, WR Sharp and PV Ammirato eds. Macmillan New York: 287

Evans, DA, Sharp, WR, Ammirato, PV eds. (1986) "Handbook of Plant Cell Culture" Vol.4 Macmillan New York.

Evans, PK and Cocking, EC (1977) in "Plant Tissue and Cell Culture" HE Street ed. Blackwell Oxford: 103.

Fowler, MW (1983) in "Plant Biotechnology" SH Mantell and H Smith eds. Cambridge Univ. press Cambridge: 3.

Furuya, T (1978) in "Frontiers of Plant Tissue Culture" ed. TA Thorpe Calgary Univ. Press Calgary: 191.

Grierson, D and Covery, S (1984) "Plant Molecular Biology" Blackie Glasgow.

Hall, JL, Flowers, TF and Roberts, RM (1981) Plant Cell Structure and Metabolism 2nd ed. Longman London.

Hamill, JD, Parr, AJ, Robins, RJ and Rhodes, MJC (1986) Plant Cell Rep. 5: 111 (1986).

Hu, C and Wang, P (1986) in "Handbook of Plant Cell Culture" DR Evans, WR Sharp, PV Ammirato eds. Macmillan New York: 43.

Hussey, G in (1983) "Plant Biotechnology" SH Mantell and H Smith, eds. Cambridge Univ. Press Cambridge: 111.

Lindsey, K, Yeoman, MM, Black, GM and Mavituna, F (1983) FEBS Lett. 115: 143.

Lindsey, K and Yeoman, MM (1984) Planta 162: 495.

Mavituna, F, Park JM, Wilkinson, AK and Williams, PD (1987a) in "Process Possibilities for Plant and Animal Cell Cultures", C Webb and F Mavituna eds. Ellis Horwood Publ: 92.

Mavituna, F, Wilkinson, AK and Williams, PD (1987b) in Process Possibilities for Plant and Animal Cell Cultures" C Webb and F Mavituna eds. Ellis Horwood Publ: 262.
Mavituna, F, Park, JM and Gardner, D (1987c) The Chemical Engineering Journal 34: B1.
Mavituna, F and Park JM, Biotechnol. Letts. (1985) 7(9): 637.
Murashige, T (1984) Ann. Rev. of Plant Physiol. 25: 135.
Reinhard, E and Alfermann, AW (1980) Adv. in Biochem. Eng. 16: 49(1980).
Reinert, J and Yeoman, MM (1982) "Plant Cell and Tissue Culture Springer-Verlag Berlin (1982).
Roken, SJ and Goldberg, I (1985) Adv. in Biotechnol. Proc. 4: 241.
Staba, EJ (1980) "Plant Tissue Culture as a Source of CRC Biochemicals" Press Florida.
Whitaker, RJ and Evans, DA (1986) in "Handbook of Plant Cell Culture" DA Evans WR Sharp PV Ammirato eds. Macmillan New York 172.
Whitaker, RJ and Hashimoto, T (1986) in "Handbook of Plant Cell Culture" DA Evans WR Sharp and PV Ammirato eds. Macmillan New York 264.
Wilkinson, AK, Williams, PD and Mavituna, F Poster Pater, NATO ASI on Plant Cell Biotechnology, Portugal, 29 March - 10 April 1987.
Wullems, GJ, Krens, FA and Schilperoort, RA (1986) in "Handbook of Plant Cell Culture" DA Evans WR Sharp and PV Ammirato eds. Macmillan New York 197.
Yamada, Y and Fujita, Y (1983) in "Handbook of Plant Cell Culture Vol.1 DA Evans WR Sharp PV Ammirato and Y. Yamada eds. Macmillan New York 717.
Yeoman, MM, Lindsey, K, Miedzybrodzka, MB and McLauchlan, WR (1982) in "Differentiation in Vitro" MM Yeoman and DES Truman eds. Cambridge Univ. press 65.

THE CONTRIBUTIONS AND PROSPECTS OF PLANT BIOTECHNOLOGY - AN ASSESSMENT

Indra K. Vasil
Laboratory of Plant Cell and Molecular Biology
Institute of Food and Agricultural Sciences
University of Florida
Gainesville, Florida 32611
United States of America

ABSTRACT

The power, promise and potential benefits of plant biotechnology in producing better and improved plants and plant products are widely recognised and appreciated. There are, however, only a few selected instances of limited value where these benefits have actually been realised. Applications of plant biotechnology to the improvement of major crop species must await intensive further research in the basic understanding of plant growth and development, and the molecular biology of plant gene structure and function, in addition to adaptation of the appropriate technologies to crops of interest. A plea is made to tone down the rhetoric and the hype of plant biotechnology and to concentrate more on developing the basic scientific competence to achieve the stated goals.

KEY WORDS: Agriculture, Plant Biotechnology.

INTRODUCTION

Although the application of the modern methods of biotechnology has led to the development and commercial availability of several bio-medical products such as vaccines, drugs and animal growth hormones, it is generally recognised that the economic value and social and political impact of agricultural biotechnology products will eventually be far greater and of worldwide significance. It is not surprising, therefore, that there are major national and international efforts - both in the

NATO ASI Series, Vol. H18
Plant Cell Biotechnology. Edited by M. S. S. Pais et al.

public and in the private sector - to support the development of plant biotechnology.

The euphoria and enthusiasm created by significant new developments in plant biotechnology has often led to very unrealistic and wide ranging forecasts of the benefits of plant biotechnology. In reality, there are very few benefits that can be presently realised, or real products that can be commercially and economically produced. Most of the stated potentials and promises of plant biotechnology remain exactly that, and will require extensive research toward the understanding of the fundamental processes of plant growth, development and molecular biology. Premature, speculative and sometimes unsubstantiated claims of the benefits and impact of biotechnology on agriculture are counterproductive and bring into question the credibility of scientists in general, and many biotechnology companies in particular (see also chapter by Dr E. Magnien, this proceedings). Many of these pronouncements and claims are unrealistic and uncritical in nature, and are often made in the form of news releases, press conferences or interviews, or reports in the popular press. They provide little or no scientific information or evidence in support of the claims, and appear to be made principally for attracting investment capital and other support. These practices cause serious harm as they unrealistically raise expectations in the public mind, and can cause backlash and serious long-lasting harm. It is, therefore, important that the widely prevalent rhetoric and hype of plant biotechnology is substantially toned down, if not entirely eliminated.

SUCCESS STORIES

The following are some of the more important current achievements and applications of plant biotechnology:

1. Clonal Propagation - The most widely used and important application of plant biotechnology is in the rapid and large scale clonal propagation of many plant species. These include numerous horticultural plants, fruit and vegetable species, oil and medicinal plants, woody tree species, etc. The currently available techniques are being exploited to their maximum efficiency in terms of plant production. Therefore, serious attempts are being made to automate and computerise the process so that much larger populations of plants, with better quality control, can be produced at

reduced cost per plant/propagule. This will extend the commercial feasibility of rapid propagation techniques to many transplant species, like the vegetables, and woody tree species.

2. Virus Elimination - The technique of meristem culture, which is often used in the clonal propagation of plants, is also useful in the elimination of certain viruses. Thus large numbers of virus free plants, which have enhanced value and quality, are produced in many important horticultural and agricultural species.

3. Embryo Rescue - The technique of embryo rescue, which involves the excision and culture of young embryos, has been used successfully to obtain many agriculturally important hybrids that can not be obtained by conventional techniques.

4. Mutant Selection - Application of selective pressure in cell cultures has been used to isolate stable mutant cell lines and plants which are resistant to certain diseases and herbicides, or are of improved nutritional quality. This has been possible only in those instances where appropriate selection pressure can be applied, and selection is for simple single gene traits. It is doubtful that such methods can be used to select for multigenic traits such as tolerance to stress (salt, drought, frost, etc.) or for agronomically important traits such as improved yield, etc.

Similarly, although there has been much speculation about the percieved uses of somaclonal variation in plant breeding and improvement, the fact remains that thus far there is not a single example of any major new variety that has been released in any of the major crop species. There is a growing feeling that much of the variation recovered from cell cultures by this method is neither novel nor of significant benefit in plant breeding and variety development.

5. Haploids - Anther and ovary culture techniques have been used to obtain haploid and doubled homozygous diploid plants. New cultivars of rice, wheat, tobacco, oil seed rape, etc. have been obtained by this method, and are being commercially grown, although not on a very large scale, in some countries.

6. Secondary Metabolites - Recent advances in fermentation technology and the synthesis of secondary metabolites in plant cell cultures have encouraged attempts to commercially produce plant products from cell cultures. Although this field of research has attracted much attention, only a few products of commercial significance, such as shikonin, are currently being produced by this method.

7. Genetic Engineering - In the long-term the novel techniques of cell fusion and genetic transformation hold the most promise. Success thus far, though encouraging, is modest. The best examples, where commercial and practical uses are imminent in some secondary crops, are the transfer of cytoplasmic male sterility by cell fusion, and the use of gene manipulation techniques for modifying bacteria to prevent frost formation, transfer of Bt genes for protection from insect pests, and conferring virus cross protection or resistance to specific herbicides.

FUTURE

The future success and applications of plant biotechnology to produce better and improved crops depends on sustained research on developmental and molecular biology of plants, and on enhanced focus and attention to the following:

1. Develop efficient regeneration protocols, particulalry from single cells or protoplasts, for species of economic importance such as cereals and grasses, legumes and vegetables, and fruit and tree species.
2. Investigate the causes and factors causing genetic instability in cultured cells and develop procedures that help maintain the genetic fidelity of cell cultures and yield true clonal populations.
3. Identify and characterize agronomically important genes for use in genetic transformation experiments.
4. Develop efficient genetic transformation systems for limited gene transfer to both dicotyledonous and monocotyledonous species.
5. Attract and train young scientists in basic plant biology - including plant physiology and biochemistry, developmental and reproductive botany, and molecular biology and genetics - in order to overcome the shortage of trained personnel.

CONCLUSIONS

Although the potential benefits of plant biotechnology are many and of wide-ranging impact, much remains to be done to achieve these objectives. Only sustained development of more efficient techniques of plant biotechnology and better understanding of plant biology will help in the exploitation of these novel technologies for the production of better and improved plants and plant products. Finally, much harm is being done by irresponsible and unrealistic forecasts of the impending benefits of plant biotechnology. The best that can be done to achieve the stated objectives is to enhance and concentrate the efforts toward developing the necessary science and technology for crop improvement and to tone down the rhetoric and speculation. It is also important to state firmly and unequivocally, that biotechnology will only supplement and complement, but will not replace, the enormously successful traditional methods of plant improvement.

SOMATIC HYBRIDIZATION OF PLANTS AND ITS USE IN AGRICULTURE

John D. Hamill and Edward C. Cocking[1]
Plant Biotechnology Group, Institute of Food Research, (Norwich Laboratory), Norwich NR4 7UA, U.K.

ABSTRACT

Since Carlson et al. (1972) first reported the successful recovery of somatic hybrid plants, very many protoplast fusion experiments have been done. These have ranged from experiments aimed at answering fundamental questions on the interaction of different plant genomes following fusion to practical questions regarding the improvement of crop species by fusion with wild species possessing desirable agronomic traits. The agricultural applications of protoplast fusion are now beginning to be realised and although the recent successes in plant transformation offer many new opportunities in plant genetic engineering for crop improvement, it is clear that protoplast fusion does have an important role to play in practical plant biotechnology.

In this article we discuss the methodologies needed to recover somatic hybrids, their key features and some examples of somatic hybrids which are proving useful in agriculture. It should be noted that there exists a huge literature on protoplast fusion and only a limited number of examples are quoted in each section.

[1]Dept. of Botany, University of Nottingham, Nottingham NG7 2RD, U.K.

NATO ASI Series, Vol. H18
Plant Cell Biotechnology. Edited by M. S. S. Pais et al.

OBTAINING SOMATIC HYBRIDS AND THEIR FEATURES

There are many technical obstacles which must be overcome before somatic hybrid plants can be obtained and then evaluated. These problems are briefly described before considering the features of somatic hybrids.

PROTOPLAST ISOLATION, CULTURE AND PLANT REGENERATION

For many years this essential prerequisite for somatic hybridization was confined to a small number of species - especially species in the *Nicotiana* and *Petunia* genera. However, in recent years there have been reports of successful and reproducible plant regeneration from protoplasts in a number of crop species, including potato, tomato, alfalfa and oilseed rape. More recently, success has been achieved with very difficult species such as rice (cereals) (Yamada *et al.*, 1986; Abdullah *et al.*, 1986; Thompson *et al.*, 1986). The factors affecting these encouraging results will not be discussed in detail here, but it is clear that growth conditions of plants and tissue type used as a source of protoplasts, genotype and precise culture conditions are all very important factors.

PROTOPLAST FUSION AND HETEROKARYON FORMATION

Protoplast fusion may be successfully accomplished by chemical fusion or electrofusion methodologies. Fusion is normally induced chemically by treatment with polyethylene glycol (Kao and Michayluk, 1974; Wallin *et al.*, 1974) or calcium ions at high pH (Keller and Melchers, 1973), although other methods have been used. Electrofusion methods have developed rapidly from initial reports in which only a limited number of protoplasts could be fused at any one time (Zimmermann and Sheurich, 1981; Vienken *et al.*, 1981), to more recent reports where large numbers of protoplasts can be efficiently fused in any one experiment (Watts and King, 1984; Negrutiu et al., 1986).

As with protoplast culture, there is no one fusion protocol which works with equal efficiency for all situations. In most cases a compromise must be reached between cell lysis, or loss of viability caused by fusion conditions, and the numbers of heterokaryons which are obtained. Generally cell suspension protoplasts are much more resistant to chemical fusion damage than protoplasts which are derived from leaf mesophyll tissue. Frequently, heterokaryons are stabilised by the cell suspension parent, presumably due to rapid mixing of the two plasma membranes (Ward et al., 1980), and can often be seen intact among lysed or collapsed leaf mesophyll protoplasts. Recently, improvements in PEG induced fusion have been reported which reduced damage to mesophyll protoplasts (Kao and Saleem, 1986). A detailed assessment of the benefits of electrofusion, if any, compared to chemical fusion is needed but initial results suggest that it may have advantages over chemical induced fusion (Negrutiu et al., 1986). Certainly electrofusion does seem to be an effective way of recovering somatic hybrid genomes (Kohn et al., 1985; Watts et al., 1985) and has been shown to be suitable for recovering interspecific somatic hybrids colonies and plants after fusion of mesophyll with mesophyll protoplasts (Hamill et al., 1987).

It is desirable to fuse protoplasts which are directly isolated from plants, if possible, with subsequent growth in the disorganised phase for as short a period as possible. In this way, mutations induced by growth as callus or cell suspension (D'Amato, 1985) may be minimised. Protoplasts isolated from hypocotyl tissue of dark grown plants may be useful to provide visual markers for selection (see later). There is also evidence that hypocotyl protoplasts may be more tolerant of fusion damage than are leaf mesophyll protoplasts (Sundberg and Glimelius, 1986).

SELECTION

This is a crucial factor in a somatic hybridization programme, not only because it is necessary to recover a limited number of heterokaryons from a larger number of parental protoplasts, but also because the selection regime can influence the nature of the hybrid cells which are recovered.

Selection regimes fall into two broad classes: i) Genomic complementation, where the hybrid genome has growth characteristics separate from either parent and ii) Heterokaryon isolation, where the heterokaryons are physically separated from the parental protoplasts and then cultured.

GENOMIC COMPLEMENTATION

There are many examples of a complementation approach to hybrid selection and only a few examples will be mentioned here. More detailed descriptions are described (Keller *et al.*, 1982; Pelletier and Chapeau, 1984).

Use has been made of complementing auxotrophs [e.g. chlorophyll deficiency (Douglas *et al.*, 1981) and nitrate reductase deficiency (Glimelius *et al.*, 1978)] drug resistant parents (White and Vasil, 1979), combinations of auxotrophy and drug resistant markers (Pental *et al.*, 1984,1986) growth conditions suitable for growth and recognition of hybrids (Kinsara *et al.*, 1986), and a large number of permentations of these approaches to provide a means of selection for hybrid cell or plant development.

Mutants are time consuming to isolate and characterize, but have been of great value in studying somatic hybridization and defining it's limits. For example, we combined the streptomycin resistant plastid of the SR1 mutant of *N. tabacum* (Maliga *et al.*, 1973) with the nitrate reductase deficient nucleus of the nia-30 mutant of *N.*

tabacum (Muller, 1983) to generate a double mutant plant (Hamill *et al.*, 1983,1984a). This produced an extremely useful selection regime whereby any wild type species could be fused with *N. tabacum* and hybrids then recovered in selection medium containing nitrate as the only nitrogen source (eliminating the *N. tabacum* parent) and also containing streptomycin (eliminating the wild type). Such a procedure was shown to be effective (Pental *et al.*, 1984) and the double mutant has been useful in studying the stability of intergeneric compared to interspecific somatic hybrids (Pental *et al.*, 1986) and also electrofusion procedures involving only mesophyll protoplasts (Hamill *et al.*, 1987). The combination of any resistance (positive selection) and auxotrophic (negative selection) characters in one genome, of an agronomic plant species central to a somatic hybridization programme, would facilitate the development of a selection regime such that all potentially useful species could be assessed regarding the production and analysis of interspecific somatic hybrids.

HETEROKARYON ISOLATION

This method of selection is potentially of greater applicability than complementation as it does not require the use of mutants and also does not necessarily prejudice the type of hybrid genome which will be recovered. In addition it allows the early stages of somatic hybridization to be closely monitored (Hamill *et al.*, 1984b). Heterokaryons are normally isolated by a micropipette procedure (Patnaik *et al.*, 1982; Hein *et al.*, 1983) and then cultured - often among cells of a nurse culture from which the hybrid cells may be later distinguished. For example, an albino nurse culture of *Petunia hybrida* cells was used to recover somatic hybrids derived from heterokaryons between *Nicotiana rustica* and *N. tabacum* (Hamill *et al.*, 1984a). Similarly, an albino nurse culture of *Nicotiana tabacum* cells was used in the recovery of cybrid plants derived from heterokaryons between different varieties of <u>Brassica napus</u> (Yarrow <u>et al</u>., 1986). Recently the use of membrane chambers has been shown to be an effective way of recovering small numbers (<50) of protoplast derived calli of <u>Medicago sativa</u>. These chambers, which were permeable to metabolites, allowed the conditioning effects of a nurse culture to sustain cell division among the small number of protoplasts in the chamber while keeping them physically separated from the nurse culture (Gilmour <u>et al</u>., 1987). Such a technique could be used to culture small numbers of heterokaryons.

Heterokaryons may be detected visually by natural differences between parents - e.g. chloroplasts (from a leaf mesophyll protoplast parent) and cytoplasmic strands (from a cell suspension protoplast parent) are often sufficient. However, fluorescent dyes [such as Fluoroscein Isothiocyanate (FITC), or Fluorscein Diacetate (FDA) - both of which fluoresce green under ultraviolet light] may be used to label one parent (normally these cells contain little chlorophyll - e.g. cell suspension protoplasts or dark grown hypocotyl protoplasts). If the other parent contains chlorophyll, then heterokaryons can be easily detected by the presence of red and green fluorescence (Galbraith and Mauch, 1980; Patnaik <u>et al</u>., 1982; Yarrow <u>et al</u>., 1986). Other non-toxic dyes may also be of value in allowing heterokaryon detection (Kanchanapoom and Boss, 1986).

Manual isolation of heterokaryons is a tedious and time consuming procedure and normally yields a limited number of heterokaryons, although there is a report of several hundred heterokaryons being isolated at one session by this method (Sundberg and Glimelius, 1986). The recent arrival of Fluorescent Activated Cell Sorting (FACS) however may revolutionise the field of somatic hybridization as potentially thousands of heterokaryons may be easily isolated. As will be discussed later, large numbers of somatic hybrids are desirable from an agricultural point of view. Early attempts to isolate heterokaryons by FACS were disappointing, because of severe damage to heterokaryons by the sorting process. However, recently the recovery of viable heterokaryons and somatic hybrids has been reported using this method (Afonso *et al.*, 1985).

CONSEQUENCES OF FUSION

In the early days of protoplast fusion some attention was given to the production of heterokaryons and hybrids between extremely divergent species. Indeed, early work showed that not only was there no barrier to protoplast fusion of evolutionary divergent species but that the nuclei of species as distant as carrot and barley fused rapidly after protoplast fusion and some of the heterokaryons divided (Dudits *et al.*, 1976). However it is now clear that the potential of the technique is not in creating hybrids between very distant species. In addition it is now recognised that the chloroplast and mitocondrial genomes may behave independently from the nuclear genomes in somatic hybrids. These organelle genomes are of great significance when considering the agronomic potential of protoplast fusion.

NUCLEAR HYBRIDITY

Use has been made of the nitrate reductase deficient, streptomycin resistant double mutant of *N. tabacum* (discussed earlier) in studying the nuclear stability of interspecific and intergeneric somatic hybrids. In fusions with another (sexually, almost incompatible) species of the *Nicotiana* genus, *N. rustica*, we found it relatively easy to recover somatic hybrid cells and plants - it was estimated that about 12% of heterokaryons (between cell suspension protoplasts of *N. rustica* and leaf mesophyll protoplasts of *N. tabacum*) gave rise to nuclear hybrids which, although aneuploid, were apparently genetically stable. Isozyme analysis, morphology and DNA hybridization experiments showed that the plants possessed genetic characters of the nuclear genomes of both parents (Pental *et al.*, 1984, unpublished observations). Indeed, both nuclear genomes remained stably together after two meiotic cycles in these somatic hybrids (Hamill *et al.*, 1985). However in protoplast fusion experiments with *N. tabacum* and species of other genera, a much lower success rate was achieved regarding the recovery of nuclear somatic hybrids. We studied in great detail the consequences of fusing *N. tabacum* protoplasts (leaf mesophyll) with protoplasts (cell suspension) of *Petunia hybrida*. Previous attempts to sexually cross these species were unsuccessful with post-zygotic incompatibility causing breakdown of the embroyo at an early stage (Zenkteler and Melchers, 1978). We recovered only eleven somatic hybrid colonies after many fusion experiments were carried out (Pental *et al.*, 1986). Even in the experiment which produced somatic hybrids, we estimated that the success rate of this intergeneric combination was at least two orders of magnitude lower than in the intrageneric somatic combination of *N. tabacum* with *N. rustica* (Pental *et al.*, 1986). Six of these lines were monitored for over a year whereupon we found all lines showed nuclear instability with the loss of the nuclear genome of one of the parents - *N. tabacum* in five of the lines and *P. hybrida* in the other line.

Other workers have reported intergeneric somatic hybrid combinations with varying degrees of nuclear stability. Fusions between members of the *Solanum* and *Lycopersicon* genera have yielded apparently stable, and intermediate intergeneric somatic hybrids (Melchers *et al.*, 1978; Shepard *et al.*, 1983; Handley *et al.*, 1986). However these genera are closely related (Rick, 1979) and thus nuclear incompatibility appears not to be so great as to cause breakdown of the hybrid. Reports of other intergeneric somatic hybrids, possessing evidence of nuclear instability have been reported between *Aribidopsis thaliana* and *Brassica campestris* (Gleba and Hoffmann, 1980), *Datura innoxia* and

Atropa belladonna (Krumbiegel and Schieder, 1981), *Lycopersicon esculentum* and *Petunia hybrida* (Tabaeizadeh *et al*., 1985), *Nicotiana tabacum* and *Atropa belladonna* (Gleba *et al*., 1986) and *Nicotiana tabacum* and *Hyoscyamus muticus* (Potrykus *et al*., 1984). Interfamily somatic hybrids have also been reported between *Glycine max* and various *Nicotiana* species. These lines also showed nuclear instability (Kao, 1977; Wetter and Kao, 1980; Chien *et al*., 1982). Thus it seems likely that interspecific somatic hybridization will yield genetically stable nuclear hybrids only when the species are from the same genus, or closely related genera. However, as is noted later, such knowledge may be very valuable as crossing sexually incompatible species, within a genus, remains an important factor in many breeding programmes

ORGANELLE GENOMES

Generally chloroplasts, in somatic hybrids, do not remain together in one cell but usually sort out to leave one or other as the remaining type (Chen *et al*., 1977; Douglas *et al*., 1981; Scowcroft and Larkin, 1981; Hamill *et al*., 1984b; Kemble *et al*., 1986). In addition recombination between chloroplast genomes has also been reported although this is a very rare phenomenon (Medgyesy *et al*., 1985).

In contrast to the chloroplast genome, the mitochondrial genome has been reported to undergo recombination in a large number of somatic hybrid combinations (Belliard *et al*., 1979; Nagy *et al*., 1981; Kemble *et al*., 1986). Evidence of independent segregation of chloroplasts and mitochondria, in somatic hybrids has also been reported (Galun *et al*., 1982; Pelletier *et al*., 1983; Clark *et al*., 1985).

There is also evidence that chloroplasts can be functional in intergeneric nuclear-cytoplasmic combinations (cybrids) which do not exhibit stability at the nuclear level. For example in our analysis of *Nicotiana* x *Petunia* somatic

hybrids (noted previously) we found that although the two nuclei did not remain together in the same cell, we were able to recover plants which possessed the nucleus of *Petunia hybrida* and the streptomycin resistant chloroplast of *Nicotiana tabacum* (Pental *et al.*, 1986). In analgous experiments, Glimeluis and Bonnett (1986) fused gamma irradiated protoplasts of *Petunia hybrida* with protoplasts from a chloroplast defective cultivar of *Nicotiana tabacum*. (This technique was previously shown to be effective in interspecific protoplast fusion experiments within the *Nicotiana* genus. The radiation causes elimination of the nucleus of the irradiated parent but allows the chloroplasts to proliferate (Zelcher *et al.*, 1978, Menczel *et al.*, 1982)). Over 100 photoautotrophic plants were ultimately recovered by Glimelius and Bonnett (1986) and the vast majority consisted of fertile plants possessing the chloroplast of *P. hybrida* and the nucleus of *N. tabacum*. The mitochondrial compositions of the plants produced by Pental *et al.* (1986), or Glimelius and Bonnett (1986) have not yet been reported.

Organelle exchange and recombination, via protoplast fusion, is potentially of great significance in agriculture. Male sterility is often associated with particular nuclear-mitochondrial combinations. Resistance to various herbicides may be due to chloroplast encoded mutations. In addition, the involvement of the chloroplast and mitochondria with the energy balance of the plant may be of significance when the productivity of somatic hybrids and cybrids is being assessed. These points are discussed later.

USE OF SOMATIC HYBRIDS IN AGRICULTURE

NUCLEAR HYBRIDS

As was noted earlier, widely divergent combinations are unlikely to be of great significance in plant breeding due to their genetic instability. However, most breeding programmes

do not require hybridization with distant genera, but rather some characters of a related species to be introduced into a cultivated species or variety. For integression of foreign genes into a cultivated genome it is necessary for the hybrids to possess some fertility. This aspect of somatic hybrids has not yet received a detailed analysis although some attention has recently been given to the problem.

It has been proposed by Pental and Cocking (1985) that the synthesis of triploid plants may be valuable for introgression of genes into a cultivated genome from a wild species. In their model a haploid genome of the wild species is fused with a diploid genome of the cultivated species. At meiosis most of the genome of the wild species would be lost, due to lack of chromosome pairing, but some trivalents might be formed if sufficient homology exists between some chromosomes of the wild species and the cultivated species. Introgression of alien genes might then occur and be recovered after selfing and backcrossing of the projeny of the somatic hybrid. Progress towards the testing of this model has been made recently by Pirrie and Power (1986) who fused haploid gametic protoplasts of _Nicotiana glutinosa_ with diploid leaf mesophyll protoplasts of _N. tabacum_ and recovered several triploid plants. These plants were reported to possess high levels of self fertility.

Homology between chromosomes of different species, in somatic hybrids, may lead to non-disjunction at meiosis with subsequent reduced fertility and aneuploidy in the offspring. These problems are well known in autotetraploids (Doyle, 1986). However, it is possible that aneuploidy, which is observed in many somatic hybrids, will influence the fertility of a somatic hybrid combination as homeolgous chromosomes may sometimes not be present and thus will not cause problems at meiosis. Often, limited numbers of interspecific somatic hybrids are available and thus firm conclusions can not be drawn regarding the potential fertility of the hybrid combination. We found that of 17 somatic hybrids produced between _N. rustica_ and _N. tabacum_ (all analysed were aneuploid), only three possessed any self-fertility and even then it was low (Hamill *et al.*, 1985). However, we did find that the self-fertility of this hybrid combination could be substantially increased by selection over successive generations, without any loss of hybrid characteristics among the offspring. Thus it may be premature to make statements upon the fertility of a somatic hybrid combination based upon only a small number of hybrids.

Other factors, besides aneuploidy and homology between chromosomes may be important regarding somatic hybrid fertility. Ploidy levels, of the parent species within a somatic hybrid, may also affect fertility. For example, Ehlenfeldt and Helgeson (1987) have evaluated two sets of somatic hybrids between the sexually incompatible combination *Solanum brevidens* and *Solanum tuberosum* (potato). One set was tetraploid (*S. brevidens* 2x=2n=24 and *S. tuberosum* 2x=2n=24) and the other was hexaploid (*S. brevidens* 2x=2n=24 and *S. tuberosum* 4x=4n=48). These authors found the hexploid plants possessed greater fertility that the tetraploids when backcrossed to *S. tuberosum*. They suggested that the ratio of maternal to paternal chromosomes (Endosperm Balance Number) was important in allowing the endosperm of the offspring of the backcrossed somatic hybrid to develop. This factor is known to be important in the sexual crossing of *Solanum* species (Johnston and Hanneman, 1982) and may be important in determining the fertility of somatic hybrids.

In fusions between *Lycopersicon esculentum* (diploid mesophyll protoplasts) and *Lycopersicon peruvianum* (tetraploid cell suspension protoplasts), hexaploid somatic hybrid plants were recovered which were self-fertile (Kinsara *et al.*, 1986). This situation contrasts with sexual hybrids between the two species which are self-sterile. It is possible that the Endosperm Balance Number hypothesis, noted above, in addition to the extent of homology between chromosomes from the different species, may be important regarding the self-fertility of this somatic hybrid combination in contrast to the self-sterility of the sexual hybrids. However, this proposal requires testing.

It is clear therefore that the subject of fertility of somatic hybrids requires more intensive research. With the successful recovery of somatic hybrids by FACS it may be possible, in future, to generate hundreds or possibly thousands of somatic hybrids, between two species. Thus the factors affecting somatic hybrid fertility can be determined with more certainty than is possible at present.

Although somatic hybrids with self-fertility may hasten the introgression of characters from a somatic hybrid into a crop species, self-fertility of somatic hybrids it is not an absolute prerequisite for the use of a plant in a breeding programme. Backcrossing of *N. tabacum* and *N. nesophilia* somatic hybrids, with limited fertility, to the *N. tabacum* parent has been carried out with the aim of transferring disease resistance into tobacco (Evans *et al.*, 1981).

In Canada, *N. rustica* and *N. tabacum* somatic hybrids have been used in a tobacco breeding programme. Backcrossing of somatic hybrids with a high nicotine content (with little or no self-fertility) to a commercial cultivar of *N. tabacum* produced, after several backcrossings, acceptable levels of seed upon selfing (Pandeya *et al.*, 1986). These authors have reported the recovery of offspring possessing significantly elevated levels of nicotine compared to the parent species. In addition, some of these lines possessed immunity or high levels of resistance to blue mold (a serious disease caused by *Peronospora tabacina* and one which neither parent was resistant to) and also black root rot resistance (caused by *Thielaviopsis basicola*) which was inherited from the *N. rustica* parent. These results demonstrate that somatic hybridization can be of great benefit when combined with a breeding programme. Somatic hybrids between the sexually incompatible species *Solanum tuberosum* (potato) and *S. brevidans* have been reported and have been found to possess resistance to potato leaf roll virus and late blight (Austin *et al.*, 1985; Helgeson *et al.*, 1986). As has been noted, studies on the fertility of these hybrids have been carried out to identify those individuals most suited to be included in a conventional breeding programme (Ehlenfeldt and Helgeson, 1987).

Organelle exchange, via protoplast fusion, is also proving to be valuable in agriculture. As was discussed previously, inheritence of chloroplasts and mitochondria may occur independently in somatic hybrid cells. Pelletier _et al._ (1983) combined an atrazine resistant chloroplast (from _Brassica campestris_), and a cytoplasmic male sterility mitochondrial trait (from _Raphanus sativus_) with the nucleus of _Brassica napus_. These novel plants were obtained via two steps. The first involved protoplast fusion of _B. napus_ and _R. sativus_ and ultimately selection of plants resembling _B. napus_ but possessing male sterility. The second step involved protoplast fusion between the cytoplasmic male sterile _B. napus_ and _B. campestris_ with subsequent selection of atrazine resistant _B. napus_ plants.

Fusion of gamma irradiated (to cause nuclear instability) protoplasts of triazine resistant (chloroplast encoded resistance) _Brassica napus_ were fused with a cytoplasmic male sterile (mitochondrial encoded) variety and heterokaryons cultured in an albino nurse culture of _N. tabacum_ (Yarrow _et al._, 1986). Whole plants were ultimately recovered which were triazine resistant and possessed cytoplasmic male sterility. This organelle combination could not be synthesized by conventional breeding. In other experiments a terbutryn resistant plastid was transferred from a resistant mutant of _Nicotiana plumbaginifolia_ to _N. tabacum_. Plants, resistant to the herbicide, were recovered after regeneration (Menczel _et al._, 1986).

In similar experiments, but with different aims, the cytoplasm of _N. debenyi_ was transferred to _N. tabacum_ by protoplast fusion to induce male sterility (Kumashiro and Kubo, 1986a). Agronomic evaluation of the male sterile cybrids showed they were suitable for incorporation into a practical breeding programme (Kumashiro and Kubo, 1986b). Such transfers are possible sexually, using _N. debenyi_ as the female parent, but repeated backcrossing to the N. tabacum parent over many generations is required to eliminate the N. debenyi nuclear genes. Protoplast fusion offers a method of transfer in one step.

Thus, somatic hybridization represents a range of techniques which are available to the plant breeder as part of an overall breeding strategy. Of course, in future, somatic hybridization, as a gene transfer technqiue will have to compete with novel sexual hybridization techniques on the one hand (such as ovule or embroyo culture, e.g. Iwai et al., 1986; Belivanis and Doré, 1986) and Agrobacterium mediated or protoplast mediated gene transfer on the other (Fraley et al., 1983; Bevan, 1984; Paszkowski et al., 1984;). However, it is clear that valuable agricultural applications of protoplast fusion are possible when biotechnologists and plant breeders work in close co-operation to determine the desired combinations and how best to achieve them. In addition, novel techniques such as ovule culture may be valuable for introgression of genes from a somatic hybrid into a cultivated plant species. Alternatively, Agrobacterium mediated gene transfer techniques may be used to transfer drug resistant markers into a species so enabling somatic hybrid selection regimes to be devised (Ye et al., 1987; Brunold et al., 1987).

In the field of secondary products, somatic hybrids may also have a novel biotechnological application. The elevated chromosome number of somatic hybrids has been a disadvantage as cell suspensions (normally used for secondary product study) are genetically unstable (D'Amato, 1985). However, hairy roots, caused by Agrobacterium rhizogenes, have recently been shown to be a genetically and biochemically stable tissue for the in vitro study and biotechnological exploitation of plant secondary products (Hamill et al., 1986; Rhodes et al., 1986; Aird et al., 1987; Hamill et al - this conference). As was demonstrated by Keller et al., 1982, somatic hybrids may, possibly due to their novel genetic constitution, synthesise greater levels of secondary products than their parental species. It may be possible to exploit this synthetic potential using hairy root technology.

Thus it is clear that somatic hybridization will continue to play an important and useful role in plant biotechnology, especially if it is not regarded as an isolated technique which is useful only to combine sexually incompatible species.

References

Abduallah R, Cocking EC, Thompson JA (1986) Efficient plant regeneration from rice protoplasts through somatic embryogenesis. Biotechnology 4: 1087-1090

Afonso CL, Harkins KR, Thomas-Compton MA, Krejci AE, Gabraith DW (1985) Selection of somatic hybrid plants in Nicotiana through fluorescence activated cell sorting. Biotechnol 3: 811-816

Aird ELH, Hamill JD, Rhodes MJC (1987) Cytogenetic analysis of a number of plant species transformed by Agrobacterium rhizogenes. Theor Appl Genet (submitted)

Austin S, Baer MA, Helgeson JP (1985) Transfer of resistance to potato leaf roll virus from Solanum brevidens into Solanum tuberosum by protoplast fusion. Plant Sci 39: 75-82

Belivanis T, Dore C (1986) Interspecific hybridization of Phaseolus vulgaris L. and Phaseolus argustissimus A. Gray using in vitro embryo culture. Plant Cell Rep 5: 329-331

Belliard G, Vedel F, Pelletier G (1979) Mitochondrial recombination in cytoplasmic hybrids of Nicotiana tabacum by protoplast fusion. Nature 281: 401-403

Bevan M (1984) Binary Agrobacterium vectors for plant transformation. Nuc Acid Res 12: 8711-8721

Bonnett HT, Glimeluis K (1983) Somatic hybridization in Nicotiana: behaviour of organelles after fusion of protoplasts from male fertile + male sterile cultivars. Theor App Genet 65: 213-217

Brunold C, Kruger-Lebus S, Saul M, Wegmuller S, Potrykus (1987) Combination of kanamycin resistance and nitrate reductase deficiency as selectable markers in one nuclear genome provides a universal somatic hybridizer in plants. Mol Gen Genet 208: 469-473

Carlson P, Smith HH, Dearing RD (1972) Parasexual interspecific plant hybridization. Proc Natl Acad Sci USA, 69: 2292-2294

Chien YC, Kao KN, Wetter LR (1982) Chromosomal and isozyme studies of Nicotiana tabacum - Glycine max hybrid cell lines. Theor Appl Genet 62: 301-304

Chen K, Wildman SG, Smith HH (1977) Chloroplasts DNA distribution in parasexual hybrids as shown by polypeptide composition of fraction 1 protein. 74: 5109-5112

Clark EM, Izhar S, Hanson MR (1985) Independent segregation of the plastid genome and cytoplasmic male sterility in Petunia Somatic hybrids. 199: 440-445

D'Amato F (1985) Cytogenetics of plant cell and tissue cultures and their regenerants. CRC Crit Rev Plant Sci 3: 73-112

Douglas GC, Keller WA, Setterfield G (1981) Somatic hybridization between Nicotiana rustica and N. tabacum. Can J Bot 59: 220-227

Doyle GG (1986) Aneuploidy and inbreeding depression in random mating and self-fertilizing autotetraploid populations. Theor Appl Genet 72: 799-806

Dudits D, Kao KN, Constabel F, Gamborg OL (1976) Fusion of carrot and barley protoplasts and division of heterokaryocytes. Can J Genet Cytol 18: 263-269

Ehlenfeldt MK, Helgeson JP (1987) Fertility of somatic hybrids from protoplast fusions of Solanum brevidens and S. tuberosum. Theor Appl Genet 73: 395-402

Fraley RT, Rogers SG, Horsch RB, Sanders P, Flick R, Adams JS, Bittner ML, Brand LA, Fink CL, Fry JS, Galluppi GR, Goldberg SB, Hoffmann NL, Woo SC (1983) Expression of bacterial genes in plant cells. Proc Natl Acad Sci USA, 80: 4803-4807

Galbraith DW, Mauch TJ (1980) Identification of fusion of plant protoplasts II: Conditions for the reproducible fluorescence labelling of protoplasts derived from mesophyll tissue. Z Pflanzenphysiol 98: 129-140

Galun E, Arzee-Gonen P, Fluhr R, Edelman M, Aviv D (1982) Cytoplasmic hybridization in Nicotiana. Mol Gen Genet 186: 50-56

Gilmour DM, Davey MR, Cocking EC, Pental D (1987) Culture of low numbers of forage legume protoplasts in membrane chambers. J Plant Physiol 126: 457-465

Gleba YY, Hoffmann F (1980) "Arabidobrassica" - A novel plant obtained by protoplast fusion. Planta 149: 112-117

Gleba YY, Kanevsky IF, Skarzhynskaya MV, Komarnitsky IK, Cherep NN (1986) Hybrids between tobacco crown gall cells and normal somatic cells of Atropa belladonna. Plant Cell Rep 5: 394-397

Glimelius K, Bonnett HT (1986) Nicotiana cybrids with Petunia chloroplasts. Theor Appl Genet 72: 794-798

Glimelius K, Eriksson T, Grafe R, Muller AJ (1978) Somatic hybridization of nitrate reductase deficient mutants of Nicotiana tabacum by protoplast fusion. Physiol Plant 44: 273-277

Hamill JD, Pental D, Cocking EC, Müller AJ (1983) Production of a nitrate reductase deficient streptomycin resistant mutant of Nicotiana tabacum for somatic hybridization studies. Heredity 50: 197-200

Hamill JD, Pental D, Cocking EC (1984) The combination of a nitrate reductase deficient nuclear genome with a streptomycin resistant chloroplast genome, in Nicotiana tabacum, by protoplast fusion. J Plant Physiol 115: 253-261

Hamill JD, Patnaik G, Pental D, Cocking EC (1984) The culture of manually isolated heterokaryons of Nicotiana tabacum and Nicotiana rustica. Proc Ind Acad Sci (Plant Sci) 93: 317-327

Hamill JD, Pental D, Cocking EC (1985) Analysis of fertility in somatic hybrids of Nicotiana rustica and N. tabacum and projeny over two sexual generations. Theor Appl Genet 71: 486-490

Hamill JD, Parr AJ, Robins RJ, Rhodes MJC (1986) Secondary product formation by cultures of Beta vulgaris and Nicotiana rustica transformed with Agrobacterium rhizogenes. Plant Cell Reports 5: 111-114

Hamill JD, King JM, Watts JW (1987) Somatic hybridization between Nicotiana tabacum and Nicotiana plumbaginifolia by electrofusion of mesophyll protoplasts. J Plant Physiol 129: 111-118

Handley LW, Nickels RL, Cameron MW, Moore PP, Sink KC (1986) Somatic hybrid plants between Lycopersicon esculentum and Solanum lycopersicoides. Theor Appl Genet 71: 691-697

Hein T, Przewozny T, Schieder O (1983) Culture and selection of somatic hybrids using an auxotrophic cell line. Theor Appl Genet 64: 119-122

Helgeson JP, Hunt GJ, Haberlach GT, Austin S (1986) Somatic hybrids between Solanum brevidens and Solanum tuberosum: Expression of a late blight resistance gene and potato leaf roll resistance. Plant Cell Rep 5: 212-214

Iwai S, Kishi C, Nakata K, Kawashima N (1986) Production of Nicotiana tabacum x Nicotiana acuminata hybrids by ovule culture. Plant Cell Rep 5: 403-404

Johnston SA, Hanneman RE (1982) Manipulations of endosperm balance number overcome crossing barriers between diploid Solanum species. Science 217: 446-448

Kanchanapoom K, Boss WF (1986) The effect of fluorescent labeling on calcium induced fusion of fusogenic carrot protoplasts. Plant Cell Rep 5: 252-255

Kao KN (1977) Chromosomal behaviour in somatic hybrids of soybean - Nicotiana glauca. Mol Gen Genet 150: 225-230

Kao KN, Michayluk MR (1974) A method for high frequency intergeneric fusion of plant protoplasts. Planta 115: 355-367

Kao KN, Saleem M (1986) Improved fusion of mesophyll and cotyledon protoplasts with PEG and high pH Ca^{++} solutions. J Plant Physiol 122: 217-225

Keller WA, Melchers G (1973) The effect of high pH and calcium on tobacco leaf protoplast fusion. Z Naturforsch 28C: 737-741

Keller WA, Setterfield G, Douglas GC, Gleddie S, Nakamura C (1982) Production, characterization and utilization of somatic hybrids of higher plants, In Tomes DT, Ellis BE, Harney PM, Kasha KJ and Peterson RL (eds) Application of Plant Cell and Tissue culture to Agriculture and Industry, Guelph 81-114

Kemble RJ, Barsby TL, Wong RSC, Shepard JF (1986) Mitochondrial DNA rearrangements in somatic hybrids of Solanum tuberosum and Solanum brevidens. Theor Appl Genet 72: 787-793

Kinsara A, Patnaik SN, Cocking EC, Power JB (1986) Somatic hybrid plants of Lycopersicon esculentum Mill. and Lycopersicon peruvianum Mill. J Plant Physiol 125: 225-234

Kohn H, Schieder R, Schieder O (1985) Somatic hybrids in tobacco mediated by electrofusion. Plant Sci 38: 121-128

Krumbiegel G, Schieder O (1979) Selection of somatic hybrids after fusion of protoplats from Datura innoxia and Atropa belladonna. Planta 145: 371-375

Kumashiro T, Kubo T (1986)a Cytoplasm transfer of Nicotiana deberyii to N. tabacum by protoplast fusion. Jap J Breed 36: 39-48

Kumashiro T, Kubo T (1986)b Stability of agronomic traits in cytoplasmic male sterile tobacco obtained by protoplast fusion. Jap J Breed 36: 284-290

Maliga P, Sz-Breznovits A, Marton L (1973) Streptomycin resistant plant from callus culture of haploid tobacco. Nature New Biol 244: 29-30

Medgyesy P, Fejes E, Maliga P (1985) Interspecific chloroplast recombination in a Nicotiana somatic hybrid. Proc Natl Acad Sci 82: 6960-6964

Melchers G, Sacristan MD, Holder AA (1978) Somatic hybrid plants of potato and tomato regenerated from fused protoplasts. Carlsberg Res Commun 43: 203-218

Menczel L, Galiba G, Nagy F, Maliga P (1982) Effects of radiation dosage on efficiency of chloroplast transfer by protoplast fusion in Nicotiana. Genetics 100: 487-495

Menczel L, Polsby LS, Steinback KE, Maliga P (1986) Fusion mediated transfer of triazine resistant chloroplasts: Characterization of Nicotiana tabacum cybrid plants. Mol Gen Genet 205: 201-205

Muller AJ (1983) Genetic analysis of nitrate reductase deficient tobacco plants regenerated from mutant cells. Evidence for duplicate structured genes. Mol Gen Genet 192: 253-261

Negrutiu I, DeBrouwer D, Watts JW, Sidorov VI, Dirks R, Jacobs M (1986) Fusion of plant protoplasts: a study using auxotrophic mutants of Nicotiana plumbaginifolia, viviani. Theor Appl Genet 72: 279-286

Nagy F, Torok I, Maliga P (1981) Extensive rearrangements in the mitochondrail DNA in somatic hybrids of Nicotiana tabacum and Nicotiana knightiana. Mol Gen Genet 183: 437-439

Pandeya RS, Douglas GC, Keller WA, Setterfield G, Patrick ZA (1986) Somatic hybridization between Nicotiana rustica and N. tabacum: Development of tobacco breeding strains with disease resistance and elevated nicotine content. Z Pflanzenzuchtg 96: 346-352

Patnaik G, Cocking EC, Hamill JD, Pental D (1982) A simple procedure for the manual isolation and identification of plant heterokaryons. Plant Sci Lett 24: 105-110

Paszkowski J, Shillito RD, Saul M, Mandak V, Hohn T, Hohn B, Potrykus I (1984) Direct gene transfer to plants EMBO J 3: 2717-2722

Pelletier G, Chupeau Y (1984) Plant protoplast fusion and somatic cell genetics. Physiol Veg 22: 377-399

Pelletier G, Primard C, Vedel F, Chetrit P, Remy R, Rousselle P, Renard M (1983) Intergeneric cytoplasmic hybridization in Cruciferae by protoplast fusion. Mol Gen Genet 191: 244-250

Pental D, Cocking EC (1985) Some theoretical and practical possibilities of plant genetic manipulation using plant protoplasts. Hereditas Suppl 3: 83-92

Pental D, Hamill JD, Cocking EC (1984) Somatic hybridization using a double mutant of Nicotiana tabacum. Heredity 53: 79-83

Pental D, Hamill JD, Pirrie A, Cocking EC (1986) Somatic hybridization of Nicotiana tabacum and Petunia hybrida. Mol Gen Genet 202: 342-347

Pirrie A, Power JB (1986) The production of fertile, triploid somatic hybrid plants (Nicotiana glutinosa (n) + N. tabacum (2n)) via gametic somatic protoplast fusion. Theor Appl Genet 72: 48-52

Rhodes MJC, Robins RJ, Hamill JD, Parr AJ (1986) Potential for the production of biochemicals by plant cell cultures. N Zealand J Technol 2: 59-70

Schiller B, Herrmann RG, Melchers G (1982) Restriction endonuclease analysis of plastid DNA from tomato, potato and some of their somatic hybrids. Mol Gen Genet 186: 453-459

Scowcroft WR, Larkin PJ (1981) Chloroplast DNA assorts randomly in intraspecific somatic hybrids of Nicotiana debenyii. Theor Appl Genet 60: 179-184

Shepard JF, Bidney D, Barsby T, Kemble R (1983) Genetic transfer in plants through interspecific protoplast fusion. Science 219: 683-688

Sundberg F, Glimelius K (1986) A method for production of interspecific hybrids within Brassiceae via somatic hybridization, using resynthesis of Brassica napus as a model. Plant Sci 43: 155-162

Tabaeizadeh Z, Perennes C, Bergounioux C (1985) Increasing the variability of Lycopersican peruvianum Mill. by protoplast fusion with Petunia hybrida L.. Plant Cell Rep 4: 7-11

Thompson JA, Abdullah R, Cocking EC (1986) Protoplast culture of rice (Oryza sativa) using media solidified with agarose. Plant Sci 47: 123-133

Vienken J, Ganser R, Haupp R, Zimmermann U (1981) Electric field induced fusion of isolated vacuoles and protoplasts of different developmental and metabolic provenience. Physiol Plant 53: 64-70

Wallin A, Glimelius K, Eriksson T (1974) The induction of aggregation and fusion of Daucus carota protoplasts by polyethylene glycol. Z Pflanzenphysiol 74: 64-80

Ward M, Davey MR, Cocking EC, Balls M, Clothier RH, Lucy JA (1980) Animal specific membrane components visualized on the surface of Animal/Plant heterokaryons. Protoplasma 104: 75-80

Watts JW, King JM (1984) A simple method for large scale electrofusion and culture of plant protoplasts. Bioscience Rep 4: 335-342

Wetter LR, Kao KN (1980) Chromosomes and isoenzyme studies on cells derived from protoplast fusion of *Nicotiana glauca* with *Glycine max* - *Nicotiana glauca* cell hybrids. Theor Appl Genet 57: 273-276

White DWR, Vasil IK (1979) Use of amino acid analogue resistant cell lines for selection of *Nicotiana sylvestris* somatic cell hybrids. Theor Appl Genet 55: 107-112

Watts JW, Doonan J, Cove DJ, King JM (1985) Production of somatic hybrids of moss by electrofusion. Mol Gen Genet 199: 349-351

Yarrow SA, Wu SC, Barsby TL, Kemble RJ, Shepard JF The introduction of CMS mitochondria to triazine tolerant *Brassica napus* L. var Regent by micromanipulation of individual heterokaryons. Plant Cell Rep 5: 415-418

Yamada Y, Yank ZQ, Tang DT (1986) Plant regeneration from protoplast derived callus of rice (*Oryza sativa* L.). Plant Cell Rep 5: 85-88

Ye J, Hauptmann RM, Smith AG, Widholm JM (1987) Selection of a *Nicotiana plumbaginifolia* universal hybridizer and its use in intergeneric somatic hybrid formation. Mol Gen Genet 208: 474-480

Zelcer A, Aviv D, Galun E (1978) Interspecific transfer of cytoplasmic male sterility by fusion between protoplasts of normal *Nicotiana sylvestris* and X-ray irradiated protoplasts of male sterile *N. tabacum*. Z Pflanzenphysiol 90: 397-407

Zenkteler M, Melchers G (1978) *In vitro* hybridization by sexual methods and by fusion of protoplasts. Theor Appl Genet 52: 81-90

Zimmermann U, Scheurich P (1981) High frequency fusion of plant protoplasts by electrical fields. Planta 151: 26-32

BIOTECHNOLOGY FOR THE IMPROVEMENT OF CEREAL AND OTHER GRASS CROPS

Indra K. Vasil
Laboratory of Plant Cell and Molecular Biology
Institute of Food and Agricultural Sciences
University of Florida
Gainesville, Florida 32611
United States of America

ABSTRACT

This paper briefly summaries the research carried out in the author's laboratory on the cell culture of gramineous species, and highlights the importance of the developmental and physiological state of the explant, the role of endogenous plant growth regulators in the control of morphogenic competence of the explants, and the early recognition and selection of competent cells in obtaining efficient regeneration of plants by somatic embryogenesis. The significance of embryogenic cell cultures in the recovery of normal plants, and in the establishment of regenerable cell suspension and protoplast systems is discussed. The successful use of gramineous protoplasts in obtaining somatic hybrids, and genetic transformation is also described.

KEY WORDS: Cereals and Grasses, Genetic Transformation, Gramineae, Plant Growth Regulators, Somatic Embryogenesis, Somatic Hybrids.

The use of physiologically suitable explants, excised and cultured at precise developmental stages, has helped develop reliable and efficient procedures for the regeneration of plants from tissue cultures of all of the important species of cereals and grasses (for survey of lietrature see IK Vasil and Vasil 1986; Morrish et al 1987; IK Vasil 1987). Early recognition and selective maintenance of embryogenic cultures has been

NATO ASI Series, Vol. H18
Plant Cell Biotechnology. Edited by M. S. S. Pais et al.

found to be critical in retaining long term competence for somatic embryogenesis and plant regeneration. Immature embryos, young inflorescences and bases of young leaves have proved to be most suitable for obtaining regenerable cultures. There is rapid and progressive loss of morphogenic competence during plant development and cellular differentiation so that mature and differentiated tissues are incapable of forming morphogenic cultures. This loss of competence is not caused by any loss or degradation of nuclear DNA (Taylor and Vasil 1987). Rather, developmentally young explants, which are embryogenically competent and comprised of meristematic and undifferentiated cells, characteristically contain higher amounts of endogenous plant growth regulators such as abscisic acid and indole-acetic acid, than older explants which have lost embryogenic competence (Rajasekaran et al 1987a,b).

In cell and tissue cultures of the Gramineae, plant regeneration generally takes place by the formation of discrete somatic embryos. In some instances development of preexisting buds as well as de novo formation of shoot meristems have been described. The somatic embryos arise - directly or indirectly - from single cells (V Vasil and Vasil 1982; Ho and Vasil 1983; Lu and Vasil 1985; V Vasil et al 1985). Cytologically the embryogenic cultures are relatively stable, and there is strong selection in favor of cytologically normal cells at the time of formation of somatic embryos (Hanna et al 1984; Swedlund and Vasil 1985; Karlsson and Vasil 1986; Rajasekaran et al 1986). Consequently, the plants recovered from embryogenic cultures are non-chimeral and cytlogically normal. Since extensive efforts to induce cell divisions in mesophyll protoplasts of grasses have not been successful, considerable attention has been directed toward the use of fast-growing, finely dispersed, embryogenic cell suspension cultures as a source of dividing and totipotent protoplasts. This approach has led to the recovery of somatic embryos (V Vasil and Vasil 1987), plantlets (V Vasil and Vasil 1980; Lu et al 1981; V Vasil et al 1983) and mature plants (Srinivasan and Vasil 1986) from cultured protoplasts. Similar cell lines have been used to obtain somatic hybrids of _Pennisetum americanum_ + _Panicum maximum_ (Ozias-Akins et al 1986), _P. americanum_ + _Saccharum officinarum_ (Tabaeizadeh et al 1986) and _P. americanum_ + _Triticum monococcum_ (V Vasil et al 1987). Examination of the mitochondrial DNA of several somatic hybrids has provided evidence of rearrangements in the mitochondrial genome indicating intergeneric recombination and amplification (Ozias-Akins et al 1987a,b; Tabaeizadeh et et 1987).

Direct delivery of DNA was achieved by electroporation and confirmed by transient expression assays in protoplasts isolated from cell lines of several gramineous species (Hauptmann et al 1987a). Furthermore, stably transformed cell lines of Panicum maximum and Triticum monococcum were recovered following delivery of appropriate resistance vectors by electroporation (Hauptmann et al 1987b). Selection was based on resistance to either kanamycin, hygromycin or methotrexate.

ACKNOWLEDGEMENTS

The research described in this report was carried out in collaboration with a number of graduate students, post-doctoral associates and colleagues. Supported by the Monsanto Company (St. Louis, MO) and by a cooperative agreement between the Gas Research Institute (Chicago, IL) and the Institute of Food and Agricultural Sciences, University of Florida.

REFERENCES

Hanna WW, Lu C and Vasil IK (1984) Uniformity of plants regenerated from somatic embryos of Panicum maximum Jacq. (Guines grass). Theoret Appl Genet 67: 155-159

Hauptmann RM, Ozias-Akins P, Vasil V, Tabaeizadeh Z, Rogers SG, Horsch RB, Vasil IK and Fraley RT (1987a) Transient expression of electroporated DNA in monocotyledonous and dicotyledonous species. Plant Cell Rep 6: 265-270

Hauptmann RM, Vasil V, Ozias-Akins P, Tabaeizadeh Z, Rogers SG, Fraley RT, Horsch RB and Vasil IK (1987b) Evaluation of selectable markers for obtaining stable transformants in the Gramineae. Plant Physiol (In Press)

Ho W and Vasil IK (1983) Somatic embryogenesis in sugarcane (Saccharum officinarum L.). I. The morphology and physiology of callus formation and the ontogeny of somatic embryos. Protoplasma 118: 169-180

Karlsson SB and Vasil IK (1986) Growth, cytology and flow cytometry of embryogenic cell suspension cultures of Panicum maximum Jacq. (Guinea grass) and Pennisetum purpureum Schum. (Napier grass). J Plant

Physiol 123: 211-227

Lu C and Vasil IK (1985) Histology of somatic embryogenesis in Panicum maximum Jacq. (Guinea grass). Amer J Bot 72: 1908-1913

Lu C, Vasil, V and Vasil IK (1981) Isolation and culture of protoplasts of Panicum maximum Jacq. (Guinea grass): somatic embryogenesis and plantlet formation. Z Pflanzenphysiol 104: 311-318

Morrish F, Vasil V and Vasil IK (1987) Developmental morphogenesis and genetic manipulation in tissue and cell cultures of the Gramineae. Adv Genet 25: 431-499

Ozias-Akins P, Ferl RJ, Vasil IK (1986) Somatic hybridization in the Gramineae: Pennisetum americanum (L.) K. Schum. + Panicum maximum Jacq. (Guinea grass). Molec Gen Genet 203: 365-370

Ozias-Akins, P, Pring DR and Vasil IK (1987a) Rearrangements in the mitochondrial genome of somatic hybrid cell lines of Pennisetum americanum (L.) K. Schum. + Panicum maximum Jacq. Theoret Appl Genet 74: 15-20

Ozias-Akins P, Tabaeizadeh Z, Pring DR and Vasil IK (1987b) Mitochondrial DNA amplification in intergeneric somatic hybrids of the Gramineae. Submitted.

Rajasekaran K, Schank SC and Vasil IK (1986) Characterization of biomass production, cytology and phenotypes of plants regenerated from embryogenic callus cultures of Pennisetum americanum x P. purpureum (hybrid Napier grass). Theoret Appl Genet 73: 4-10

Rajasekaran, K, Hein MB, Davis, GC, Carnes MG and Vasil IK (1987a) Endogenous growth regulators in leaves and tissue cultures of Napier grass (Pennisetum purpureum Schum.) J Plant Physiol 130: 13-25

Rajasekaran K, Hein MB and Vasil IK (1987b) Endogenous abscisic acid and indole-3-acetic acid and somatic embryogenesis in cultured leaf explants of Pennisetum purpureum Schum.: effects in vivo and in vitro of glyphosate, fluridone and paclobutrazol. Plant Physdiol 84: 47-51

Srinivasan C and Vasil IK (1986) Plant regeneration from protoplasts of sugarcane (Saccharum officinarum L.). J Plant Physiol 126: 41-48

Swedlund B and Vasil IK (1985) Cytogenetic characterization of embryogenic callus and regenerated plants of Pennisetum americanum (L.) K. Schum. Theoret Appl Genet 69: 575-581

Tabaeizadeh Z and Vasil IK (1986) Somatic hybridization in the Gramineae: Saccharum officinarum L. (sugarcane) and Pennisetum americanum (L.) K. Schum. (pearl millet). Proc Nat Acad Sci USA 83: 5616-5619

Tabaeizadeh Z, Pring DR and Vasil IK (1987) Analysis of mitochondrial DNA from somatic hybrid cell lines of Saccharum officinarum (sugarcane) and Pennisetum americanum (pearl millet). Plant Molec Biol 8: 509-513

Taylor MG and Vasil IK (1987) Analysis of DNA size, content and cell cycle in leaves of Napiergrass (Pennisetum purpureum Schum.). Theoret Appl Genet (In Press)

Vasil IK (1987) Developing cell and tissue culture systems for the improvement of cereal and grass crops. J Plant Physiol 128: 193-218

Vasil IK and Vasil V (1986) Regeneration in cereal and other grass crops. In: Vasil IK (ed). Cell Culture and Somatic Cell Genetics of Plants, Volume 3, Plant Regeneration and Genetic Variability. Academic Press 1986 Orlando:121-150

Vasil V and Vasil IK (1980) Isolation and culture of cereal protoplasts. II. Embryogenesis and plantlet formation from protoplasts of Pennisetum americanum. Theoret Appl Genet 56: 97-99

Vasil V and Vasil IK (1982) The ontogeny of somatic embryos of Pennisetum americanum (L.) K. Schum.: in cultured immature embryos. Bot Gaz 143: 454-465

Vasil V and Vasil IK (1987) Formation of callus and somatic embryos from protoplasts of a commercial hybrid of maize (Zea mays L.). Theoret Appl Genet 73: 793-798

Vasil V, Wang D and Vasil IK (1983) Plant regeneration from protoplasts of Pennisetum purpureum Schum. (Napier grass). Z Pflanzenphysiol 111: 319-325

Vasil V, Lu C and Vasil IK (1985) Histology of somatic embryogenesis in cultured embryos of maize (Zea mays L.). Protoplasma 127: 1-8

Vasil V, Ferl RJ and Vasil IK (1987) Somatic hybridization in the Gramineae: Triticum monococcum L. (Einkorn) + Pennisetum americanum (L.) K. Schum. (Pearl millet). J Plant Physiol (In Press)

CELL CULTURE OF THE POACEAE (GRAMINEAE)

B. V. Conger, R. N. Trigiano,[a] and D. J. Gray[b]
Department of Plant and Soil Science
University of Tennessee
P.O. Box 1071
Knoxville, Tennessee 37901
USA

INTRODUCTION

Plant cell and tissue culture constitutes an important and integral part of modern plant biotechnology including in the areas of application such as mass propagation, mass production of chemicals, and genetic engineering which will be emphasized during the course of this meeting. Most gene transfer systems, either in practice or proposed, rely on the regeneration or recovery of whole plants from single cells. It is well known that the development of in vitro culture systems for the Poaceae (Gramineae) are much less advanced than for certain other plant families, especially, the Solonaceae. However, the regeneration of rice from protoplasts, as reported recently by groups in Japan, France, and Britain represents a long-awaited breakthrough and the conquering of a major barrier in cereal cell culture technology (Coulibaly and Demarly 1986; Fujimura et al 1986; Yamada et al 1986; Abdullah et al 1986).

In terms of providing food for the world's human population, no botanical family approaches the Gramineae. Approximately half of the calories from plants consumed directly by man come from only three species wheat, rice, and maize (Wilkes 1983). If the other major cereals, barley, oats, rye, etc., are included, this proportion increases to about 65%. These

[a]Department of Ornamental Horticulture and Landscape Design, University of Tennessee, P.O. Box 1071, Knoxville, Tennessee 37901, USA

[b]AREC, IFAS, University of Florida, P.O. Box 388, Leesburg, Florida 32749, USA

NATO ASI Series, Vol. H18
Plant Cell Biotechnology. Edited by M.S.S. Pais et al.

figures do not include the indirect contribution of forage grasses through meat and milk products of the ruminant animal which are especially important in developed countries. For example, in cash receipts for agricultural products in the United States, beef cattle is the top commodity in 21 states and ranks among the top 5 in 47 states while dairy cattle and products are the top commodities in 9 states and rank among the top 5 in 39 states (Hodgson 1976). Thus, the vast attention given to species in this family in many research areas, both basic and applied and including biotechnology, is understandable.

Even though recent progress in grass and cereal biotechnology is encouraging, there is a continuing need to advance our abilities to regenerate plants from cultured cells and tissues, especially across a wide range of species and genotypes. In vitro culture systems have been developed for the forage grass _Dactylis glomerata_ L. (orchardgrass or cocksfoot) that are among the most advanced for grasses and cereals. These include: (1) direct embryogenesis, without an intervening callus, from cultured explants, including from mesophyll cells and (2) the development of embryos to a germinable stage in suspension cultures entirely within one medium. This presentation will focus on these systems, particularly on somatic embryo ontogeny from mesophyll cells and the regulation of embryogenesis in suspension cultures, and their relevance in biotechnology of the Gramineae. The extension of experimental protocols to _Zea mays_ will also be briefly mentioned.

REGENERATION

The two primary modes of regeneration from plant cell and tissue cultures are organogenesis and somatic embryogenesis. Reports of regeneration in cereals and grasses prior to 1980 were limited and in most cases probably involved organogenesis. Regeneration frequencies were usually low and decreased further with increasing time in culture and number of subcultures. These characteristics were clearly shown in our early experiments with tall fescue (_Festuca arundinacea_ Schreb.) (Lowe and

Conger 1979). During this period it was also suggested that cells of Gramineae were not totipotent and that plant regeneration in tissue cultures of these species was not de novo but rather a derepression of preexisting shoot primordia which proliferated adventitiously in culture (King et al 1978).

Plant regeneration via somatic embryogenesis has now been reported in numerous cereal and grass species (Conger and Gray 1984; Vasil 1985). In most cases somatic embryos are derived from calli produced from cultured meristematic tissues such as immature embryos, inflorescences, and basal portions of leaves. In *D. glomerata* embryos are produced directly from cells of the explant in cultured leaves (Hanning and Conger 1982; Conger et al 1983) and anthers and pistils (Songstad and Conger 1986). The basic protocols used to induce direct embryogenesis include explanting onto SH medium (Schenk and Hildebrandt 1972) with 30 µM 3,6-dichloro-o-anisic acid (dicamba) and culturing in the dark at 25°C. After embryo development (3-6 weeks), cultures are transferred to SH medium without auxin and placed in the light at 25°C for embryo germination and seedling development. The substitution of 10 µM 2,4-dichlorophenoxyacetic acid for dicamba provides equally good results for somatic embryo induction in leaf cultures (Hanning and Conger 1986).

EMBRYOGENESIS FROM MESOPHYLL CELLS IN *D. GLOMERATA*

The leaf culture system has received primary emphasis in our studies. The basal portions of the two innermost leaves of mature greenhouse plants are split at the midvein, cut transversely into segments approximately 3 mm square, plated serially, and cultured as stated above. After 4-6 weeks in culture the gradient response typical for in vitro culture of grass and cereal leaves and first reported in *Sorghum* seedlings (Wernicke and Brettell 1980) is apparent. However, in addition to the formation of embryogenic callus on the most basal segments, embryos form directly from the leaf tissue in segments beginning 6-9 mm and 3-6 mm from the bases of the innermost and next leaf outward, respectively. An explanation for the gradient response has been provided (Conger et al 1983).

Of primary interest is the cellular origin of embryos, i.e., whether they have a single or multicellular origin. Since the embryos develop without obscuring floral parts and tissues, lemma and palea, endosperm, etc., the process can be monitored at both the macroscopic and microscopic levels. A time lapse photomicrographic and scanning electron microscopy study of somatic embryo development from cultured leaves has been published (Gray and Conger 1985a). More recently we have concentrated on studying embryogenesis at the microscopic level.

Haccius (1978) defines a plant embryo, either zygotic or nonzygotic, as a new individual arising from a single cell and having no vascular connection with maternal tissues. Even those embryos having a broad based attachment to underlying callus tissue are considered to have a single cell origin. In this case the embryogenesis is analogous to that occurring in gymnosperms and certain primitive angiosperms in which embryos are derived from proembryonal cell complexes which in turn originate from single cells.

We have conducted extensive histological analyses on cultured leaf sections of _D. glomerata_. Within one week of culture single cells become densely cytoplasmic and begin to divide. More than 90% of the first divisions are periclinal. Subsequent divisions may be: (1) periclinal producing an embryo with a uniseriate suspensor, (2) both anticlinal and periclinal to produce a multiseriate suspensor, or (3) oblique as well as anticlinal and periclinal to produce a proembryonal cell complex similar to that described by Haccius (1978) and mentioned above. Embryos may arise with either uni or multiseriate suspensors from the complex.

A microscopic examination of 317 cells showed that more than 90% of the initial division sites were from mesophyll cells. Most of these were within one or two cell layers of the epidermis and were nearly equally distributed between the abaxial and adaxial portions. Slightly less than 5% of the initial sites were from cells of the outer bundle sheath and less than 2% were from the epidermis. No other tissues gave rise to initial cell divisions.

POTENTIAL APPLICATIONS OF LEAF CULTURE SYSTEMS

Applications of leaf culture systems, such as developed for D. glomerata, have potential in areas such as mutagenesis and gene transfer. Theoretically mutagen treatment of leaf tissue could induce mutations in single mesophyll cells that could give rise to mutations which would be pure (nonchimeric) for the mutant trait. Mutations induced in heterozygous loci would be expressed immediately.

Gene transfer systems are presently limited for cereals and grasses and most of those proposed rely on direct uptake of DNA by protoplasts. Purified exogenous DNA was shown to be taken up, integrated, and expressed in isolated protoplasts of Lolium multiflorum (Potrykus et al 1985), Triticum monococcum (Lörz et al 1985), and Zea mays (Fromm et al 1986). It was possible to grow the protoplasts into organized calli but plants were not regenerated. The more recent success with rice protoplast culture (mentioned above) adds new hope for this approach in grass and cereal transformation.

The discovery or development of either bacterial or viral vectors for monocot hosts could allow for effective gene transfer to vegetative cells analogous to that in dicot species with the Agrobacterium Ti plasmid system. The leaf-disc transformation systems developed for tobacco, petunia, and tomato are highly efficient and have largely obviated the need for protoplast regeneration (Horsch et al 1985; McCormick et al 1986). Leaf culture systems in cereals and grasses offer a similar potential for gene transfer in those species. Potential virus vectors already exist (Ahlquist et al 1984; Lörz et al 1985) and recently Agrobacterium mediated maize streak virus (MSV) infection was accomplished when whole maize plants were inoculated with strains of the bacterium carrying tandemly repeated copies of MSV genome in their T-DNA (Grimsley et al 1987).

CELL SUSPENSION CULTURES

Reports of somatic embryogenesis from suspension cultures of Gramineae are much fewer than from callus cultures. In most experiments, the cultures are initiated from calli derived from immature embryos, young leaves, or inflorescences. Somatic embryos develop only to a young proembryo stage in liquid and the suspension must be plated onto solid medium for further embryo development and maturation (Green et al 1983; Vasil 1985; Vasil and Vasil 1986). As mentioned above, D. glomerata is the only Gramineae species in which embryo development has progressed beyond the proembryo stage in liquid culture. Embryos develop in a single liquid medium to a stage capable of germination (Gray et al 1984; Gray and Conger 1985b) and therefore represents a system analogous to that which has existed in certain dicot species for many years.

Suspension cultures in D. glomerata are usually initiated by placing highly developed embryos from leaf cultures onto solid SH medium with 30 µM dicamba (SH-30)at 25°C. After 7-10 days and callus initiation, embryos and the proliferated calli are placed in multiwell plates with SH-30 liquid medium containing 0.3% filter sterilized casein hydrolysate. Approximately 5 embryos are placed in 1 ml of medium per well. After 10-14 days, 1 ml of fresh medium is added to each well. When the wells are reasonably full (approximately 3 weeks), the resulting cultures are transferred to flasks containing 5 ml of SH-30 medium. The cultures are progressively transferred to 10, 20, and 30 ml volumes. Cultures are transferred weekly and fully developed embryos result in about 12 weeks.

To study the effects of different treatment regimes on embryogenesis, the cultures are resuspended in 25 ml of SH-30 without casein hydrolysate to inhibit embryogenesis and then transferred weekly for 5 weeks. One day after the final transfer, the contents are homogenized in a blender using high speed pulses (one pulse per second for 30 seconds). Two milliliters of suspension are pipetted into flasks containing 8 ml of SH-30 medium amended with medium components or combinations of components which we wish to study. At this point, the suspension is finely divided and consists of single cells and small groups of cells.

In following embryo development in suspension cultures, we have documented two-cell, four-cell, and multicellular stages to the progression of fully differentiated embryos. To produce embryos to a germinable stage, we found that casein hydrolysate (optimum concentration 3-4 grams per liter) was a necessary medium addendum (Gray et al 1984; Gray and Conger 1985b). More recently, we have discovered that proline-serine combinations of 12.5 to 25.0 mM will also result in full embryo development (Trigiano and Conger 1987). The proline-serine combinations provide the advantage of a defined culture medium. Also, the suspension is composed primarily of small aggregates of embryogenic cells and single cells and the embryos form free of the cell masses. Embryos formimg in SH-30 with casein hydrolysate are often embedded in large cell masses. Embryos forming freely in the medium would be preferable in propagation systems since no physical separation of the embryos from cell masses would be required.

POTENTIAL PROPAGATION OF SOMATIC EMBRYOS

The mass production of fully developed embryos directly in liquid culture and the ability of these to germinate and grow into plants is of potential high significance for naturally outcrossing, seed propagated, perennial crops which do not produce plant parts readily available for vegetative propagation. In the synthetic cultivar development procedures used for crops such as _D. glomerata_, outstanding characters in individual plants are often lost in the 3-4 generations of genetic recombination required for production of seed utilized by the farmer. Direct propagation of outstanding individual or F_1 hybrid plants has the potential for producing very large gains in yield and improvement of other characters.

Undoubtedly, much refinement in the technology is necessary to make it economically feasible for crops of relatively low unit value such as forage crops. Factors that are necessary for mass in vitro propagation include: (1) the production of large numbers of embryos, (2) the synchronization of embryo development, (3) the development of methods to easily package and sow embryos on a large scale, (4) the presence of the embryogenic trait in a wide range of "desirable" species and genotypes, and (5) the maintenance of phenotypic and genotypic stability in plants produced from the embryos. The ability to induce dormancy in embryos may be a useful additional component.

We have produced up to 2000 embryos per 35 ml of suspension (Gray and Conger 1985b). With existing fermentation and bioreactor technology it should be theoretically possible to "scale up" to produce thousands or perhaps even millions of embryos (Styer 1985; Tanaka et al 1983). Working with both carrot and caraway, Ammirato (1983) reported that a abscisic acid (ABA) selectively inhibited certain aspects of embryo development without totally inhibiting growth and development. He was able to inhibit precocious germination and obtain over 90% normal embryos. By using appropriate combinations of dicamba (15.0 to 30.0 μM) and ABA (12.5 to 25.0 μM) we were also able to reduce precocious germination in suspension cultures from approximately 40% to an average of less than 10% (Table 1). Most proposed delivery systems for somatic embryos rely on embryo encapsulation (Redenbaugh et al 1986). Since

TABLE 1. Influence of dicamba (3,6-dichloro-o-anisic acid) and ABA (abscisic acid) on precocious germination in Dactylis glomerata suspension cultures. (Numbers are per 35 ml of liquid)

DICAMBA μM	ABA μM	NUMBER OF GERMINATED EMBRYOS[a]	PERCENT GERMINATED
15.0	0.0	1205 A	42.8
30.0	0.0	1003 A	39.0
15.0	12.5	77 B	6.8
30.0	12.5	160 B	13.6
15.0	25.0	15 B	2.4
30.0	25.0	175 B	14.7

[a]Numbers followed by the same letter are not significantly different at $\alpha = 0.05$ according to Duncan's New Multiple Range Test.

our embryos will readily germinate and produce plantlets, even under nonsterile conditions, the technique for establishing grass seedlings by an automatically-fed "bandoleer" transplanter (Hauser 1983) may be appropriate for our system.

Genotypic differences for in vitro culture response have been reported for numerous plant species, including grasses and cereals (Abe and Futsuhara 1986; Foroughi-Wehr et al 1982; Lazar et al 1983; Maddock et al 1983; Duncan et al 1985; Hodges et al 1986). We previously reported differences in in vitro culture response of D. glomerata between open-pollinated populations of germplasm collections and plants within the cultivar 'Potomac' (Hanning and Conger 1986). More recently, data obtained from crossing embryogenic and nonembryogenic genotypes indicated that the embryogenic trait can be transmitted through both female and male gametes (Lujan 1987). These results are of potential great significance in the utilization of in vitro technology for low or nonresponding genotypes which otherwise possess outstanding agronomic characters. It has been stated that plants regenerated by somatic embryogenesis do not exhibit phenotypic and genotypic abnormalities that are often induced by tissue culture cycles (Vasil 1983). Numerous other published reports are in disagreement with this premise (see Evans and Sharp 1986 for a review). We have not observed chlorophyll deficiency or other morphological abnormalities among our regenerated plants but an extensive cytogenetical or genetical analysis has not been performed.

Recently, we reported that somatic embryos of D. glomerata become quiescent upon desiccation to 13% water (Gray et al 1987). After 3 weeks storage at 23°C, 12% germinated and 4% developed into green plants. Although refinement of the technique and improvement of percentage of plant recovery are needed, the technology offers a potential for embryo dormancy and storage analogous to that which exists for seeds. Embryos could be sown at a time when convenience or conditions are most favorable.

APPLICATION OF PROTOCOLS TO ZEA MAYS

Results of recent experiments indicate that protocols used to induce direct embryogenesis in cultured leaf segments of D. glomerata also have application to Zea mays L. (Conger et al 1987). Basal leaf segments of 3-4 week-old maize seedlings of the inbred lines CHI31, S615, and S7 plated onto SH medium with 30 µM dicamba and incubated in the dark at 25°C produced embryogenic callus and embryos directly from the explant in a gradient response similar to that obtained for D. glomerata. When leaf segments with embryos, or isolated embryos, were transferred to regeneration medium and placed in the light, some of the embryos germinated and produced plants which could be established in soil.

SUMMARY AND CONCLUSIONS

The foregoing has focused on tissue and cell culture of the Gramineae using in vitro systems developed in the forage grass Dactylis glomerata for exemplification. Evidence exists that somatic embryos arising from mesophyll cells of cultured leaves have a single cell origin. The development of somatic embryos to a germinable stage in suspension culture depends on proper concentrations of casein hydrolysate or combinations of the amino acids proline and serine along with the synthetic auxin dicamba in SH medium. The systems have potential applications in other areas of biotechnology such as genetic engineering and mass propagation. Results of recent experiments indicate that protocols used to induce direct embryo formation from cultured explants of D. glomerata also have application to leaf cultures of Zea mays.

ACKNOWLEDGMENTS

Preparation of this manuscript, in Vienna, Austria, was partially supported by a professional leave award to the senior author by the University of Tennessee, Knoxville, Tennessee 37901, USA. The research was supported in part by USDA/CRGO Grants Nos. 82-CRCR-1-1086 and 86-CRCR-1-2215.

REFERENCES

Abdullah R, Cocking EC and Thompson JA (1986) Efficient plant regeneration from rice protoplasts through somatic embryogenesis. Bio/Technology 4: 1087-1090

Abe T and Futsuhara Y (1986) Genotypic variability for callus formation and plant regeneration in rice (*Oryza sativa* L.). Theor Appl Genet 72: 3-10

Ahlquist P, French R, Janda M and Loesch-Fries SL (1984) Multicomponent RNA plant virus infection derived from cloned viral cDNA. Proc Natl Acad Sci U.S.A. 81: 7066-7070

Ammirato PV (1983) The regulation of somatic embryo development in plant cell cultures: suspension culture techniques and hormone requirements. Bio/Technology 1: 68-74

Conger BV and Gray DJ (1984) In vitro culture in forage grass improvement. In: Barker RE and Burson BL (eds). Proc 28th grass breeders work planning conf USDA/ARS: 50-64

Conger BV, Hanning GE, Gray DJ and McDaniel JK (1983) Direct embryogenesis from mesophyll cells of orchardgrass. Science 221: 850-851

Conger BV, Novak FJ, Afza R and Erdelsky K (1987) Somatic embryogenesis from cultured leaf segments of *Zea mays*. Plant Cell Rep (in press)

Coulibaly MY and Demarly Y (1986) Regeneration of plantlets from protoplasts of rice, *Oryza sativa* L. Z Pflanzenzüchtg 96: 79-81

Duncan DR, Williams ME, Zehr BE and Widholm JM (1985) The production of callus capable of plant regeneration from immature embryos of numerous *Zea mays* genotypes. Planta 165: 322-332

Evans DA and Sharp WR (1986) Somaclonal and gametoclonal variation. In: Evans DA, Sharp WR and Amirato PV (eds). Handbook of plant cell culture v.4 Techniques and applications. MacMillan Pub Co New York: 97-132

Foroughi-Wehr B, Friedt W and Wenzel G (1982) On the genetic improvement of androgenetic haploid formation in *Hordeum vulgare* L. Theor Appl Genet 62: 233-239

Fromm ME, Taylor LP and Walbot V (1986) Stable transformation of maize after gene transfer by electroporation. Nature 319: 791-793

Fujimura T, Sakurai M, Akagi H, Negishi T and Hirose A (1986) Regeneration of rice plants from protoplasts. Plant Tissue Culture Lett 2: 74-75

Gray DJ and Conger BV (1985a) Time lapse light photomicrography and scanning electron microscopy of somatic embryo ontogeny from cultured leaves of *Dactylis glomerata* (Gramineae). Trans Am Microsc Soc 104: 395-399

Gray DJ and Conger BV (1985b) Influence of dicamba and casein hydrolysate on somatic embryo number and culture quality in cell suspensions of *Dactylis glomerata* (Gramineae). Plant Cell Tissue Organ Culture 4: 123-133

Gray DJ, Conger BV and Hanning GE (1984) Somatic embryogenesis in suspension and suspension-derived callus cultures of *Dactylis glomerata*. Protoplasma 122: 196-202

Gray DJ, Conger BV and Songstad DD (1987) Desiccated quiescent embryos of orchardgrass for use as synthetic seeds. In Vitro Cellular Devel Biol 23: 29-33

Green CE, Armstrong CL and Anderson PC (1983) Somatic cell genetic systems in corn. In: Downey K, Voellmy RW, Ahmad F and Schultz J (eds). Advances in gene technology: Molecular genetics of plants and animals. Academic Press New York: 147-157

Grimsley N, Hohn T, Davies JW and Hohn B (1987) Agrobacterium-mediated delivery of infectious maize streak virus into maize plants. Nature 325: 177-179

Haccius B (1978) Question of unicellular origin of nonzygotic embryos in callus cultures. Phytomorphology 28: 74-81

Hanning GE and Conger BV (1982) Embryoid and plantlet formation from leaf segments of Dactylis glomerata L. Theor Appl Genet 63: 155-159

Hanning GE and Conger BV (1986) Factors influencng somatic embryogenesis from cultured leaf segments of Dactylis glomerata. J Plant Physiol 123: 23-29

Hauser VL (1983) Grass establishment by bandoleers, transplants and germinated seeds. Trans ASAE 26: 74-78

Hodges TK, Kamo KK, Inbrie CW and Becwar MR (1986) Genotype specificity of somatic embryogenesis and regeneration in maize. Bio/Technology 4: 219-223

Hodgson HJ (1976) Forage crops. Sci Amer 234(2): 61-75

Horsch RB, Fry JE, Hoffman N, Eichholtz D, Rogers SG and Fraley RT (1985) A simple and general method for transferring genes into plants. Science 227: 1229-1231

King PJ, Potrykus I and Thomas E (1978) In vitro genetics of cereals: problems and perspectives. Physiol Veg 16: 381-399

Lazar MD, Collins GB and Vian WE (1983) Genetic and environmental effects on the growth and differentiation of wheat somatic cell cultures. J Hered 74: 353-357

Lörz H, Baker B, Junker B, Matzeit V and Göbel E (1986) Studies on cereal transformation. In: Nuclear techniques and in vitro culture for plant improvement. Intern Atomic Energy Agency Vienna: 505-509

Lörz H, Baker B and Schell J (1985) Gene transfer to cereal cells mediated by protoplast transformation. Molec Gen Genet 199: 178-182

Lowe KW and Conger BV (1979) Root and shoot formation from callus cultures of tall fescue. Crop Sci 19: 397-400

Lujan AM (1987) Transmission of somatic embryogenesis via sexual hybridization in Dactylis glomerata L. M.S. Thesis University of Tennessee Knoxville: 58 p

Maddock SE, Lancaster VA, Rissiott R and Franklin J (1983) Plant regeneration from cultured immature embryos and inflorescences of 25 cultivars of wheat (Triticum aestivum). J Exp Bot 34: 915-926

McCormick S, Niedermeyer J, Fry J, Barnason A, Horsch R and Fraley R (1986) Leaf-disc transformation of cultivated tomato (L. esculentum) using Agrobacteriuim tumefaciens. Plant Cell Rep 5: 81-84

Potrykus I, Saul MW, Petruska J, Paszkowski J and Shillito RD (1985) Direct gene transfer to cells of a graminaeous monocot. Molec Gen Genet 199: 183-188

Redenbaugh K, Paasch BD, Nichol JW, Kossler ME, Viss PR and Walker KW (1986) Somatic seeds: encapsulation of asexual plant embryos. Bio/Technology 4: 797-801

Schenk RU and Hildebrandt AC (1972) Medium and techniques for induction and growth of monocotyledonous and dicotyledonous plant cell cultures. Can J Bot 50: 199-204

Songstad DD and Conger BV (1986) Direct embryogenesis from cultured anthers and pistils of *Dactylis glomerata*. Amer J Bot 73: 989-992

Styer DJ (1985) Bioreactor technology for plant propagation. In: Henke RR, Hughes KW, Constantin MJ and Hollaender A (eds). Tissue culture in forestry and agriculture. Plenum Press New York: 117-130

Tanaka H, Nishijima F, Suwa M and Iwamota T (1983) Rotating drum fermentor for plant suspension cultures. Biotechnol Bioeng 25: 2359-2370

Trigiano RN and Conger BV (1987) Regulation of growth and somatic embryogenesis by proline and serine in suspension cultures of *Dactylis glomerata*. J Plant Physiol (in press)

Vasil IK (1983) Regeneration of plants from single cells of cereals and grasses. In: Lurquin PF and Kleinhofs A (eds). Genetic engineering in eukaryotes. Plenum Press New York: 233-252

Vasil IK (1985) Somatic embryogenesis and its consequences in the Gramineae. In: Henke RR, Hughes KW, Constantin MJ and Hollaender A (eds). Tissue culture in forestry and agriculture. Plenum Press New York: 31-47

Vasil V and Vasil IK (1986) Plant regeneration from friable embryogenic callus and cell suspension cultures of *Zea mays* L. J Plant Physiol 124: 399-408

Wernicke W and Brettell R (1980) Somatic embryogenesis from *Sorghum bicolor* leaves. Nature 287: 138-139

Wilkes G (1983) Current status of crop plant germplasm. CRC Crit Rev Plant Sci 1: 133-181

Yamad Y, Yang Z-Q and Tang D-T (1986) Plant regeneration from protoplast derived callus of rice (*Oryza sativa* L.). Plant Cell Rep 5: 85-88

CALLUS INDUCTION IN ENDOSPERMS FROM MAIZE MUTANTS WITH DIFFERENT IAA CONTENT

Silvana Castelli, Lucia A. Manzocchi and Giuseppe Torti

Istituto Biosintesi Vegetali, C.N.R.
Via Bassini, 15, 20133 Milano, Italy

INTRODUCTION

From the first successful in vitro culture of maize endosperm tissue, reported in 1949 (La Rue 1949), long-term cultures of callus derived from immature endosperms have been established from sweet corn varieties (Straus and La Rue 1954; Tamaoki and Ullstrup 1958), from starchy inbreds varieties (Shannon and Batey 1973) and from several endosperm mutants: ae, su2 (Shannon and Batey 1973), wx (Reddy and Peterson 1977), o2 (Manzocchi and Racchi 1986); endosperm cultures have been demonstrated suitable for metabolic studies, and have been employed for research on cell division and chromosome constitution (Boyer and Shannon 1974), starch synthesis (Chu and Shannon 1975) and anthocyanins biosynthesis (Straus 1959; Racchi 1985).

A large number of studies have dealt with nutritional requirements of callus induction and proliferation (reviewed in Shannon 1982): maize endosperm cultures can be grown in simple media containing salts, sucrose and thiamine (Shannon and Liu 1977). As far as auxins are concerned (which are required for callus initiation in other maize tissues as scutellum: Green et al. 1974), most authors report that auxins are not required either for endosperm callus initiation (Straus 1960) or for subsequent culture proliferation (Tamaoki and Ullstrup 1958; Straus 1960; Shannon and Liu 1977), with the exception of the work of Reddy and Peterson (1977), who report stimulation by 2,4 D in callus induction for the inbred variety R 168. In other cases, added auxins seem to be detrimental to cell proliferation: the high endogenous content of IAA of the endosperm (Bandurski and Shulze 1977) and of endosperm cultures (Stowe et al. 1956)

NATO ASI Series, Vol. H18
Plant Cell Biotechnology. Edited by M. S. S. Pais et al.

have been interpreted as the cause of the insensitivity of endosperm cultures to auxin addition.

RESULTS AND DISCUSSION

We have investigated the initial callus proliferation on a group of viable and defective endosperm mutants from the collection of the Istituto Sperimentale per la Cerealicoltura, Bergamo; these mutations are inherited as monogenic recessive traits and are not allelic (Manzocchi et al. 1980a,b). Although slightly reduced in the overall plant size, they are similar to the normal in ear and leaf shape: the effect of mutations is evident in the reduction in the accumulation of dry matter and protein in the seed (Manzocchi et al. 1980b). Moreover, significant differences in the levels of both free and esterified indoleacetic acid in the endosperm have been described (Torti et al. 1984), while seed germination and plant IAA content appear to be unaffected (Torti et al. 1986).

Table 1 summarizes IAA levels in the endosperms for the mutations examined in this study.

Table 1. Free and bound IAA levels in mature endosperms of de mutants and of the isogenic wild-type inbred B37+ (mg kg^{-1} dry weight; Torti et al. 1984)

Genotype	Free IAA	Bound IAA
B 37 +	1.13 ±0.35	6.72 ±0.50
B 37 de*-B22	2.71 ±0.35	16.16 ±0.24
B 37 de*-B10	0.47 ±0.22	5.32 ±0.41
B 37 de*-B18	0.13 ±0.03	0.03 ±0.01

Plants of the inbred line B37+ and of the isogenic versions of the mutant alleles were grown in the field in 1986 and selfed. Ears were collected 11 days after pollination, dehusked and surface sterilized; the kernel tops were removed with a scalpel, the endosperms were excised and placed on agar medium containing Nitsch salts (Nitsch and Nitsch 1969), 3% sucrose, 0.2% asparagine, 0.4 mg l^{-1} thiamine and the indicated amounts of 2.4 D. One hundred endosperms were plated for each test; callus initiation was scored as the number of proliferating explants 30 days after plating. Results are reported in Table 2.

Table 2. Callus induction from mutants and wild-type maize endosperms (% of proliferating endosperms)

Genotype / Medium:	No auxin	2.4 D (0.5 mg l^{-1})	2.4 D (1 mg l^{-1})
B 37 +	24	47	66
B 37 de*-B22	72	79	74
B 37 de*-B10	45	80	n.d.
B 37 de*-B18	63	77	95

n.d.= not determined

B37 is a genotype showing a very poor endosperm callus induction, which is nevertheless improved by 2.4 D addition. A fairly better induction is observed in high-IAA de*-B22 endosperms: in this case, however, tissue proliferation is not further enhanced by 2.4 D addition. Low-IAA de*-B10 and de*B18 show a stimulation of callus induction by 2,4 D supplementation, in addition to an unexpected high proliferation on auxin-free medium.

Since our mutants are highly isogenic with respect to their residual background, we can infer that the differences observed in cell proliferation can be directly ascribed to different hormone balances, resulting from interactions between the concentration of the growth substances in the culture media and the endogenous levels in the tissues placed in culture. We can conclude that in culture initiation from immature endosperms, initial cell proliferation cannot simply be related to the auxin level (as evidenced by the fairly good proliferation of low-IAA de-*B18 endosperms in hormone-free medium); stimulation by synthetic auxin is nevertheless evident in all cases, with the exception of the high-IAA endosperms.

Endosperm mutants with altered hormone content offer therefore a useful tool for investigating the mechanism of hormone control of cell proliferation.

REFERENCES

Bandurski RS, Shulze A (1977) Concentration of indole-3-acetic acid and its derivatives in plants. Plant Physiol 60:211-213

Boyer CD, Shannon JC (1974) Chromosome constitution and cell division in in vitro cultures of *Zea mays* L. endosperm. In Vitro 9:458-462

Chu LC, Shannon JC (1975) In vitro cultures of maize endosperms. A model for studying in vivo starch biosynthesis. Crop Sci 15:814-819

Green CE, Phillips RL, Kleese RA (1974) Tissue culture of maize (Zea mays L.): initiation, maintenance and organic growth factors. Crop Sci 14:54-58

La Rue CD (1949) Cultures of endosperms of maize. Am J Bot 36: 798

Manzocchi LA, Daminati G, Gentinetta E, Salamini F (1980a) Viable defective endosperm mutants in maize. I. Kernel weight, protein fractions and zein subunits in mature endosperms. Maydica 25:105-116

Manzocchi LA, Daminati MG, Gentinetta E (1980b) Viable defective endosperm mutants in maize. II. Kernel weight, nitrogen content and zein accumulation during endosperm development. Maydica 25:199-210

Manzocchi LA, Racchi ML (1986) Maize endosperm culture: in vitro expression of tissue specific proteins. 5th FESPP Congress, Hamburg Aug. 31-Sept.4 1986, Abstr 1-08

Nitsch JP, Nitsch C (1969) Aploid plants from pollen grains. Science 163:85-87

Racchi ML (1985) Effect of the genes B and Pl on anthocyanin synthesis in maize endosperm culture. Plant Cell Rep 4:184-1987

Reddy AR, Peterson PA (1977) Callus initiation from waxy endosperm of various genotypes of maize. Maydica 22:125-130

Shannon JC (1982) Maize endosperm cultures. In: Sheridan WF (ed) Maize for biological research, Plant Molecular Biology Association, p 397-400

Shannon JC, Batey JW (1973) Inbred and hybrid effect on establishment of in vitro cultures of Zea mays endosperm. Crop Sci 13:491-493

Shannon JC, Liu JW (1977) A simple medium for the growth of maize (Zea mays) endosperm tissue in suspension culture. Physiol Plant 40:285-291

Stowe BB, Thiman KV, Kelford NP (1956) Further studies of some plant indoles and auxins by paper chromatography. Plant Physiol 31:162-165

Straus J (1959) Anthocyanin synthesis in corn endosperm tissue cultures. I. Identitiy of the pigments and general factors. Plant Physiol 34:536-541

Straus J (1960) Maize endosperm grown in vitro. III. Development of a synthetic medium. Am J Bot 47:641-647

Straus J, La Rue CD (1954) Maize endosperm tissue grown in vitro. I. Culture requirements. Am J Bot 41:687-694

Tamaoki T, Ullstrup AJ (1958) Cultivation in vitro of excised endosperm and meristem tissues of corn. Bull Torrey Bot Club 85:260-272

Torti G, Lombardi L, Manzocchi LA, Salamini F (1984) Indole-3-acetic acid content in viable defective endosperm mutants of maize. Maydica 29:335-343

Torti G, Manzocchi L, Salamini F (1986) Free and bound indole acetic acid is low in the endosperm of the maize mutant defective endosperm _B-18_. Theor Appl Genet 72:602-605

SOMATIC EMBRYOGENESIS IN *Hordeum vulgare* AND *Secale vavilovii*

A.M. Vázquez, M. Candela, J. Espino, J. Rueda[1], M.L. Ruiz and M.I. Peláez[2]

[1]Dpto. Genética, Fac. Biología, Univ. Complutense, 28040 Madrid, Spain. [2]Dpto. Genética, Fac. Biología, Univ. León, 24071 León, Spain.

INTRODUCTION

Somatic embryogenesis has been reported in several species of Gramineae. Different explants have been used and immature embryos, inflorescences and leaves have been cultured successfully in some of these species. Here, we report on the obtainment of somatic embryogenesis in *Hordeum vulgare* and *Secale vavilovii*.

MATERIAL AND METHODS

Immature kernels were sterilized by rapid immersion in ethanol (70%) followed by sodium hypochlorite within 20-30 min. After rinsing in distilled, sterile water the embryos were aseptically removed and cultured. Embryos of different sizes were used, from 0.1 to 1.5 mm in length in the case of *Hordeum vulgare*, and from 1 to 2 mm in the case of *Secale vavilovii*.

Young leaves were taken from 1-month-old plants growing in a basic Murashige and Skoog medium. The leaves were cut into 2 mm pieces, and were considered as an explant (Linacero and Vázquez 1986).

In the case of *Secale vavilovii*, immature inflorescences were also cultured. The inflorescences, with the leaves which surrounded them, were surface sterilized in the same manner as immature embryos. Afterwards, the leaves were removed and the inflorescences cut into 2-mm pieces. Each inflorescence was scored as an explant, their sizes ranged from 5 to 32 mm.

The media used were: MS: Murashige and Skoog media supplemented with different concentrations of 2,4-D (from 0.5 to 5 mg l^{-1}). It was used for leaves, immature inflorescences and immature embryos of rye; L3: Murashige and Skoog salts with

NATO ASI Series, Vol. H18
Plant Cell Biotechnology. Edited by M. S. S. Pais et al.

Gamborg B5 vitamins supplemented with 2 mg l^{-1} 2,4-D. It was used in the case of immature embryos of barley and rye.

RESULTS AND DISCUSSION

Hordeum vulgare

In the case of immature embryos the results obtained are shown in Table 1. The most noticeable feature was the differences in the percentage of embryogenic calluses between the cultivars tested. In some of them no embryogenic calluses were formed at all, while others presented a fairly good response. The size of the embryos in some cases seemed to be important but the best size varied depending upon the cultivar used. Thus, for Hellas small embryos appeared to give a better response, but in other cases, as in Ingrid, for example, the embryos with sizes between 1.1 and 1.5 mm proved to be the best.

Table 1. Immature embryo cultures of *Hordeum vulgare*

Size	Cultivar	Germination (%)	Callus formation (%)	Embryogenic calluses (%)
0.1-0.5	Albacete	0	44.35	7.84
	Ingrid	1.61	63.98	33.91
	Hellas	0	53.31	33.33
	Porthos	6.09	65.65	9.27
0.6-1.0	Albacete	26.66	90.25	8.70
	Ingrid	20.17	81.97	27.56
	Hellas	38.81	75.24	30.92
	Porthos	20.78	85.71	13.36
	Beka	21.25	97.50	55.12
	Dissa	30.00	62.00	38.70
1.1-1.5	Albacete	66.66	95.65	6.56
	Ingrid	83.93	94.22	42.86
	Hellas	82.01	97.84	6.62
	Porthos	56.79	86.41	7.14
	Astrix	63.57	0	0
	Almunia	90.00	0	0

The results obtained in the case of leaves are summarized in Table 2. In these cases two factors are implicated in the embryogenic callus formation. First, it was noted that the concentration of 2,4-D affected the response, the best results being obtained in the media with 2 mg l^{-1}. In the 1 mg l^{-1} media the calluses grew smaller and produced roots. The second factor is again the cultivar, different responses being obtained between them. It could be observed that some of the cultivars did not produce calluses. In other cases, even when calluses were produced, they were not embryogenic.

Table 2. Leaf cultures of *Hordeum vulgare*

Medium	Cultivar	Leaves forming calluses (%)	Embryogenic calluses (%)
1 mg l^{-1}	Albacete	0	0
2,4-D	Perfumada	50.00	9.25
	Ingrid	81.25	12.50
	Astrix	0	0
2 mg l^{-1}	Albacete	0	0
2,4-D	Perfumada	80.00	13.33
	Ingrid	91.30	40.47
	Astrix	0	0
	Porthos	50.00	3.85
	Golden Promise	100	31.57
	Dissa	100	60.00

Comparing the results obtained in the case of leaves with those for immature embryos, it can be noted that when a cultivar had a good response in one of the explants the same occurred in the others. As cultivars represent different genotypes we conclude that the genotype is implicated in the "in vitro" behaviour and so, have a strong influence on the frequency of the appearance of embryogenic calluses.

Our data are in agreement with that reported by Goldstein and Kronstad (1986) who observed the same kind of genotypic influence in barley.

Secale vavilovii

The results obtained for leaves and immature embryos are summarized in Tables 3 and 4 respectively.

Table 3. Leaf cultures of Secale vavilovii

	2,4-D (mg l^{-1})	Leaves forming calluses (%)	Embryogenic calluses (%)
Darkness	1	100	4.67
	2	100	4.55
	5	100	4.76
Light	1	96.55	14.29
	2	100	25.64
	5	100	0

In the case of leaves it can be noted that the percentage of embryogenic calluses was enhanced in the cultures maintained in light. The concentration of 2,4-D influenced the embryogenic callus formation, 2 mg l^{-1} being the best. However, the response was low when compared with the ones reported for leaves of Secale cereale (Linacero and Vázquez 1986).

In the case of embryos the response varied depending upon the culture medium and the size of the embryo, attaining around 50% under the best conditions. These frequencies are much higher than those previously reported for the same species (Rybcynsky and Zdunczyk 1986).

The best results were obtained for the immature inflorescences. Independently of the size (from 0.5 to 32 mm) or the 2.4-D concentration (0.5 and 2 mg l^{-1}), all the explants produced embryogenic calluses. However, in the medium with 0.5 mg l^{-1}, the calluses had small developed roots, and it was impossible to maintain them. Embryogenic calluses have been reported to be produced from immature inflorescence of Secale cereale (Krumbiegel-Schroeren et al. 1984), its frequency being low.

Acknowledgements. This research has been supported by grant No. 2093-83 from the Comisión Asesora de Investigación Científica y Técnica of Spain.

Table 4. Immature embryo cultures of Secale vavilovii

Medium	Size (mm)	Germination (%)	Embryogenic calluses (%)
L3	1	1.83	49.54
	1.1-1.5	25.40	21.31
	1.6-2.0	30.30	27.27
	2	72.41	0
MS+1	1	76.47	11.76
	1.1-1.5	82.92	26.83
	1.6-2.0	100	33.33
	2	100	12.82
MS+2	1	76.92	17.94
	1.1-1.5	87.80	24.39
	1.6-2.0	57.14	14.28
	2	100	15.65

REFERENCES

Goldstein CS and Kronstadt WE (1986) Tissue culture and plant regeneration from immature embryo explants of barley, Hordeum vulgare. Theor Appl Genet 71:631-636

Krumbiegel-Schroeren G, Finger J, Schroeren V and Binding H (1984) Embryoid formation and plant regeneration from callus of Secale cereale. Z Pflanzenzüchtg 92:89-94

Linacero R and Vázquez AM (1986) Somatic embryogenesis and plant regeneration from leaf tissues of rye (Secale cereale L.). Plant Science 44:219-222

Rrybczynski JJ and Zdunczyc W (1986) Somatic embryogenesis and plantlet regeneration in the genus Secale. 1. Somatic embryogenesis and organogenesis from cultured immature embryos of five wild species of rye. Theor Appl Genet 73:267-271

SOMATIC EMBRYOGENESIS IN CONIFERS

Larry Fowke and Inger Hakman
Department of Biology, University of Saskatchewan,
Saskatoon, Saskatchewan,
Canada, S7N 0W0

INTRODUCTION

There is currently considerable interest in developing techniques for producing somatic embryos from conifers. Somatic embryos could be used for micropropagation of trees with desirable characteristics such as improved dimension increment, better quality wood and disease resistance. The availability of somatic embryos would also provide excellent experimental material for basic studies of conifer embryo development. Finally, embryogenic callus and suspension cultures should facilitate the development of a protoplast regeneration system to be used in future genetic manipulation studies. The use of embryogenic material has previously proved successful for developing protoplast systems with other recalcitrant species (e.g. Vasil and Vasil 1986).

The first report of successful somatic embryogenesis was published by Hakman et al. in 1985 for *Picea abies* (Norway spruce). Cultured immature zygotic embryos produced a translucent callus containing numerous somatic embryos. Plantlets were regenerated from these somatic embryos. In the last couple of years there has been an explosion of interest in this area of research and successful somatic embryogenesis has been achieved with a number of conifers (Table 1). Despite the difference in technique used in these studies, the resulting calli and somatic embryos are remarkably similar.

WHITE SPRUCE SOMATIC EMBRYOS

This chapter examines the methodology used and results obtained for somatic embryogenesis in *Picea glauca* (white spruce). Cones collected at Christopher Lake, Saskatchewan

NATO ASI Series, Vol. H18
Plant Cell Biotechnology. Edited by M.S.S. Pais et al.

Table. 1 Conifer Somatic Embryogenesis

Species	Source	Reference
Picea abies (Norway spruce)	imm embryos	Hakman et al. 1985
		Hakman and von Arnold 1985
	mat embryos	von Arnold and Hakman 1986
		Gupta and Durzan 1986a
		Krogstrup 1986
Larix decidua (larch)	♀ gametophyte	Nagmani and Bonga 1985
Picea glauca (white spruce)	imm embryos	Hakman and Fowke 1987a
	cell suspension	Hakman and Fowke 1987b
Picea mariana (black spruce)	imm embryos	Hakman and Fowke 1987a
Pinus lambertiana (sugar pine)	imm embryos	Gupta and Durzan 1986b
Pinus taeda (loblolly pine)	imm embryos	Gupta and Durzan 1987

imm = immature
mat = mature

were used as source material for the initiation of callus (Hakman and Fowke 1987a). Seeds were surface sterilized and the embryos isolated and cultured on agar solidified LP medium containing 2,4-D (1 X 10^{-5}), BA (5 X 10^{-6}) and 1% sucrose. Immature embryos with tiny cotyledons proved to be the most successful. Within 6-8 weeks of culture, calli formed on the zygotic embryos. Embryogenic callus was easily recognized by its translucent whitish appearance. Cones stored at 4°C for up to three months yielded embryogenic callus but after further storage yields dropped dramatically. Embryogenic calli have been maintained for over 2 years. By manipulating the culture conditions, plantlets have been obtained from these calli but in very low frequencies. Considerable effort is needed to improve regeneration frequencies. A thorough investigation of explant age and culture conditions will likely result in significant improvement of plantlet formation.

Suspension cultures can be established from embryogenic callus by growing small aliquots of callus in liquid LP medium on a rotary shaker at 150 rpm under continuous light (Hakman and Fowke 1987b). Such cultures have been maintained for approximately 1 year. They contain somatic embryos of various stages, elongate cells and small cell clusters. Embryogenic callus can easily be re-established from the suspension cultures, however, plantlets have not been regenerated from them.

Numerous somatic embryos at different developmental stages are observed in both callus and suspension cultures (Figs. 1-4). They are characterized by an embryonal region consisting of meristematic cells, subtended by a suspensor containing elongate highly vacuolated cells. Mitotic figures are frequently observed in the embryonal regions of embryos at all stages. Cell divisions increase the size of the embryonal region and provide cells which expand to form new suspensor cells.

Fine structural studies (Figs. 5,6 and Hakman et al. 1987) indicate that the embryonal region contains typical meristmatic cells with thin cell walls, small vacuoles and a distribution of organelles reflecting rapid growth. The upper suspensor cells while being highly vacuolated, are also apparently metabolically active. These cells exhibit active cytoplasmic streaming and contain a network of actin cables (Hakman et al. 1987). In contrast, the lower suspensor cells are apparently senescing.

Current studies are directed towards the development of cryopreservation techniques for storing somatic embryos. Suspension cultures have been cryopreserved and successfully regenerated to form embryogenic callus after storage in liquid nitrogen (Kartha et al. - unpublished). The isolation, culture and regeneration of protoplasts of white spruce is also under investigation.

ACKNOWLEDGEMENTS

We wish to express our appreciation to Pat Rennie for her excellent technical assistance. Financial support from the Program of Research by Universities in Forestry, Canadian Forestry Service and the Natural Sciences and Engineering Research Council of Canada are greatly appreciated.

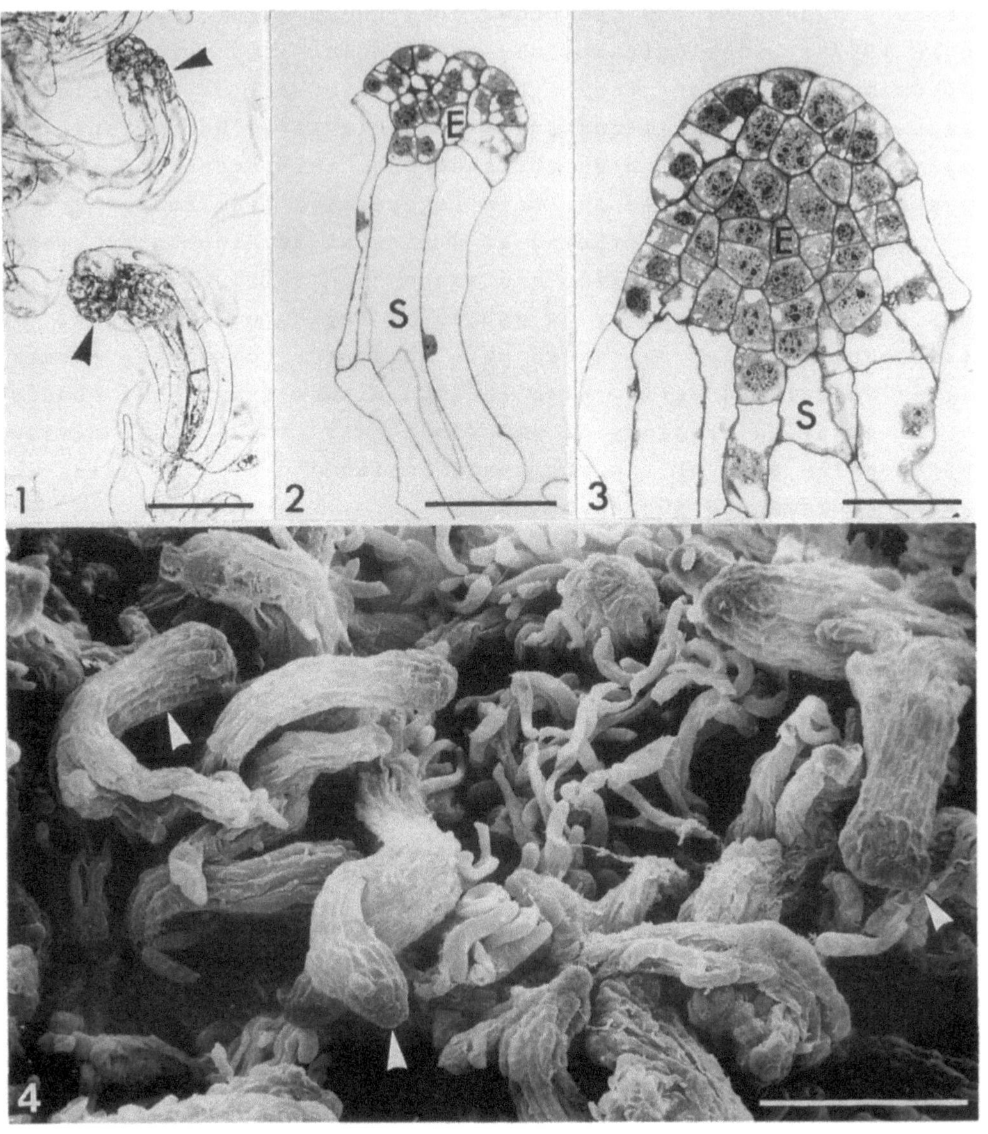

Fig. 1. Light micrograph showing somatic embryos (arrows) of white spruce. Bar = 200 µm.
Fig. 2. Light micrograph of sectioned white spruce somatic embryo from an embryogenic suspension culture. Note the embryonal (E) and suspensor (S) regions. Bar = 100 µm.
Fig. 3. Light micrograph of sectioned white spruce somatic embryo from an embryogenic callus culture showing meristematic cells of the embryonal (E) region and highly vacuolated suspensor (S) cells. Bar = 50 µm.
Fig. 4. Scanning electron micrograph showing callus culture containing numerous white spruce somatic embryos (Arrows). Bar = 0.5 mm.

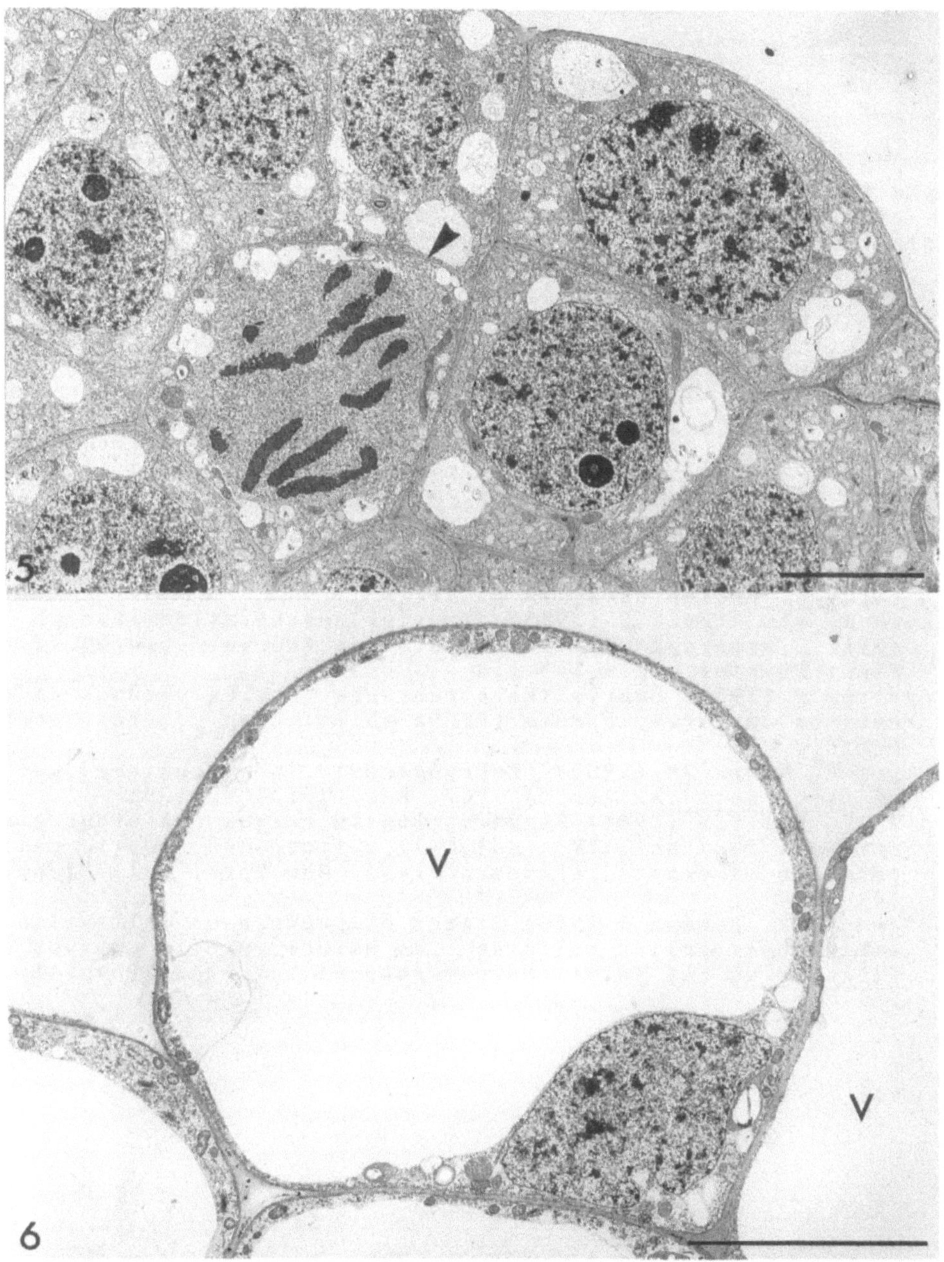

Fig. 5. Transmission electron micrograph showing meristematic cells in the embryonal region of a white spruce somatic embryo. The cells have thin cell walls and small vacuoles. Note the cell in mitosis (arrow). Bar = 10 µm.

Fig. 6. Transmission electron micrograph showing a cross section of suspensor cells from a somatic embryo. Note the large central vacuoles (V). Bar = 10 µm.

REFERENCES

Gupta PK, Durzan DJ (1986a) Plantlet regeneration via somatic embryogenesis from subcultured callus of mature embryos of Picea abies (Norway spruce). In Vitro Cell. Develop. Biol. 22: 685-688

Gupta PK, Durzan DJ (1986b) Somatic polyembryogenesis from callus of mature sugar pine embryos. Biotechnology 4: 643-645

Gupta PK, Durzan DJ (1987) Biotechnology of somatic polyembryogenesis and plantlet regeneration of loblolly pine. Biotechnology 5: 147-151

Hakman I, Fowke LC (1987a) Somatic embryogenesis in Picea glauca (white spruce) and Picea mariana (black spruce). Can. J. Bot. 65: 656-659

Hakman I, Fowke LC (1987b) An embryogenic cell suspension culture of Picea glauca (white spruce). Plant Cell Reports 6: 20-22

Hakman I, Fowke LC, von Arnold S, Eriksson T (1985) The development of somatic embryos in tissue cultures initiated from immature embryos of Picea abies (Norway spruce) Plant Science 38: 53-59

Hakman I, Rennie P, Fowke LC (1987) A light and electron microscope study of Picea glauca (white spruce) somatic embryos. Protoplasma, in press

Hakman I, von Arnold S (1985) Plantlet regeneration through somatic embryogenesis in Picea abies (Norway spruce). J. Plant Physiol. 121: 149-158

Krogstrup P (1986) Embryolike structures from cotyledons and ripe embryos of Norway spruce (Picea abies). Can. J. For. Res. 16: 664-668

Nagmani R, Bonga JM (1985) Embryogenesis in subcultured callus of Larix decidua. Can. J. For. Res. 15: 1088-1091

Vasil IK, Vasil V (1986) Regeneration in cereal and other grass species. In: Vasil IK (ed) Cell culture and somatic cell genetics of plants, Academic Press, New York, Vol. 3, pp. 121-150

von Arnold S, Hakman I 1986) Effect of sucrose on initiation of embryogenic callus cultures from mature zygotic embryos of Picea abies (L) Karst (Norway spruce). J. Plant Physiol. 122: 261-265

REGENERATION OF THALLICLONES[1] FROM LAURENCIA SP. (RHODOPHYTA)

Guillermo Garcia-Reina, Rafael R. Romero and Angel Luque

Universidad Politecnica Canarias, Departamento de Biologia, Box 550, Las Palmas de Gran Canaria, Spain

INTRODUCTION

Tissue culture techniques could be applied to improve the genetic qualities of seaweeds. In order to apply those techniques, efficient methods for obtaining and regenerating calli are needed. The term "callus" or "calluslike" has been applied to define very different structures in seaweeds (Dixon 1963; Fries 1980; Chen 1982; Saga et al. 1982; Tsekos 1982; Polne-Fuller et al. 1984; Yan 1984). Histological studies have been performed on "tumourlike" growths induced by bacteria in Gigartina teedii (Tsekos 1982), and to our knowledge nothing is known on the cytological events preceding differentiation from true callus. In a previous paper (Garcia-Reina et al. 1987), we reported the spontaneous formation of morphogenetic calli, the different calligenic potentials among Laurencia species and primary explants, and the lack of necessity for axenicity. Callus growth was drastically reduced after its organogenetic trigger, and the organogenetic potential seemed to decrease with time in culture.

The objective of this paper was: (1) to study the effect of different physical states of the culture media and its supplementation with several hormones to determine their effect on Laurencia callus induction, growth and in vitro regeneration, and (2) the histological study of the callus, in an attempt to clarify the callus versus calluslike composition.

---

[1]Thalliclones = thalli regenerated from callus.

NATO ASI Series, Vol. H18
Plant Cell Biotechnology. Edited by M.S.S. Pais et al.

MATERIAL AND METHODS

Apical branched segments (5 mm long) from Laurencia sp. were used as the primary explant. Two hours after collection they were subjected to the "sonication - 1% Betadine - GAN" treatment described previously (Garcia-Rheina et al. 1987) and cultured in petri dishes with 20 ml Provasoli-enriched seawater media (PES) (Provasoli 1968). Seawater (not the enrichment) was diluted 70 and 50% with distilled water in the "seawater assay". PES was supplemented with GEO_2 (0.5 mg l^{-1}), Ampicillin (10 mg l^{-1}), Nyastatin (2 mg l^{-1}), (GAN) and agar (Bacto Difco, 8 g l^{-1} agar were tested; also 45 explants were cultured in PES ± GAN liquid media. Two or 5 mg l^{-1} kinetin, BA or 2,4-D was added to the media in the "hormone assay". The effect of 5 mg l^{-1} Na-Naphtenate (Pfalz and Bauer, Inc., N00910) was also tested.

The cultures were incubated in a growth chamber adjusted to 21 + 2°C, 18 h light and 1500 lx. Transfers of cultures to fresh media were done at 15-day intervals. After 45, 60, 75 and 90 days the number of buds arising from the calli were evaluated. Calli with a homogeneous phenotype were transferred to an aerated liquid medium (PES + GAN) after 60 days in culture. The histological study of calli was performed following previously described methods (Devilopoulos and Tsekos 1986). Semi-thin sections (0.5 μ) were stained with toluidine blue and post-stained with lugol.

RESULTS

After 45 days in culture (uni-algal, non-axenic), we found different types of growth: (1) callus arising from the cut surface or from the disorganization of the apical neogrowth, with or without thalliclonal regeneration; (2) finger like growths of buds from the cut or apical zone; (3) thallus arising from the cut end or developing from the buds of the primary explant.

Reducing seawater concentration (osmotic potential) to 70 and 50% halted development of any type of growth (Table 1).

Reducing agar concentration (matric potential) produced an increase in the number of viable explants, but a decrease in the number of explants producing callus (Table 1) None of the explants in liquid media produced callus.

Table 1. Effects of the different assays on the type of growth after 45 days in culture

Assay[b]		N	V	Type of growth[a] C	F	T
Seawater	50%	45	0	-	-	-
	70%	45	2	0	0	2
	100% (*)	45	16	10	2	4
Agar (g l^{-1})	3	45	43	0	0	43
	8 (*)	45	17	7	4	6
	20	45	17	17	0	0
2 mg l^{-1} Kinetin		30	21	15	4	2
2 mg l^{-1} BA		30	17	8	6	3
2 mg l^{-1} 2,4-D		30	28	18	7	3
5 mg l^{-1} Kinetine		30	0	-	-	-
5 mg l^{-1} BA		30	0	-	-	-
5 mg l^{-1} 2,4-D		30	0	-	-	-
5 mg l^{-1} Na-Naph.		30	0	-	-	-
Control (*)		30	11	7	0	4

[a]V= Viability (number of explants showing any type of growth after 45 days), C= callus, F= finger, T= tallus. [b](*)= control = PES + GAN + 8 g l^{-1} agar.

Two mg l^{-1} kinetin, BA or 2,4-D increases viability and number of explants forming callus. Five mg l^{-1} kinetin, BA, 2,4-D or Na-Naphtenate has an inhibitory effect upon any type of growth (Table 1). Morphogenetic calli have an average of three to four buds after 45 days. No differences in callus growth and number of regenerating buds were observed among agar assay, hormone assay and the control. Differences were among individual calli. Later controls (60, 75 days) of the highest morphogenetic calli (at 45 days) show a decrease in the number of buds. The decrease is related to the fusion of thalliclones. Calli can be cultured for more than 5 months in agarized media, thus retaining their morphogenetic potential, and producing "air-growing" thalliclones.

In liquid media some calli developed many fast-growing thalliclones around the callus, while others developed a few slow-growing thalliclones, despite showing a homogeneous phenotype at the transfer. Fast- and slow-growing lines have retained their phenotype for at least 4 months in culture.

Histological study under light microscopy shows a clear difference between the cells from the explant and the callus. Callus cells are small, meristematic and filled with florid granules, which became brown-black after being post-stained with lugol. Enlarged cells and wide intercellular spaces seem to be oriented to the regenerating areas on the surface of the calli. Toluidine blue stained the calli metachromatic.

DISCUSSION

The physical state of the media (agarized) seems to be a key factor in the induction of callus in seaweed (Fries 1980; Chen 1982; Saga et al. 1982; Saga and Sakai 1983; Neushul 1984; Polne-Fuller et al. 1984; Lee 1985). Our results show that the decrease of the hydric potential (more negative) enhances callus formation related to the matric and osmotic potentials (percent agar and seawater respectively). However, Polne et al. (Polne-Fuller et al. 1984) reported callus formation in several species of Porphyra in PES supplemented with 0.3% agar, and callus from protoplasts of several green species when grown on 1% agar, while reporting only thalli when grown on 0.5% agar (Polne-Fuller et al. 1986). Differences between species (and their habitats) may be involved.

The effect of the addition of hormones to seaweed media in order to obtain or regenerate callus is confusing with regards to the effects and type of hormones and concentrations (Bradley and Cheney 1986; Chen 1982). Na-Naphtenate (5 mg l^{-1}) has been described as a "crucial hormone" for inducing callus in Laminaria japonica (Yan 1984). Our results suggest that great differences in the in vitro uptake, metabolism and effect of plant hormones may exist between seaweed species.

The high facility of the buds arising from the calli cultured in agarized media to fuse together, and of the apex of the thalliclones to differentiate into a disc, indicates the high platicity of the apex of Laurencia sp. The fusion among regenerated buds shows that no loss of the morphogenetic potential, nor a mechanism of dominance, is involved in the apparent morphogenetic potential decline. High plasticity and ano dominance seem to be desirable characteristics for selecting a primary explant to start tissue cultures.

The spontaneous apparition of fast- and slow-growing lines and its apparent stability indicate the usefulness of seaweed tissue culture for selecting superior strains.

The dimensions and high concentration of starch grains in callus cells resemble apical cells and carpospores (Devilopoulos and Tsekos 1986) of tip regions of Laurencia and other tubular branched Rhodophyta. Thalliclones show the normal histologly of _Laurencia_ thalli. Copious deposition of starch in plastids has been described as the first visible change accompanying callus differentiation of the moss _Psicomitrium coorgense_ (Lal and Narang 1985). _Laurencia_ calli are cell systems closely related to the calli of higher plants, but different from structures referred to as calli in brown seaweeds.

REFERENCES

Bradley PM and Cheney D (1985) Stimulation of cell division in a marine red alga using plant growth regulating substances. Plant Physiol Suppl 77:69

Chen LCM (1982) Callus-like formations from Irish moss. Biol Bull Nat Taiwan Nor Univ 17:63-68

Devilopoulos SG and Tsekos I (1986) Ultrastructure of carposporogenesis in the red alga _Gracilaria verrucosa_. Bot Mar 29:27-35

Dixon PG (1963) Biology of the Rhodophyta. Oliver & Boyd, Edinburgh

Fries L (1980) Axenic tissue cultures from the sporophytes of _Laminaria digitata_ and _Laminaria hyperborea_. J Phycol 16: 475-477

Garcia-Reina G, Romero RR and Brinkhuis B (1987) Induction of morphogenetic callus from _Laurencia_ (Rhodophyta). _J Phycol_ (submitted)

Lal M and Narang M (1985) Ultrastructural and histochemical studies of transfer cells in the callus and apogamous sporophytes of _Psicomitrium coorgense_ Broth. New Phytol 100:225-231

Lee TF (1985) Aposporous gametophyte formation in stipe explants from _Laminaria saccharina_. Bot Mar 28:179-185

Neushul M (1984) Marine biomass program plant breeding and genetics. Gas Res Inst Rep N. 84/0076

Polne-Fuller M, Biniaminov M and Gibor A (1984) Vegetative propagation of Phorphyra perforata. Hydrobiologia 116/117:308-313

Polne-Fuller M, Saga N and Gibor A (1986) Algal cell, callus and tissue cultures and selection of algal strains. Beihefte zur Nova Hedwigia 83:30-36

Provasoli L (1968) Media and prospects for the cultivation of marine algae. In Watanabe A, Hatori A (eds) Cultures and col lection of marine algae. Jap Soc Plant Physiol, p 3075

Saga N, Motomura T and Sakai Y (1982) Induction of callus from the marine brown alga Dyctiosiphon foeniculaceus. Plant and Cell Physiol 23:727-730

Saga N and Sakai Y (1983) Axenic tissue culture and callus formation of the marine brown alga Laminaria angustata. Bull Jap Soc Sci Fish 49:1561-1563

Tsekos I (1982) Tumour-like growths induced by bacteria in the thallus of a red alga, Gigartina teedii. Ann Bot 49:123-126

Yan ZM (1984) Studies on tissue culture of Laminaria japonica and Undaria pinnatifida. Hydrobiologia 116/117:314-316

GENETIC INSTABILITY IN PLANT CELL CULTURES : UTILIZATION IN PLANT BREEDING AND GENETIC STUDIES

Bruce I. Reisch

Department of Horticultural Sciences, New York State Agricultural Experiment Station, Cornell University, Geneva, New York 14456, USA

INTRODUCTION

Plant cell cultures are now acknowledged as a rich source of genetic variation. Numerous recent reviews have summarized observations on the genetic instability of cultured plant cells and the associated variability among regenerated plants (Evans and Sharp 1985; Larkin et al. 1985; Reisch 1983). Variation associated with culture conditions has also permitted the selection of numerous types of biochemical and agronomic mutants at the cellular level (Chaleff 1983). The list of agriculturally useful *in vitro* selected mutant plants is growing rapidly. In light of the amount of culture induced variability, it is not surprising that the addition of mutagens has not often resulted in significant increases in mutant recovery. Yet, not all cultures are inherently variable (Krikorian et al. 1983). Before the notion that variability is inherent in the tissue culture process can be broadly accepted, a wide number of species, genotypes and culture conditions must be surveyed for conditions associated with variability.

This chapter will discuss recent developments in areas related to genetic variability produced by cultured cells, and present some thoughts on the use of this variability in crop

NATO ASI Series, Vol. H18
Plant Cell Biotechnology. Edited by M.S.S. Pais et al.

improvement and in studies on plant genetics.

Culture-associated variability has been termed "somaclonal variation", defined as the variation among plants derived from any form of cell culture (Larkin and Scowcroft 1981). Meristem culture and its state of tissue organization is associated with high levels of genetic stability. Variants derived from *in vitro* proliferated shoot tips are usually the result of chimeral rearrangements, or epigenetic changes (Reisch 1983; Swartz et al 1981). Yet the process of growing de-differentiated tissue, such as callus, protoplasts or cell suspensions, is often associated with genetic instability (Evans and Sharp 1986; Reisch 1983).

Somaclonal variants have been regenerated from more than 25 plant genera, and are not just a recent phenomena, having been observed among some of the first plants regenerated from plant cell cultures (reviewed in Evans and Sharp 1986; Reisch 1983). Where no intentional selection has occurred *in vitro*, regenerated plants (1-50%) have varied for numerous traits when compared to the original plant. Varying traits include flower type, flower color, waxiness, plant vigor, herbage yield, tuber production, tuber quality, leaf color, plant height and disease resistance. Where inheritance has been examined, both mendelian and non mendelian inheritance patterns have been observed (Reisch 1983).

The underlying causes of somaclonal variation have evoked much speculation. Variation might arise from mutant cells which pre-exist in the explant tissue or from variation that arises during the culture process itself. Much is already known about chromosomal changes *in vitro*. Cultured cells may vary in chromosome number for a variety of reasons: 1. Pre-existing chromosome variability in the original explant sources (D'Amato 1964); 2. Nucleus fragmentation during the first *in vitro* cell division (Cionini et al 1978); 3. Endomitosis during culture initiation (D'Amato 1978); and 4. Mitotic abnormality resulting in aneuploid cells. On the other hand, by adjusting subculture intervals, and with the use of an appropriate culture medium, chromosome stability *in vitro* can be achieved (Nandi et al 1978; Evans and Gamborg 1982). Virtually all possible types of genetic changes have been found to occur in cultured cells and regenerated plants (Larkin et al 1985).

VALUE OF SOMACLONAL VARIATION FOR GENETIC STUDIES

A number of unusual genetic events have been observed among regenerated plants. These include, aside from more common events like chromosome doubling, point mutations, and aneuploidy, such occurrences as alterations in cytoplasmic genomes, movement of transposable elements, homozygous mutations, gene amplification and increased genetic distances between linked loci (Larkin et al 1985). Some of these events may be much more common, but, while causing DNA level changes, there may be no easily observed phenotypic change. These silent events may not be observable unless searched for in a systematic fashion.

Somaclonal variation has been shown in several cases to cause changes in mitochondrial DNA and chloroplast DNA. Numerous investigations have shown that, whether *in vitro* selection for T-toxin resistance takes place or not, regenerated maize plants have altered T-toxin resistance, male fertility and changes in mtDNA restriction fragment patterns (Gengenbach et al 1979 , 1981; Brettell et al 1980; Umbeck and Gengenbach 1983). Changes in chloroplast DNA restriction patterns have been revealed in tomato (Evans et al 1984). Some maternally inherited morphological variants of lettuce have also been reported (Sibi 1976).

The activation of transposable elements may play a considerable role in the production of somaclonal variation. The first reported case of the activation of previously silent controlling elements is that of a high-spotting variant of *Nicotiana tabacum* among protoclones of SU/su (sulfur mutant) light green plants (Lorz and Scowcroft 1983). Two of 120 protoclones had a high frequency of leaf spotting, indicating activation of a mutator gene. In a study of tetraploid alfalfa (*Medicago sativa* L.), a somaclonal variant for flower color was reported (Groose and Bingham 1984). The donor genotype was purple flowered, simplex (Cccc) and the mutant was white flowered (cccc). In planta, the recessive allele was unstable, giving 2.2% purple flowered shoots among 180 shoots. Yet the *in vitro* reversion rate was 23% purple flowered regenerants among 1356 plants derived from cultures of the original white flowered variant. Culture may have not only activated a mutator gene, but also enabled a high frequency of reversion in the second culture cycle.

An unusual study on recombination frequencies was carried out on tomato (Sibi et al 1984). Tomato hybrids heterozygous for 2 loci on chromosome I and 2 loci on chromosome II were produced. Each seedling was grown to maturity while donating a single cotyledon to initiate a culture. F_2 plants from control and cultured F_1's were analyzed for rate of recombination between the markers genes. About half of the cultured F_1's had increased distances between markers. On chromosome I the mean map distance in the control was 20.25 and in the F_2 plants from culture the distance was 26.06 units. On chromosome II the mean was 25.81 in the controls and 31.29 from the cultured plants. This important phenomenon deserves further study to elucidate the mechanism by which it may occur and to examine its role in other species and genotypes. As a means to increase genetic map distance, culture produced plants may be useful in breaking undesirable linkage groups.

Somaclonal variants will undoubtedly lead to useful genetic studies. Probing into the origins of somaclonal variation will lead to a better understanding of plant genetic make-up. Some understanding of the phenomena that induce gene amplification, transposable element activation and increased genetic distance will be of importance in understanding phenotypic responses to genotypic changes and will lead to a more basic understanding of plant genetics. Somaclonal variants will also be of use in clarifying the genetic control of altered traits. Aneuploids could be of use in localizing genes to chromosomes. Transposable element activation could be of great utility for gene tagging and subsequent cloning of genes with unknown gene products.

UTILIZATION FOR CROP IMPROVEMENT

Genetic variability derived from cultured cells and its impact upon crop improvement strategies has been reviewed (Larkin et al 1985; Chaleff 1983). Though viewed as a rich source of genetic variability, certain requirements must be met before variants can be used efficiently for crop improvement. The first requirement is to have available an improved variant, as many somaclonal variants are less fit than their donor genotypes. The positive trait must also be stable and heritable. Furthermore, alteration should take place in important lines or be easily transferred to advanced breeding lines. Entirely new phenotypes not available or not readily transferred sexually into elite lines are especially desirable. Finally, the considerable effort in producing useful variants should not exceed the efforts put forth for similar ojectives in conventional breeding programs.

The end result of somaclonal variation efforts is akin to the results obtained following mutation breeding and backcross breeding efforts (Reisch 1983) in that the genotype produced resembles the original genotype in most respects but with the addition of an altered trait. As compared to these efforts, somaclonal variation has some advantages and disadvantages (Table 1). Following careful consideration of the pros and cons, somaclonal variation related efforts will certainly find a place as an adjunct to conventional crop improvement efforts in certain species.

Somaclonal variation could be directly useful, especially where an improved variant form is produced in an elite vegetatively propagated genotype. There is then no need to transfer the variant trait to another genetic background, and the variant can be propagated and used directly. Also, variants of self-pollinated crops as well as of inbreds of cross-pollinated crops could be directly improved via somaclonal variation. On the other hand, certain forms of somaclonal variation would have to be manipulated by plant breeding approches in order to achieve desirable horticultural types. Backcross breeding is the method of choice for the management of variability among regenerated plants. This method would be especially useful in cases where positive variation is obtained jointly with negative variation, or where positive variation is obtained in a line used for culture investigations but lacking agricultural merit. Useful variation will be of little value if not separated from negative variation, and if not transferred to an advanced breeding line of cultivar.

METHOD	Backcross Breeding	Mutation Breeding	Somaclonal Variation
Source of variation	Natural	Induced	Spontaneous/ Induced
Success	Almost guaranteed	Undirected variation	Undirected variation
Quantitative traits ?	No	Possible	Possible
Rate of progress	Slow	Fast	Fast
Chimerism	None	Possible	Possible
Species limits	Sexually reproducing species	All species	Regenerable species

Table 1. Comparison of somaclonal variation with backcross and mutation breeding methods.

CONCLUSIONS

There has been considerable progress in understanding the nature and origin of somaclonal variation and utilizing it for plant improvement and genetic studies. Yet considerable goals still remain to be met. As the causes of somaclonal variation become better understood, it may become possible to gain exacting control over stability and variation in culture. A tighter link, through interdisciplinary efforts, between culturists, physiologists, geneticists and breeders is imperative to this process. There is also renewed interest in selecting somaclonal variants as early as possible, and even in returning to the cultured cell level for selection prior to plant regeneration. Where cellular and whole plant traits fail to correlate, early screening of regenerated plants would be most desirable. Since somaclonal variation is at present an uncontrollable process, variants of a desired type can not be obtained on demand, thus renewing interest in *in vitro* mutant selection.

A wide variety of genetic changes have been demonstrated to occur within cell cultures as well as among regenerated plants. This variation should prove to be of widespread use in understanding the nature of control over heritable traits in plants, and in manipulations for the purpose of crop improvement.

ACKNOWLEDGEMENTS

The contributions of E.T. Bingham are gratefully acknowledged.

REFERENCES

Brettell RIS, Thomas E and Ingram DS (1980) Theor Appl Genet **58**:55
Chaleff RS (1983) Science **219**: 676
Cionini A, Bennici A and D'Amato F (1978) Protoplasma **96**: 101
D'Amato F (1964) Caryologia **14**: 41
D'Amato F (1978) In Thorpe TA Ed, Frontiers of Plant Tissue Culture, IAPTC/University of Calgary, Canada p.287

Evans DA and Gamborg OL (1982) Plant Cell Reports **1**:104
Evans DA and Sharp WR (1986) In Evans DA, Sharp WR and Ammirato PV Eds. Handbook of Plant Cell Culture , MacMillan New York **4**: 97
Evans DA, Sharp WR and Medina-Filho HP (1984) Amer J Bot **71**: 759
Gengenbach BG, Green CE, and Donovan CM (1979) Proc Natl Acad Sci, USA 74:5113
Gengenbach BG, Connelly JA, Pring DR and Conde MF (1981) Theor Appl Genet 59:161
Groose RW and Bingham ET (1984) Crop Sci **24**: 655
Krikorian AD, O'Connor SA and Fitter MS (1983) In Evans DA, Sharp WR, Ammirato PV and Yamada Y Eds.Handbook of Plant Cell Culture, MacMillan, New York 4:541L
Larkin PJ, Brettell RIS, Ryan SA, Davies PA, Pallota MA and Scowcroft WR (1985) In Zaitlin M, Day P and Hollaender A Eds, Biotechnology in Plant Science, Academic Press, New York, p. 83
Larkin PJ and Scowcroft WR (1981) Theor Appl Genet **60**: 197
Lorz H and Scowcroft WR (1983) Theor Appl Genet **66**: 67
Nandi S, Fridborg G and Eriksson T (1977) Hereditas **85**: 57
Reisch B (1983) In Evans DA, Sharp WR, Ammirato PV and Yamada Y Eds, Handbook of Plant Cell Culture, MacMillan, New York 1:748
Sibi M (1976) Ann Amelior Plant **26**: 523
Sibi M, Biglary M and Demarly Y (1984) Theor Appl Genet **68**: 317
Swartz HJ, Galletta GJ and Zimmerman RH (1981) J Amer Soc Hort Sci **106**: 667
Umbeck PR and Gengenbach BG (1983) Crop Sci **23**: 584

GENE CHARACTERIZATION IN HIGHER PLANTS

J.H. Weil

Institut de Biologie Moléculaire et Cellulaire, Université Louis Pasteur, 15 rue Descartes, 67084 Strasbourg, France

INTRODUCTION

Higher plants possess three different genomes, located respectively in the nucleus, in the chloroplasts and in the mitochondria.

The smallest of these three genomes is the chloroplast genome, which, in most species studied, consists of a circular DNA molecule of about 150 kilobases (kb). The pastid genome is also the best known, as the complete nucletotide sequence of the chloroplast genome of Nicotiana tabacum and that of Marchantia polymorpha have recently been determined (Shinozaki et al. 1986; Oyama et al. 1986). In most plants, the circular chloroplast DNA molecule contains two inverted repeats (10-25 kb long) separated by a large and a small single copy region, but there are a few exceptions to this general organization, for instance in some legumes (Vicia faba, Pisum sativum) and in Euglena.

The mitochondrial genome is somewhat larger and its size varies, depending on the plant species considered, from about 200 kb in the case of Brassica to over 2500 kb in the case of watermelon. In fact, circular DNA molecules of different sizes can be found in the mitochondria of a given plant species, because the master chromosome (or master circle) contains direct or inverted repeats allowing recombination events which can generate smaller circular molecules (Lonsdale et al. 1984; Quetier et al. 1985). In addition, linear plasmid DNA molecules are often found in plant mitochondria.

The nuclear genome is considerably larger, but there are important differences, depending on the plant considered and on the ploidy level: In Nicotiana tabacum for instance there is about 3×10^6 kb of nuclear DNA, while there is only about 7×10^4 kb in Arabidopsis thaliana.

NATO ASI Series, Vol. H18
Plant Cell Biotechnology. Edited by M.S.S. Pais et al.

During the past few years, abundant information has been obtained on the organization of plant genomes due to the development of new methods in molecular biology, particularly in gene cloning.

I. METHODS USED IN GENE IDENTIFICATION AND LOCALIZAITON

1) Physical map of an organellar genome

Chloroplast genomes, which consist of relatively short circular DNA molecules (about 150 kb), can be cut into a small number of fragments by enzymes called restriction endonucleases, which cleave DNA at specific sites (consisting usually of 4 or 6 bp) and thereby generate restriction fragments (usually 10 or 20, depending on the enzyme used). These fragments can be inserted into a suitable vector (a plasmid, a phage or a cosmid) and then cloned after introduction of the recombinant plasmid (or phage, or cosmid) into *E. coli*. It is thus possible to obtain a genomic bank, or a genomic library, i.e. a collection of recombinant plasmids (or phages, or cosmids) which together contain all the sequences of a given genome.

Because of the small size of the chloroplast genome and the small number of restriction fragments generated by a given endonuclease, it is rather easy to determine the relative order of these fragments on the circular DNA molecule, usually by combining the information obtained from the study of overlapping fragments generated by different restriction endonucleases. One can therefore establish a physical map of the chloroplast genome. More recently such maps have also been established for a small number of plant mitochondrial genomes, namely those of *Brassica* (Lonsdale et al. 1984), maize (Palmer and Schields 1984) and wheat (Quetier et al. 1985).

2) Localization of ribosomal and transfer RNA genes

The identification and localization of genes coding for ribosomal and transfer RNAs is relatively easy, as the hybridization of the products of these genes to the various restriction fragments can be tested directly. Once a purified ribosomal RNA (rRNA) or transfer RNA (tRNA) has been isolated, which may require two-dimensional polyacrylamide gel electrophoresis, e.g.

to purify a given tRNA from a total tRNA preparation containing 30 or 40 different tRNA species (Burkard et al. 1982), this RNA is labeled, usually with ^{32}P, either at its 5' or 3' end, and hybridized to the genomic bank, in order to detect which restriction fragment(s) contains the complementary DNA sequence (in other words, the corresponding gene). If a more precise localization is needed, this restriction fragment can be cleaved, using other restriction enzymes, into smaller subfragments which are tested by hybridization to the labeled RNA.

3) Localization of protein-coding genes

The identification and localization of the gene(s) coding for a protein are somewhat more difficult. The usual strategy consists first in isolating and characterizing the mRNA coding for this protein. This is not an easy task, considering the fact that there are generally a few 100 (or more) different mRNA species in a cell and that they represent together only a small percentage (2-4%) of the total cellular RNAs.

A) Isolation of the mRNA

The first animal mRNAs were isolated from specialized cells synthesizing preferentially one protein, such as red blood cells (to obtain globin mRNA) or hen oviduct cells (to obtain ovalbumin mRNA), but such situations are rather exceptional.

Another approach consists in using antibodies raised against a given protein to specifically precipitate the polyomes engaged in the synthesis of this protein and thus selectively extract the mRNA coding for this protein.

The mRNAs can also be first separated from the other cellular RNAs (rRNAs, tRNAs) by chromatography on an oligo U-Sepharose or oligo dT-Sepharose column, as they will bind to such a column through the poly-A tail which they have at their 3' end (this is the case of most cytoplasmic mRNAs, but the chloroplast mRNAs do not have a polyadenylated 3' end). Then fractionation of mRNAs can be achieved by several techniques, such as electrophoresis or sucrose gradient centrifugation.

Finally, identification of a given mRNA is based on its ability to program the synthesis (in _Xenopus oocytes_ or in a cell-free protein-synthesizing system) of the corresponding polypeptide which is revealed, after polyacrylamide gel electrophoresis of the translation products, using specific antibodies.

The mRNA itself can be used as a probe and hybridized to DNA fragments, to identify and isolate the corresponding gene(s). But it is usually difficult to obtain a completely pure mRNA, with high enough radioactivity and in sufficient amounts, especially when the corresponding gene(s) is(are) expressed at low levels, so that one now prefers to use a cloned cDNA of the mRNA.

B) Synthesis of the cDNA and preparation of a cDNA library

In order to synthesize a double-stranded DNA, corresponding to a given mRNA, one first anneals a short oligo dT primer to the poly-A tail present at the 3' end of the mRNA. The reverse transcriptase then uses the 3'OH end of this primer to polymerize deoxynucleotides as directed by the mRNA template and thus synthesizes a DNA strand complementary to the mRNA. Upon reaching the end of the mRNA template, the reverse transcriptase loops back, and uses the last few bases of the newly made DNA strand as a template to direct the synthesis of a short complementary sequence, thus creating a hairpin about 10-20 bp long. Upon alkaline hydrolysis of the mRNA, the hairpin of the remaining single-stranded complementary DNA (called cDNA) provides a 3'OH end, which is used as a primer by DNA polymerase I to synthesize a DNA strand complementary to the cDNA, thus forming a double-stranded (or duplex) DNA, which still has a hairpin at one end. But this hairpin can be removed by treatment with nuclease S1, which specifically hydrolyzes single-stranded regions, generating a classical DNA duplex. This DNA duplex (sometimes also called cDNA) can be inserted into a suitable vector (phage λ, for instance) and cloned. Such cDNA clones can be prepared starting from a total cellular mRNA population (rather than from an individual purified mRNA species), yielding then a cDNA bank (or a cDNA library).

C) Screening a CDNA library prepared using an expression vector

If the cDNA library has been prepared using a so-called expression vector, namely a vector allowing expression (i.e. transcription and translation) of inserted gene(s) in E. coli, the bacterial colonies can be screened using specific antibodies in order to detect the bacteria in which the corresponding protein has been synthesized, thus allowing identification of the clone(s) containing the corresponding cDNA. It should be mentioned that in order to have an eukaryotic gene expressed in E. coli, this gene is usually placed (on the vector) under the control of a strong bacterial promoter (for instance, the lac promoter) to allow efficient transcription. To allow efficient translation, a proper ribosome binding site must be present upstream of the AUG initiation codon. Another possibility is to insert the gene shortly after the start of a sequence coding for a bacterial protein in the correct reading frame, thus producing a fusion protein, whose N-terminal region is that of a bacterial protein, while the rest of the protein is coded by the inserted DNA. One should keep in mind that bacteria are not equipped to splice out introns, so that one cannot expect the correct expression of an eukaryotic interrupted gene in E. coli; it is therefore recommended to use a cDNA (prepared from a mature mRNA template), rather than a genomic DNA fragment, to obtain expression of eukaryotic genes in bacteria.

D) Screening the cDNA library by hybrid-released or hybrid-arrested translation

Instead of identifying, in a cDNA library, the clone containing a given sequence by looking for the corresponding protein expressed in vivo in E. coli cells, two alternative in vitro techniques can be used: (1) hybrid-released translation consists in hybridizing total mRNAs to cloned DNA immobilized on nitrocellulose, washing extensively, thus releasing the mRNA from the hybrid and introducing it into a cell-free protein-synthesizing system to see whether the expected protein is made (using the corresponding antibodies) to reveal it, after polyacrylamide gel electrophoretic fractionation of the in vitro translation products). (2) Hybrid-arrested translation consists in checking whether the cloned DNA tested is able to hybridize the mRNA studied and thus inhibit the translation in vitro, so that the corresponding polypeptide is no longer present among the translation products.

E) Screening a genomic library

Once the cloned cDNA coding for a given protein is available, it can be made radioactive and used as a probe to look for the genomic clones containing the same sequences by colony hybridization. In this technique, the bacterial colonies containing the chimeric vectors are lysed on a nitrocellulose filter, their DNA is denatured, fixed on the filter and hybridized to the radioactive cloned cDNA. Upon autoradiography, the colonies containing DNA, which has hybridized to the probe, are seen as dark spots, and their chimeric DNA can be obtained from the untreated colonies in the reference library.

One has to realize that the number of random fragments which must be cloned in order to ensure a high probability that every sequence of the genome studied is present in at least one chimeric vector, increases with the genome size and the desired probability. For a probability level of 99%, about 10^6 cloned fragments are necessary in the case of a mammalian or a plant nuclear genome, whose size is about 10^9 bp (or 10^6 kb). It should be pointed out that a single gene of about 5000 bp, although rather large (if it has no intron, it codes for a protein of over 1600 amino acids), only represents 0.00005% of this nuclear DNA. One can see why, in order to identify such a small proportion, a very specific and highly radioactive probe is necessary in the hybridization experiments.

Because of the high degree of sequence homology often seen in corresponding genes from various organisms (the sequence conservation of histone genes for instance is particularly striking), it has often been possible to detect a gene in one species by using the cloned gene (or cDNA) from another species as a probe.

It is also possible to screen a library, using as a probe a radioactive, synthetic oligodeoxynucleotide, consisting of a dozen to a few dozen nucleotides, whose sequence has been established on the basis of the sequence of the corresponding protein, when this protein (either from the same organism or from related species) has been purified and completely or partially sequenced.

F) Gene identification upon sequencing and sequence comparison by computer

Finally, progress in cloning and sequencing techniques has recently culminated in the determination of the complete nucleotide sequence of the chloroplast genomes of tobacco (Shinozaki et al. 1986) and *Marchantia* (Ohyama et al. 1986), and because of the high degree of homology between chloroplast and prokaryotic genes, a simple comparison by computer of chloroplast and *E. coli* DNA sequences has allowed the identification and localization of some chloroplast genes which had not been detected before (see below).

II. GENES CHARACTERIZED

1) Chloroplast genomes

A) General organization of the genome

The chloroplast genome consists of circular DNA molecules, which all have the same length in a given plant species (this is not the case in plant mitochondria, as we shall see). This length is usually about 150 kb (it varies from 120 to 190 kb, depending on the species considered).

In most species studied so far, the chloroplast circular DNA comprises a region present twice, each in opposite orientation (inverted repeat), carrying an rDNA unit (an operon of the ribosomal RNA genes). The size of this repeated region varies from 10 kb in *Marchantia* (Ohyama et al. 1986) to 25 kb in tobacco (Shinozaki et al. 1986). The two copies of the inverted repeat are separated by: (1) a large single copy region, which is about 80 kbp long in *Marchantia* and tobacco, (2) a small single copy region, which is about 20 kbp long in *Marchantia* and tobacco.

There are a few exceptions to this general organization:

1. In certain legumes (broad bean, pea) there is only one rDNA unit.

2. In *Euglena* there are three rDNA units in the same (tandem) orientation (strains have been described which have one or five rDNA units).

B) Genes identified, localized and sequenced

In tobacco (Shinozaki et al. 1986), identified, localized and sequenced chloroplast genes code for:

1. Four rRNAs: 23S, 16S, 5SS and 4.5S.
2. Thirty different tRNAs.
3. About 40 proteins. In addition, a dozen putative genes have been located which might code for proteins, as well as approximately 40 ORFs or URFs (Open Reading Frames or Unidentified Reading Frames) comprising over 70 codons each.

The genes coding for the four rRNAs are clustered in the inverted repeat region and are thus present twice in the chloroplast genome. The genes for the 23S, 16S and 5S rRNAs are very similar to those in *E. coli*; for instance, there is about 75% homology for the 16S RNA (Schwarz and Kossel 1980) and 70% for the 23S RNA (Edwards and Kossel 1980) between maize chloroplasts and *E. coli*. The 4.5S RNA is a typical chloroplast RNA species, but it corresponds to the 5' end of the 23S RNA in *E. coli*.

Thirty different tRNA genes have been localized, and as 7 of them are located in the repeated region, there are 37 tRNA genes (Shinozaki et al. 1986). In theory, a minimum of 32 tRNAs are necessary to translate the 61 sense codons of the "universal" genetic code, but if one considers the possibility of a U-N wobble and/or a "two of three" mechanism for codon-anticodon base pairing, protein synthesis could take place in the chloroplast without import of tRNAs. These tRNA genes are scattered over the chloroplast genome in higher plants (Weil et al. 1982), but in *Euglena* chloroplasts most of them are clustered (Hallick et al. 1984).

Among the plastid genes coding for proteins, one can distinguish those coding for polypeptides found in the stroma, and those coding for polypeptides of the thylakoids.

In the first group, one finds the genes coding for the following *stroma polypeptides*:

1. The large subunit of ribulose 1,5 bisphosphate carboxylase/ oxygenase (Rubusco). This enzyme consists of eight chains of the large subunit and eight chains of the small subunit. The latter is coded in the nuclear genome, synthesized in the cytoplasm as a precursor which is imported into the chloroplast and processed to yield the mature small subunit that associates with the large subunit to form the complete enzyme (Chua and Schmidt 1978; Smith and Ellis 1979).

2. Nineteen of the 60 or so (58 to 62) chloroplast ribosomal proteins (Shinozaki et al. 1986; Ohyama et al. 1986), namely 11 r-proteins of the small particle (30S) and 8 r-proteins of the large particle (50S). The 40 or so other r-proteins are therefore imported from the cytoplasm.

There is a rather high homology between these r-proteins and the corresponding ones of E. coli. The nucleotide sequences of the genes and therefore the structures of the proteins, as revealed by cross-immunological reactions, can reach homologies as high as 75%. The clustering of some genes is also similar; for example proteins L23, L2, S19, L22, S3, L16, L14 and S8 have their genes clustered in this order in both the chloroplast genome and E. coli genome (Shinozaki et al. 1986; Oyama et al. 1986).

3. The translation initiation factor IF-1 (Shinozaki et al. 1986; Ohyama et al. 1986) and the elongation factor EF-Tu (Montandon and Stutz 1983) are coded in chloroplast DNA.

4. The determination of the complete chloroplast genome sequence has revealed in Marchantia and in tobacco (Shinozaki et al. 1986; Ohyama et al. 1986) the presence of genes homologous to those in E. coli which code for the α, β and β' subunits of RNA polymerase, but it has not been demonstrated so far that chloroplast RNA polymerase consists of these chloroplast DNA-coded polypeptides.

The genes coding for ribosomal proteins, polypeptides of the RNA polymerase, and translation factors are sometimes considered together with those coding for rRNAs and tRNAs, as forming a group of genes coding for products involved in the expression of the chloroplast genome.

The second group of plastid genes comprises those coding for <u>thylakoid proteins which participate in photosynthesis</u> (Shinozaki et al. 1986; Ohyama et al. 1986):

1. Proteins of photosystem I (PS I):
   a) the apoproteins A_1 and A_2 of P 700 (the reaction center of PS I).
2. Proteins of photosystem II (PS II):
   a) the 32-kD protein, responsible for the sensitivity to herbicides such as atrazine (in resistant mutants sequencing of this gene has shown that only one nucleotide is different);
   b) the apoprotein of P 680;
   c) the 44-kD protein;
   d) the D_2 protein;
   e) cytochrome b_{559}.
3. Proteins of the cytochrome b_6/f complex:
   a) cytochrome f;
   b) cytochrome b_6;
   c) subunit 4.
4. Proteins of the chloroplast ATPase complex:
   a) Six of the nine subunits of this complex, namely α, β, ε, I, III and a.

The proteins of the fifth thylakoid complex (the light-harvesting protein-chlorophyll complex) appear to be coded in the nuclear genome.

It should be pointed out that two "ars" (autonomously replicating sequences) have been localized (Shinozaki et al. 1986) which are capable of autonomous replication in yeast, but their role in the replication of chloroplast DNA is not known, as they are located outside the region of the origin of chloroplast DNA replication.

Among the ORFs there are six sequences homologous to those coding in the mitochondria for the subunits of NADH dehydrogenase (Shinozaki et al. 1986; Ohyama et al. 1986), so that one wonders whether a respiratory chain, including this NADH dehydrogenase, exists and functions in the chloroplasts.

In tobacco (Shinozaki et al. 1986) a small number of chloroplast genes (six tRNA genes and nine protein genes) contain introns, while in Marchantia these numbers are 6 and 12 respectively (Ohyama et al. 1986). Some of them are very long (several hundreds of base pairs), even in the case of genes coding for tRNAs (although tRNAs have only 70-80 nucleotides), as illustrated by the split $tRNA^{Lys}$ gene which has a 2526-bp intron. In Euglena chloroplasts, on the contrary, tRNA genes have no introns, whereas multiple introns are found in some protein genes which have no intron in higher plants, such as the gene coding for the large subunit of Rubisco which has nine introns, or the gene of the 32-kD protein of PS II which has four introns.

2) MITOCHONDRIAL GENOMES

A) General organization of the genome

The mitochondrial genomes of plants are much larger than those of mammals (about 16 kb) or yeast (about 78 kb, in the case of Sacharomyces cereviseae), and their size varies from about 200 kb (in Brassica) to over 2500 kb (in the case of watermelon). The fact that circular DNA molecules of different sizes are found in the mitochondria of a given plant species was puzzling, until it was shown that the master chromosome (or master circle) contains direct or inverted repeats allowing recombination events which can generate smaller circular molecules (Lonsdale et al. 1984; Quetier et al. 1985). In addition, linear plasmid DNA molecules are found in the mitochondria of fertile and male-sterile plants.

B) Genes identified, localized and sequenced

Plant mitochondrial DNA codes for three rRNAs, namely two high molecular weight rRNAs (18s and 26S) and a 5S rRNA (which is present in prokaryotes, chloroplasts and in the cytoplasm of eukaryotes, but not in fungal or mammalian mitochondria); these mitochondrial RNA genes have been localized and sequenced in several plants. A number of tRNA genes have also been mapped and sequenced, mostly in maize and wheat mitochondria, and they resemble more prokaryotic or chloroplast tRNA genes than tRNA genes found in fungal or mammalian mitochondria.

Less than a dozen protein-coding genes have been identified and sequenced so far in plant mitochondria (Leaver et al. 1985); these genes code for:

1. Subunits I, II and III of the cytochrome oxidase complex;
2. Apocytochrome b of the cytochrome bc_1 complex;
3. Subunits 6 and 9 (F_O) and subunit (F_1) of the F_O-F_1 ATPase complex;
4. Subunits 1 and 3 of the NAD-Q1 complex;
5. Ribosomal proteins S12 and S13 of the small ribosomal particle.

It is interesting to note that most polypeptides coded either in the chloroplast genome or in the mitochondrial genome and synthesized in the corresponding organelle, must become associated with polypeptides imported from the cytoplasm in order to form functional enzymes or complexes, for example Rubisco, photosystems I and II, cytochrome b_6/f (in chloroplasts) cytochrome bc_1 (in mitochondria), ATPase, ribosomes (in chloroplasts and mitochondria). The biogenesis and function of both types of organelles therefore require a control of gene expression in the organellar and nucleocytoplasmic compartments, in order to ensure a concerted synthesis of polypeptides, which must assemble in the organelles to form functional oligomeric enzymes or complexes.

It should be pointed out that in plant mitochondrial DNA, sequences homologous to chloroplast and nuclear DNA sequences have been found (for instance sequences homologous to chloroplast genes coding for 16S rRNA, tRNAs, or the large subunit of Rubisco). The origin of these sequences and their function (if any) in mitochondria are not known.

3) NUCLEAR GENOMES

A) General organization of the genome

Plant nuclear genomes are much larger than organellar genomes, but the amounts of nuclear DNA vary considerably, depending on the species considered and the ploidy level: for instance there is approximately 3 x 10^6 kb of nuclear DNA in *Nicotiana tabacum*, whereas there is only 7 x 10^4 kb in *Arabidopsis thalina*. Consequently, digestion of nuclear DNA with a restriction endonuclease does not yield discrete fragments nicely fractionated upon gel electrophoresis, but gives a continuous smear. Discrete bands, whenever visible on these restriction patterns, usually repre-ent highly repeated sequences (see below).

On the basis of renaturation kinetics studies performed on nuclear DNA, one can distinguish between single copy sequences (present only once per haploid genome) and repeated sequences, usually classified into three categories, depending upon the reiteration frequency, which is either low (dozen of copies), medium (hundreds or thousands of copies), or high (up to several millions of copies). The proportion of repeated sequences in higher plant nuclear DNA varies considerably; it is relatively low in *Arabidopsis thaliana* (37%), but it is usually higher than 50%, as for instance in *Nicotiana tabacum* (70%), *Triticum monococcum* (80%), *Vicia faba* (85%) or *Secale cereale* (90%). Although it has been suggested that these large variations in the proportion of repeated nuclear DNA are correlated with phenotypic characters, there is really no strong evidence and some people think that repeated DNA is "selfish" or "junk" DNA.

Some of the repeated sequences are clustered (satellite DNA and rRNA genes are tandemly repeated sequences), while others (such as transposons) are dispersed in various regions of the genome and interspersed with single copy sequences.

One of the first satellite DNAs studied was that of *Secale cereale*. Six different repeated sequences have been cloned (Bedbrook et al. 1980): The most reiterated unit consists of a 140-bp doublet interspersed with an unrelated 230-bp sequence, and there are several hundred thousands of copies of this unit (it represents over 5% of the genome). The functions, if any, of satellite DNA, are not known, although it has been postulated that these sequences may be involved in chromosome movements and pairing (Bennett 1984).

The rRNA genes are found in the nucleolar organizers as adjacent repeated units. Each of them consists of a transcription unit (transcribed into a common precursor of the 18S, 5.8S and 25S rRNAs) and a non-transcribed spacer located between the transcription units of adjacent repeated units. The size of these rDNA units varies between 8 and 12 kb, and their copy number varies between 1000 and 15 000 per diploid genome (for a review, see Vedel and Delseny 1987). The existence of transposable elements, suggested by studies in maize genetics performed about 40 years ago by McClintock (McClintock 1948), has been confirmed by molecular analyses performed not only on plant genomes, but also on other genomes (bacteria, yeast, Drosophila, etc.). These transposable elements (mobile genetic elements, insertion elements or transposons) are DNA fragments which can move from one site of the genome to another, either on the same or on a different chromosome. Among these elements, two have been generally defined very early and have been well characterized at the molecular level: the Ac element can promote its own transposition and that of Ds from one site to another on any chromosome. Sequence studies (Muller-Neumann et al. 1984) have shown that transposons have terminal inverted repeats (11 bp in the case of Ac), which are supposed to be the substrate for transposase activity, flanked by direct replication of the target size (8 bp in the case of Ac). A rather high proportion of the maize genome (about 30%) consists of middle-repetitive sequences, assumed to derive from transposable elements, and according to the genome stress hypothesis proposed by McClintock (McClintock 1978), this high proportion might be due to the fact that plants have evolved to adapt to genetic stress (chromosome breaks) or environmental stress (e.g. viral infection).

B) Genes identified, localized and sequenced

In addition to rRNA genes (found in repeated units, see above), and some tRNA genes (Green et al. 1986), a number of protein-coding genes have been identified and sequenced.

It has been estimated that approximately 15 000 genes exist and are transcribed into mRNAs at some point of the life cycle and in the different tissues of the plant (Goldberg et al. 1978). This represents only a very small percentage of the plant nuclear genome: In the case of Nicotiana tabacum, expressed genes would represent about 1.8×10^4 kb of DNA, i.e. less than 1% of the total 3×10^6 kb of nuclear DNA.

Thanks to progress made in gene identification, cloning and sequencing (see above), information is accumulating on an increasing number of protein-coding genes. Amont the main features, one should mention the fact that some proteins are coded by multigene families. The number of these genes can vary from a few copies to several 100 copies; sometimes they are clustered, sometimes they are dispersed over the chromosomes. Their level of expression can vary depending on the tissue considered, but some genes have been altered by mutations and have become nonfunctional (pseudogenes). Most of the protein-coding genes contain introns (histone genes have no intron).

Among the protein-coding genes studies (see Vedel and Delseny 1987, for a review), one can mention the genes coding for:

a) Seed storage proteins

In cereals:

Zein (maize), gliadin (wheat) and hordein (barley) are coded by multiple genes.

Sequence studies on these genes have revealed the presence of (sometimes imperfect) internal repeats, suggesting that these genes have evolved by rearrangements implicating these repeats.

In dicotyledons:

Glycinin and conglycinin (soybean), legumin, vicilin and convicilin (pea), cruciferin and napin (rape seed) and phaseolin (bean) are also coded by small multigenic families.

In addition to the major storage proteins, other proteins are found in seeds, but in smaller amounts, for instance lectin, phytohemagglutinin, trypsin inhibitor, which are also encoded by small multigenic families.

b) Chloroplast proteins

Among the chloroplast proteins encoded by nuclear genes, two have been particularly studied, namely the small subunit of ribulose 1,5 bisphosphate carboxylase (Rubisco) and the light-harvesting chlorophyll a/b binding protein. They are coded by small multigene families, which sometimes differ in the 5' and 3' flanking regions but not in the coding region.

c) Stress proteins

Heat-shock proteins of relatively small (15-25 kD) or larger size (65-110 kD) are also coded by multigenic families, as revealed by studies performed in *Zea mays* and in *Glycine max*.

Other stress proteins whose genes have been studied include hydroxyproline-rich proteins, endochitinase and enzymes of the phenylpropanoid pathway such as phenylalanine ammonia lyase and chalcone synthase.

d) Other proteins

Other proteins have been shown to be encoded by multigene families, for instance histones H_3 and H_4 (about 100 genes in *Zea mays*), actin, leghemoglobin and several nodulins in *Glycine max* and various enzymes, such as α-amylase, glutamine synthetase, alcohol dehydrogenase 1 and 2, sucrose synthetase, triosephosphate isomerase (for a review, see Brown 1986).

CONCLUSION

Considerable progress has been made recently in understanding the general organization of plant genomes and the structure of plant genes. Many studies, not analyzed in this review, have been devoted lately to the expression of plant genes and to the control of gene expression. However, only a few of the 15 000 genes supposed to be active during the life cycle of a plant have been studied, and one still does not know the function (if any) of most of the plant nuclear DNA. But one can be optimistic and predict that within the next years, further progress will be made in our knowledge of plant genome organization and expression, leading to a better understanding of the production of improved plants, thanks to the development of genetic engineering.

REFERENCES

Bedbrook PE, O'Dell M and Flavell RB (1980) Nature 288, 133

Bennett MD (1984) In: Vickery RK (eds). Plant biosystematics. Academic Press

Brown JWS (1986) Nucleic Acids Res 14:9549

Burkard G, Steinmetz A, Keller M, Mubumbila M, Crouse E and Weil JH (1982) In: Edelman M, Hallick R and Chua NH (eds). Methods in chloroplast molecular biology. Elsevier Biomedical Press (Amsterdam) p. 347

Chua NH and Schmidt GW (1978) Proc Nat Acad Sci USA 75:6110

Edwards K and Kossel H (1981) Nucleic Acids Res 9:2853

Goldberg RD, Hoschek G, Kamalay JC (1978) Cell 14:123

Green G, Weil JH and Steinmetz A (1986) Plant Mol Biol 7:207

Hallick RB, Hollingsworth M and Nickoloff JA (1984) Plant Mol Biol 3:169

Leaver CJ, Isaac PG, Bailey-Serres J, Small ID, Hanson DK and Fox TD (1985) In: Quagliariello et al (eds). Achievements and perspectives of mitochondrial research, Elsevier vol II, p. 111

Lonsdale DM, Hodge TP and Fauron CM (1984) Nucleic Acids Res 12:9249

McClintock B (1948) Carnegie Inst Washington Yearbook 47:155

McClintock B (1978) Stadler Genet Sympos 10,25

Montandon PE and Stutz E (1983) Nucleic Acids Res 11:5877

Muller-Neumann M, Yoder J and Starlinger P (1984) Mol Gen Genet 198:19

Palmer JD and Schields CR (1984) Nature 307:436

Quetier F, Lejeune B, Delorme S, Falconet D and Jubier MF (1985) In: van Vloten-Doting, Groot GSP and Hall TC (eds). Molecular form and function of plant genome. Plenum Press 1985 New York:413

Shinozaki K, Ohme M, Tanaka M, Wakasugi T, Hayashida N, Matsubayashi T, Zaita N, Chunwongse J, Obokata J, Yamaguchi-Shinozaki K, Ohto C, Torazawa K, Meng BY, Sugita M, Deno H, Kamogashira T, Yamada K, Kusuda J, Takaiwa F, Kato A, Tohdoh N, Shimida H and Sugiura M (1986) EMBO J 5:2043

Smith SM and Ellis RJ (1979) Nature 278:662

Ohyama K, Fukuzawa H, Kohchi T, Shirai H, Sano T, Sano S, Umesono K, Shiki Y, Takeuchi M, Chang Z, Hota S, Inokuchi H and Ozeki H (1986). Nature 322:572

Vedel F and Delseny M (1987) Plant Physiol Biochem 25:191

Weil JH, Mubumbila M, Kuntz M, Keller M, Steinmetz A, Crouse E, Burkard G, Guillemaud P, Selden R, Mc Intosh L, Bogorad L, Loffelhardt W, Mucke H and Bohnert HJ (1982) In: Akoyunoglou G et al (eds), Cell function and differentiation. Liss New York (982):321

REGULATION OF GENES INVOLVED IN T-DNA PROCESSING: AN INITIAL STEP IN THE GENETIC MODIFICATION OF PLANT CELLS

C. I. Kado, P. Rogowsky, T. J. Close[1], and T. J. A. Quayle
Davis Crown Gall Group,
Department of Plant Pathology
University of California, Davis

ABSTRACT

Directed genetic modification of plants requires the transmission of specific foreign DNA to the host plant, where the DNA is incorporated and stably maintained in the plant nuclear genome. Although there are several DNA transmission techniques, the transfer of DNA mediated through the natural transmission of T-DNA of Ti plasmids of *Agrobacterium tumefaciens* has proven relatively efficient and practical. The initial processing and transmission of the T-DNA requires genes situated on the Ti plasmid and on the bacterial chromosome. The T-DNA processing genes on the Ti plasmid are located in the virulence (Vir) region. The *vir*D2 gene of the *vir*D operon, which appears to be involved in the formation of T-DNA intermediates by encoding a specific endonuclease, is constitutively synthesized in a permissive mutant *ros* of *A. tumefaciens*. Gene fusions to reporter genes such as the luciferase cassette of *Vibrio fischeri* made it possible to measure the expression of Vir genes during the on-going interaction between *A. tumefaciens* and carrot disks. Using phenolic inducers such as acetosyringone, *vir*B, *vir*C, *vir*D and *vir*E were identified as inducible operons of the Vir region in the presence of *vir*A, *vir*G and inducer. *vir*C and *vir*D are also controlled by the *ros* gene of the chromosome as mutations in the *ros* gene results in the constitutive expression of these operons. This has enabled us to study the formation of double stranded T-DNA intermediates in the absence of acetosyringone or other phenolic inducers. Integration of the T-DNA was

[1]Present address: Division of Plant Industry, CSIRO, GPO Box 1600, Canberra, ACT 2601, Australia.

NATO ASI Series, Vol. H18
Plant Cell Biotechnology. Edited by M.S.S. Pais et al.

directly observed in _Haplopappus_ _gracilis_ chromosome A by in situ hybridization analysis.

INTRODUCTION

Efforts to expedite the incorporation of genes conferring desirable traits into existing crop plants can be categorized into three major areas. 1) The approach to modify conventional plant breeding techniques through _in vitro_ cell propagation that rely on spontaneous genetic alterations through mutations (including transposon mediated diversifications), somatic clonal variations, and through directed heterokaryosis (e. g., protoplast fusion). 2) The direct approach to either inject or incubate protoplasts with DNA containing selected gene(s) and usually a marker gene(s) into recipient cells. 3) The use of a helper organism or virion to deliver and stably incorporate selected gene(s) into plants. Each of these methods have inherent limitations (Kado and Kleinhofs, 1980; Potrykus et al., 1985), and there is clearly room for continued optimization of each procedure to fit the needs of each group of crop plants.

1. Transformation with the _Agrobacterium_ system.

As shown in Fig. 1, the genetic resource for desirable plant genes can be broad, but it is generally dependent on existing crop plant resources (either as cultivars or wild-type species). The culturing and subculturing of embryos and shoot-tips have been useful in obtaining disease free stock from infected cultivars (Fig. 1, left). On the other hand, genetic modification by chemical mutagenesis have not been highly fruitful in obtaining mutants with desirable traits. Somatic variants that are known to arise upon subculturing cells in vitro have relied on random selection, often yielding an unstable product. The consequences of prolonged _in vitro_ cell culture are well known where somaclonal variation is a genetic hazard for in vitro micropropagation and for germplasm conservation of asexually propagated species (Scowcroft et al.,

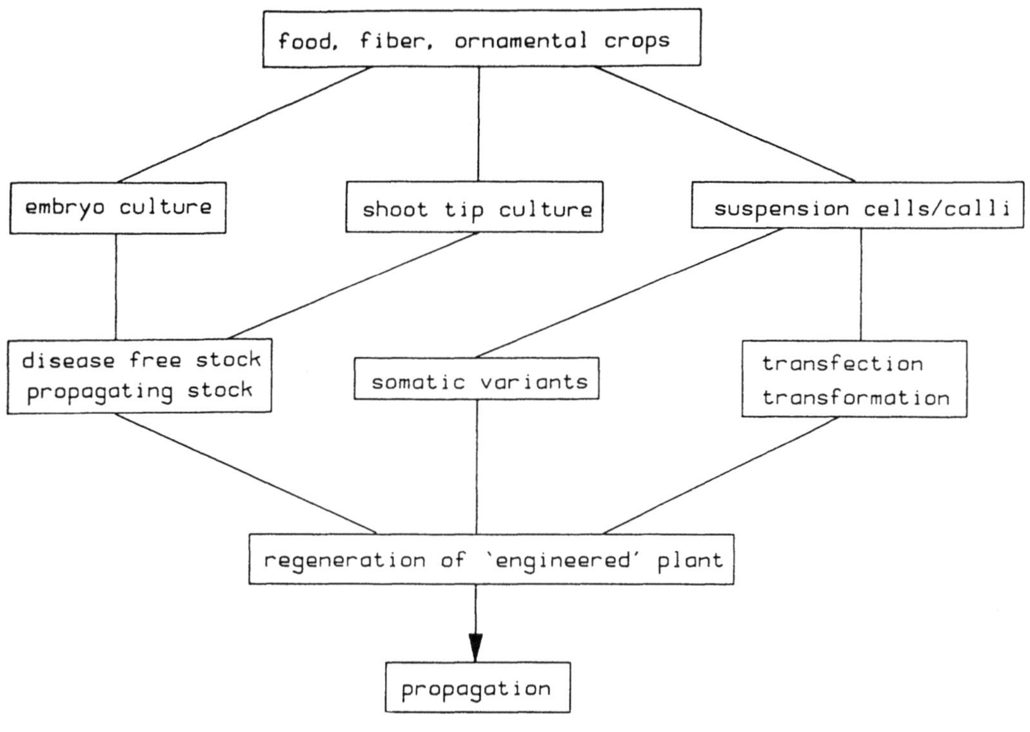

Fig. 1. Diagram of the technologies used in the development of genetically engineered plants. The source of desirable plant genes come from the indicated crops and from their wild type descendants. Embryo and shoot tip culture have been used for "escaping" from diseased stock and developing propagating stock with desirable features. Suspension cell and callus culture have provided somaclonal variants and have been used in the transformation and transfection experiments. These strategies culminate in the development of genetically engineered plants that are propagated.

1985). Therefore, it is apparent that genetically modified plant material needs to be processed quickly to progeny propagules, minimizing the lag time in tissue culture.

The approaches obviating randomized selection of genetic variants are those techniques that incorporate specific genes with functional promoters into plant cells directly. This includes transfection and transformation of DNA fragments containing the desired gene or gene set. The proven procedure of incorporating such genes in a relatively stable form into the chromosomes of plants is through the use of the natural transformation system of Agrobacterium tumefaciens. This system is reliable, efficient, and easy to perform, quickly yielding T1 regenerants from many different plant sources.

2. Direct transformation.

It has been long known that protoplasts incubated with DNA will incoporate that DNA into their nuclei (Lurquin and Kado, 1977; Fernandez et al., 1978). Like in the animal systems (Mertz and Gurdon, 1977; DeRobertis and Mertz, 1977; Wigler et al., 1979; 1980; Brinster et al., 1981; Gordon et al., 1981; Wagner et al., 1981; Constantini and Lacy, 1981; Harbers et al., 1981; Rusconi and Schaffner, 1981), virtually any DNA may be incorporated into the plant cell genomic nuclear DNA (Peerbolte et al., 1985). With further studies, it is likely that many different types of heterologous DNAs will be transcribed and translated in directly transformed plants irrespective of their promoter region as observed in Xenopus cells (Mertz and Gurdon, 1977; DeRobertis and Mertz, 1977). Part of this may be explained by DNA reorganization into expressible regions of the genome during or after integration. The scrambling of DNA as exemplified by the T-DNA in direct transformation experiments has been reported (Krens et al., 1985).

The direct incorporation of genetic material into higher cells include the use of calcium phosphate, polyethylene glycol, microinjection, liposomes, and electroporation that facilitate DNA incorporation either directly or by cell fusions

(Table 1). The DNA to be incorporated usually contains besides the gene of interest a gene with selectable traits such as aminoglycoside phosphotransferase that confers resistance to an aminoglycosidic antibiotic (e. g., kanamycin), or a screenable trait such as nopaline production catalyzed by nopaline synthase. Antibiotic resistance genes derived from bacteria are usually placed under the control of a promoter and a polyA site known to function in plants. Such genes serve as selectable markers and also reflect the efficiency of transformation.

3. Transfection.

In addition to transformation, transfection techniques have been reported (Table 2). These techniques have employed either cauliflower mosaic virus itself, or tandem arrays of the virus arranged in modified T-DNAs of the Ti plasmid of A. tumefaciens, or the cDNA of potato spindle tuber viroid, or selected regions of the genome of tobacco mosaic virus that were cloned between the T-DNA borders of modified pTi plasmids. The delivery of the virion was mediated through the natural infection process of A. tumefaciens, known as 'agroinfection' (Grimsley et al., 1986).

MATERIALS AND METHODS

Bacterial strains. The recombination deficient mutant LBA4301 (Klapwijk et al., 1979) of A. tumefaciens Ach-5 was used. Strain C58 was used to transform Haplopappus gracilis.

Plasmids. pUCD615 containing a lux cassette derived from Vibrio fischeri (Rogowsky et al., 1987) was used to determine the levels of gene expression of vir genes. Promoter active fragments were cloned into one of several unique restriction sites (BamHI, EcoRI, XcyI, XbaI) located about 600 bp upstream of the first start codon of the lux cassette. Promoter activity was measured by the amount of light produced in A. tumefaciens LBA4301(pTiC58) harboring the respective vir-lux fusion plasmids.

Plants. *H. gracilis* seeds (N = 2) were obtained from a vacant lot near the campus of the University of Arizona, Tucson. Carrot roots were purchased from a local market.

In situ hybridization. Protoplasts of *H. gracilis* were generated from suspension cells as described elsewhere (Quayle and Kado, 1987). Metaphase chromosomes were arrested with colchicine during midlog growth and used as the source of protoplasts. The protoplasts were lysed by dropping them from 1 meter onto acid washed glass slides. The preparations were air dried, treated with ribonuclease (Gerhard et al., 1981), and fixed in 5% paraformaldehyde (Haase et al., 1984). DNA was denatured in 0.15M NaOH in 70% ethanol (Landegent et al., 1984). The preparations were then sequentially dehydrated in increasing concentrations of ethanol, air dried and stored in a desicator. Denatured probe DNA (0.1 ug/ml) labelled with [^{3}H] was added immediately after carrier DNA had been denatured to a hybridization solution described elsewhere (Harper et al, 1981). The slides were then incubated for 16 h at 37°C with 20 ul of the hybridization mixture which was spread over the target area. The slides were washed in 2 X SSC (1X SSC = 0.15M NaCl, 0.015M Na citrate) for 2 h and then three times in 50% formamide in 3X SSC at 39°C for 5 min, in 2X SSC containing 0.1% Na lauryl sulfate at 37°C for 2 h, and in 0.1X SSC at room temperature for 2 h followed by washing at 4°C in fresh 0.1X SSC overnight. The slides were rinsed with distilled water and placed in increasing concentrations of ethanol. The air dried slides were dipped in nuclear track emulsion film (Kodak NTB2) and stored 5-7 days. The chromosomes were visualized by staining with azure B eosinate solution (Dvorak et al., 1984).

RESULTS

1. *Agrobacterium* mediated transformation.

Transmission of the T-DNA and chimeric T-DNA by infection of recipient plants with *A. tumefaciens* is reliable and yields stable integrations. Although rearrangements occur, especially

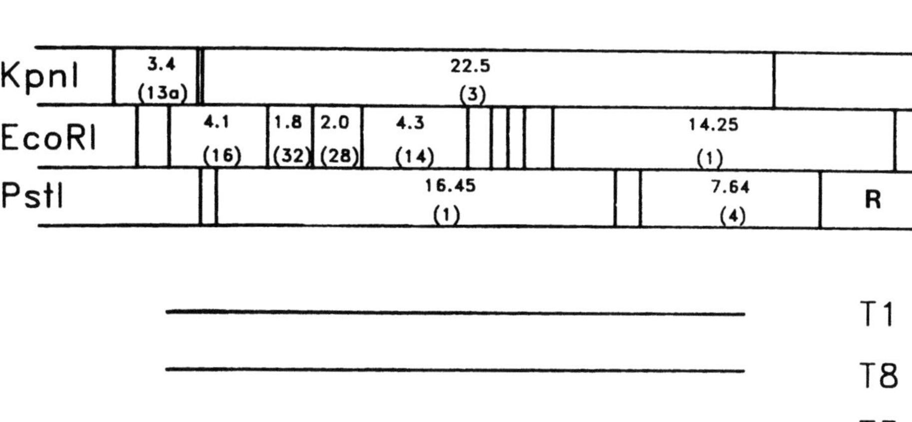

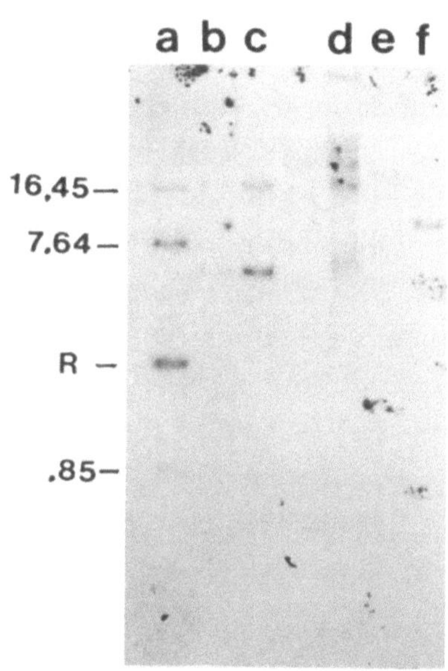

Fig. 2. Southern hybridization analysis of pTiC58 T-DNA in the nuclear DNA of transformed lines of *Haplopappus gracilis*. Genomic DNA digested with PstI, blotted, and hybridized with EcoRI fragment 1 of pTiC58. Lane a: pTiC58 (1 copy); b: T1 (0.2 copy); c: T1 (1 copy); d: T8 (1 copy); e: T8 (0.2 copy); f: T9: (1 copy).

when binary T-DNA vector systems are employed (Spielmann and Simpson, 1986), direct infection with A. tumefaciens containing chimeric T DNAs in full size Ti plasmids has been the current method of choice for dicotyledonous plants.

Southern hybridization analyses of crown gall tumor DNA from four transformed lines of H. gracilis (N =2) showed that various integration events can take place through natural infections with wild-type Ti plasmids (Fig. 2). Starting with the right border, T-DNA fragments of various sizes up to full length T-DNA had been stably inserted into the H. gracilis chromosome (Fig. 3). The integration of the T-DNA from one transformed line occurred only near the centromere of chromosome A.

2. Regulation of the vir genes during A. tumefaciens cell contact with carrot disks.

Genes in the virulence region of the pTi plasmid are required for the processsing of the T-DNA. These vir genes are virtually silent when A. tumefaciens cells are maintained in laboratory culture, but are induced when these bacterial cells come into contact with suspended plant cells and protoplasts. Phenolic compounds such as sinapinic acid involved in lignin biosynthesis, and derivatives such as acetosyringone (3',5'-dimethoxy-4'-hydroxyacetophenone) work effectively as inducers in the absence of plant cells under laboratory conditions (Stachel et al., 1985; Bolton et al., 1986).

Although vir gene induction has been shown in extracts of bacterial cells after contact with suspended plant cells or protoplasts, the induction and expression of vir genes in vivo has not been demonstrated. To determine whether the vir genes are actually induced during early tumorigenesis of wounded plant tissue, we have constructed gene fusions of the six vir operons of pTiC58 to the lux operon in pUCD615 (Rogowsky et al., 1987). The expression of each vir gene was determined by the amount of light produced by A. tumefaciens during its interaction with freshly wounded carrot disks (Table 3). It is apparent that virB, virC, virD and virE are expressed at high levels during the infection process. Unlike the virA and virG genes of pTiA6 (Stachel and Nester, 1986), some induction occurs with virA, but little if any induction is observed with virG. Aside from these differences between octopine (pTiA6) and nopaline (pTiC58) type plasmids, these data corroborate the results from the induction experiments carried out with cocultivation of tobacco suspension cells and with culture media containing acetosyringone. Although, acetosyringone was not shown to be present in freshly wounded carrot disks, the parallel induction of the vir genes suggests that either acetosyringone itself or a related or a set of related compounds serve as inducer(s).

The vir region of the octopine-type Ti plasmids pTiA6 and pTiAch5 consists of seven complementation groups: virA, virB, virC, virD, virE, virF, and virG (Hille et al., 1984; Hooykaas et al., 1984; Iyer et al., 1982; Klee et al., 1983). The vir region of the nopaline-type Ti plasmid pTiC58 consists of six complementation groups (Lundquist et al., 1984) and shares considerable DNA homologies with the octopine pTi counterparts of virB (Hirooka et al., 1987), virC (Close et al., 1987), virD (Hagiya et al., 1985), virE (Hirooka et al., 1987), and virG (Powell et al., 1987). There is no virF in pTiC58 (Otten et al., 1985; Hirooka and Kado, 1986). The expression of the vir genes of both octopine- and nopaline-type pTi plasmids is modulated by at least two mechanisms: positive regulation of virB, virC, virD, virE, and virG is dependent on plant phenolic compounds such as acetosyringone, and negative regulation of virC and virD is abolished by strains carrying a chromosomal mutation in the ros gene (Close et al., 1985; Rogowsky et al., 1987; Close et al., 1987b).

Clearly, the regulation of the vir genes is more complicated than initially viewed. There is extensive conservation of the carboxy terminal regions between products of virA of A. tumefaciens and ntrB of Bradyrhizobium parasponia and Klebsiella pneumoniae, and envZ, cpxA, and phoR of Escherichia coli (Nixon et al., 1986). There is also homology between the product of virG of A. tumefaciens and ompR, sfrA, phoB, cheY and cheB of E. coli, and ntrC of K. pneumoniae, and spoOF of B. subtilis (Winans et al., 1986; Powell, et al., 1987). In the above two component regulatory systems, virA and its homologs are thought to encoded sensors of environmental stimuli, while virG and its homologs code for positive regulatory elements. VirG is required for the positive regulation virB and virE of pTiA6NC (Winans et al., 1986), and virB, virC, virD, and virE of pTiC58 (Rogowsky et al., 1987). Superimposed on this regulation is the ros gene, which when mutated leads to the full expression of virC and virD (Close et al., 1985; Close et al., 1987b).

3. T-DNA processing.

One of the early processing steps involves the excision of the T-DNA, which is clearly distinguished by flanking bordering sequences of nearly perfect 25 bp directly repeated nucleotides. Interestingly, as shown below any DNA linked to the right border is transferred and integrated into the plant nuclear genome. The T-DNA transfer process is directed by products of the vir region, of which virD has been identified to encode a site-specific endonuclease that nicks at a unique site within each of the border repeats that flank the T-DNA (Yanofsky et al., 1986) to produce a single stranded product when A. tumefaciens cells were treated with acetosyringone (Stachel et al., 1986) or cocultivated with suspension cells of tobacco (Albright et al., 1987). On the other hand, double stranded intermediates of the T-DNA are produced if the bacterial cells are exposed to regenerating tobacco protoplasts, (Koukolikova-Nicola et al., 1985; S. Gelvin, et al., 1987). In order to resolve these differences, we have analyzed T-DNA intermediate formation in the ros mutant of A. tumefaciens that is permissive for the constitutive expression of virC and virD (Close et al., 1985) and therefore does not require induction with plant cells or with acetosyringone. Our analysis showed that double stranded intermediates are produced from the wild-type Ti plasmid in the ros mutant (Fig. 4). Mutations in the second open reading frame of virD results in the loss in the ability to produce the T-DNA intermediate,

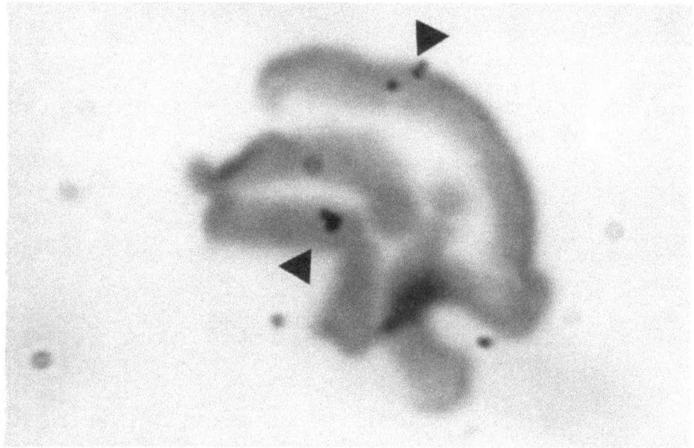

Fig. 3. In situ hybridization of the two chromosomes of transformed line T1 of tetraploid H. gracilis. Sites of T-DNA integration are indicated by the arrows.

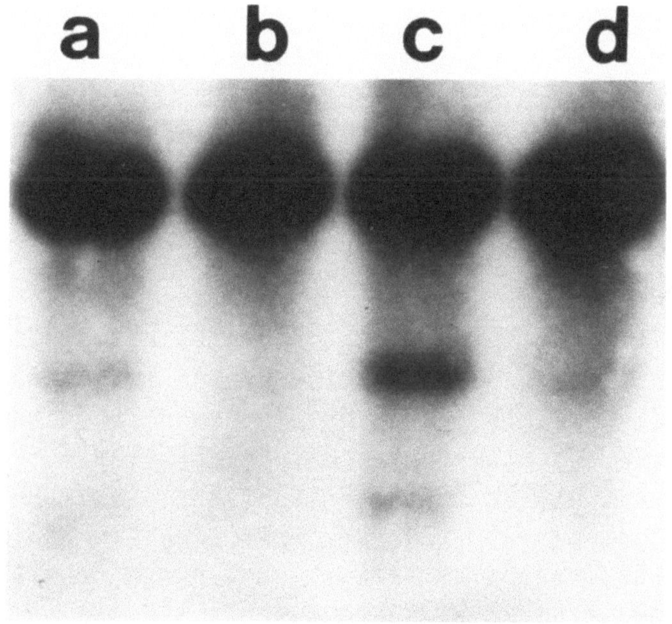

Fig. 4. Double stranded intermediate formation in LBA4301 (lanes a and b) and LBA4301ros (lanes c and d) with (lanes a and c) and without (lanes b and d) induction with acetosyringone. The hybridization probe was the 14.2 kb EcoR1 fragment 1 of pTiC58, which was labelled by primer extension with [32P]-NTP. pTi DNAs obtained from the treated bacterial cells were cleaved with EcoR1, blotted and hybridized. Double stranded intermediates of 7.3 and 6.9 kb are below the pTiDNA. Wild-type LBA4301 that was induced with acetosyringone shows intermediate formation. The same intermediate is formed in induced and uninduced LBA4301ros. The hybridization analysis was carried out on each vir gene mutant as summarized in Table 4.

suggesting the the endonuclease (Yanofsky et al., 1986) is required for its production. Mutations in *vir*C or other *vir* genes have no effect on intermediate production (Table 4).

DISCUSSION

Three major approaches have been employed in the genetic modification of plants. These are propagation of somaclonal variants, direct and helper organism mediated transformation, and transfection. The transfomation system employing *A. tumefaciens* has been the method of choice since stable transgenic progeny have been consistently obtained. However, the processing of the T-DNA and its transmission to the nuclear genome of the host plant is not well understood. Current work therefore has focused on the genes involved in T-DNA processing. These genes are located in the *vir* region of the Ti plasmid and extensive work is being done on each *vir* operon. Interestingly, the *vir*D operon contains at least one gene (*vir*D2) that encodes an endonuclease (Yanofsky et al., 1986) which is responsible for generating the T-DNA intermediates in the *ros* mutant (Fig. 4). This endonuclease is reminiscent of endonuclease A of *A. tumefaciens*, which was reported to cleave Ti plasmid DNA and superhelical DNA but not linear DNA (LeBon et al., 1978).

Elucidating the processing step will be complicated since it is not known whether the excised T-DNA molecules observed in the bacterial cell are the actual intermediates that are transferred to the plant cell, or whether single or double stranded intermediates are even required. Since only the right border is required for transfer, the major responsibility for processing the T-DNA lies with the *vir* genes. The regulation of the *vir* genes is tuned by environmental signals in the form of plant phenolics that are emmitted by wounded plant cells such as in our carrot disk experiments where the *vir* genes are induced during the on-going infection process. It is not known if these phenolics are a general signal produced by all wounded plants.

TABLE 1. TRANSFORMATION OF PLANT PROTOPLASTS BY VARIOUS METHODS

DNA	METHOD USED	RECIPIENT	REFERENCE
PBR313	ZNSO4-POLY-L-ORNITHINE	COWPEA	LURQUIN AND KADO. 1977. MOL. GEN. GENET. 154:113-121.
PBR313	ZNSO4-POLY-L-ORNITHINE PROTAMINE SULFATE	COWPEA TURNIP CARROT	FERNANDEZ ET AL. 1978. FEBS LETT. 87:277-282.
PBR322	LIPOSOMES	COWPEA CARROT	LURQUIN AND SHEEHY. 1982. PLANT SCI. LETT. 25:133-146. LURQUIN. 1979. NUCL. ACIDS RES. 6:3773-3784.
PTIACH-5	POLY-L-ORNITHINE	PETUNIA	DAVEY ET AL., 1980. PLANT SCI. LETT. 18:307-313.
PYEARG4	ZNSO4-POLY-L-ORNITHINE	CHLAMYDOMONAS REINHARDII	ROCHAIX AND DILLEWIJN. 1982. NATURE 296:70-72.
PTIACH5	POLYETHYLENE GLYCOL AND CACL2	TOBACCO	KRENS ET AL. 1982. NATURE 296: 72-74. KRENS ET AL. 1985. PLANT MOL. BIOL. 5:223-234.
PTIB6S3 PTIC58 PRI15834	SPHEROPLAST FUSION POLYETHYLENE GLYCOL	TOBACCO	HAIN ET AL. 1984. PLANT CELL REPT. 3:60-64.
PABD1	POLYETHYLENE GLYCOL	TOBACCO	PASZKOWSKI ET AL. 1984. EMBO J. 3:2717-2722.
PABD1	POLYETHYLENE GLYCOL HEAT SHOCK	ITALIAN RYEGRASS	POTRYKUS ET AL. 1985. MOL. GEN. GENET. 199:183-188.
PABD1	ELECTROPORATION-POLYETHYLENE GLYCOL-HEAT SHOCK	TOBACCO	SHILLITO ET AL. 1985. BIO-TECHNOL. 3:1099-1103
PNOSCAT4	DEAE-DEXTRAN	MAIZE	HOWARD ET AL. 1985. IN: PLANT GENETICS P. 225-234. (M. FREELING, ED.) ALAN R. LISS, INC. NY)
PBL1102-4	POLYETHYLENE GLYCOL	TRITICUM MONOCOCCUM	LORZ ET AL. 1985. MOL. GEN. GENET. 199:178-182.
PLGV23NEO	LIPOSOMES POLYETHYLENE GLYCOL	TOBACCO	DESHAYES ET AL. 1985. EMBO J. 4:2731-2737.
PBR322 WITH APTII GENE	CALCIUM PHOSPHATE	TOBACCO	HAIN ET AL. 1985. MOL. GEN. GENET. 199:161-168.
PTIA208	SPHEROPLAST FUSION	VINCA ROSEA	OKADA ET AL. 1985. PLANT CELL REPT. 4:133-136.
PTIC58	ELECTROFUSION	CARROT	LANDRIDGE ET AL. 1985. PLANT CELL REPT. 4:355-359.
PNOSCAT	ELECTROPORATION	CARROT	FROMM ET AL. 1985. PROC. NAT. ACAD. SCI. USA 83:5602-5606 FROMM ET AL. 1986. NATURE 319:791-793.
PMON200	ELECTROPORATION	TOBACCO	RIGGS AND BATES. 1986. PROC. NAT. ACAD. SCI. USA 83:5602-5606.
PCGN561	MICROINJECTION	TOBACCO	CROSSWAY ET AL. 1986. MOL. GEN. GENET. 202:179-185.
PTIA208 PTIA277	SPHEROPLAST FUSION	RICE	BABA ET AL. 1986. PLANT CELL PHYSIOL. 27:463-471.
PLGV2103NEO	CALCIUM PHOSPHATE	TOBACCO	CZERNILOFSKY ET AL. 1986. DNA 5:101-114.
PTI	MICROINJECTION	ALFALFA	REICH ET AL. 1986. BIO/TECHNOL. 4:1001-1004.

DNA	Method Used	Recipient	Reference
DIHYDROFOLATE REDUCTASE GENE OF R67	CAULIFLOWER MOSAIC VIRUS	TURNIP	BRISSON ET AL. 1984. NATURE 310:511-514. AL. 1984.
65 BP FRAGMENT OF LAC PROMOTER	CAULIFLOWER MOSAIC VIRUS	TURNIP	GRONENBORN ET AL. 1981. NATURE 294:773-776.
TANDEM CaMV	'AGROINFECTION'	TURNIP	GRIMSLEY ET AL. 1986. PROC. NAT. ACAD. SCI. USA 83:3282-3286.
SINGLE OR TANDEM TOMATO GOLDEN MOSAIC VIRUS	'AGROINFECTION'	PETUNIA	ROGERS ET AL. 1986. CELL 45: 593-600.
COAT PROTEIN cDNA GENE OF TOBACCO MOSAIC VIRUS	'AGROINFECTION'	TOBACCO	ABEL ET AL. 1986. SCIENCE 232:738-743.
POTATO SPINDLE TUBER VIROID cDNA	'AGROINFECTION'	TOMATO	GARDNER ET AL. 1986. PLANT MOL. BIOL. 6:221-228; OWENS ET AL. 1986. PLANT MOL. BIOL. 6:179-192.
CHLORAMPHENICOL ACETYLTRANSFERASE	RECONSTITUTED TMV cDNA TRANSCRIPTS	TOBACCO	TAKAMATSU 1987. EMBO J. 6:307-311.

TABLE 2. TRANSFECTION OF PLANT CELLS

Plasmid in LBA4301	Promoter fusion	Relative light produced*	
		+pTiC58	-pTiC58
pUCD1186	*virA*	++	+
pUCD1187	*virB*	++++	-
pUCD1168	*virC*	++	-
pUCD1173	*virD*	+++	-
pUCD1194	*virE*	++++	-
pUCD1195	*virG*	-	-
pUCD615	none	-	-
pUCD607	tet	++++++	++++++

Table 3. Induction of *vir* genes in *A.* tumefaciens C58 during infection of carrot disks

*Each carrot disk was inoculated with *A.*tumefeciens containing each plasmid indicated. After a 16 hour incubation at room temperature, the amount of light emmitted from each carrot disk was measured with a photometer and photographed as described elsewhere (Shaw et al., 1987).

Table 4. Vir genes involved in T-DNA intermediate formation

Gene with Tn5	T-DNA intermediate formation from pTiC58 in:	
	LBA4301	LBA4301*ros*
*vir*A	no	yes
*vir*B	no	yes
*vir*C	no	yes
*vir*D1	no	no
*vir*D2	no	no
*vir*D4	no	yes
*vir*E	no	yes
*vir*G	no	yes

ACKNOWLEDGMENTS

We thank Dr. Salome S. Pais, members of the Scientific Committee, and the NATO Scientific Affairs Division for giving us the opportunity to present our latest research, and Dr. Todd Steck for manuscript review. This work was supported by DHHS grant CA-11526 from the National Cancer Institute.

REFERENCES

Abel, P. P., R. S. Nelson, B. De, N. Hoffmann, S. G. Rogers, R. T. Fraley, and R. N. Beachy. 1986. Science 232: 738-743.
Albright, L. M., M. F. Yanofsky, B. Leroux, D. Ma, and E. W. Nester. 1987. J. Bacteriol. 169: 1046-1055.
Alt-Moerbe, J., B. Rak, and J. Schroder. 1986. EMBO J. 5: 1129-1135.
Baba, A., S. Hasesawa, and K. Syono. 1986. Plant Cell Physiol. 27: 463-471.
Bolton, G. W., E. W. Nester, and M. P. Gordon. 1986. Science 232: 983-985.
Brinster, R. L., H. Y. Chen, M. Trumbauer, A. W. Senear, R. Warren, and R. D. Palmiter. 1981. Cell 27: 223-231.
Brisson, N., J. Paszkowski, J. R. Penswick, B. Gronenborn, I. Potrykus, and T. Hohn. 1984. Nature 310: 511-514.
Close, T. J., R. C. Tait, and C. I. Kado. 1985. J. Bacteriol. 164: 774-781.
Close, T. J., R. C. Tait, H. Rempel, T. Hirooka, L. Kim., C. I. Kado. 1987a. J. Bacteriol. in press
Close, T. J., P. M. Rogowsky, C. I. Kado., S. Winans, M. F. Yanofsky, and E. W. Nester. 1987b. J. Bacteriol. in press.
Constantini, F., and E. Lacy. 1981. Nature 294: 92-94.
DeRobertis, E. M., and J. E. Mertz. 1977. Cell 12: 175-182.
Crossway, A., J. V. Oakes, J. M. Irvine, B. Ward, V. C. Knauf, and C. K. Shewmaker. 1986. Mol. Gen. Genet. 202: 179-185.
Czernilofsky, A. P., R. Hain, L. Herrera-Estrella, H. Lorz, E. Goyvaerts, B. J. Baker, and J. Schell. 1986. DNA 5: 101-113.
Davey, M. R., E. C. Cocking, J. Freeman, N. Pearce, and I. Tudor. 1980. Plant Sci. Lett. 18: 307-313.
Deshayes, A., L. Herrera-Estrella, and M. Caboche. 1985. EMBO J. 4: 2731-2737.
Dvorak, J., M. W. Lassner, R. S. Kota, and K. C. Chen. 1984. Can. J. Genet. Cytol. 26: 628-632.
Fernandez, S. M., P. F. Lurquin, and C. I. Kado. 1978. FEBS Letters 87: 277-282.
Fromm, M., L. P. Taylor, and V. Walbot. 1985. Proc. Natl. Acad. Sci. USA 82: 5824-5828.
Fromm, M., L. P. Taylor, and V. Walbot. 1986. Nature 319: 791-793.
Gardner, R. C., K. R. Chonoles, and R. A. Owens. 1986. Plant Mol Biol. 6: 221-228.
Gerhard, D. S., E. S. Kawasaki, F. C. Bancroft, and P. Szabo.

1981. Proc. Natl. Acad. Sci. USA 78: 3755-3759.
Gordon, J. W., G. A. Scangos, D. J. Plotkin, J. A. Barbosa, and F. H. Ruddle. 1980. Proc. Natl. Acad. Sci. USA 77: 7380-7384.
Grimsley, N., B. Hohn, T. Hohn, and R. Walden. 1986. Proc. Natl. Acad. Sci. USA 83: 3282-3286.
Gronenborn, B., R. C. Gardner, S. Schaefer, and R. J. Shepherd. 1981. Nature 294: 773-776.
Haase, A., M. Brahic, L. Stowring, and H. Blum. 1984. Methods in Virology 7: 189-226.
Hagiya, M., T. J. Close, R. C. Tait, and C. I. Kado. 1985. Proc. Natl. Acad. Sci. USA 82: 2669-2673.
Hain, R., P. Stabel, A. P. Czernilofsky, H. H. Steinbiss, L. Herrera-Estrella, and J. Schell. 1985. Mol. Gen. Genet. 199: 161-168.
Harbers, K., D. Jahner., and R. Jaenisch. 1981. Nature 293: 540-542.
Harper, M. E., A. Ullrich, and G. F. Saunders. 1981. Proc. Natl. Acad. Sci. USA 78: 4458-4560.
Hepburn, A. G., L. E. Clarke, K. S. Blundy, and J. White. 1983. J. Mol. Appl. Genet. 2: 211-224.
Hille, J., J. Van Kan, and R. Schilperoort. 1984. J. Bacteriol. 158: 754-756.
Hirooka, T., and C. I. Kado. 1986. J. Bacteriol. 168: 237-243.
Hirooka, T., P. M. Rogowsky, and C. I. Kado. 1987. J. Bacteriol. 169: 1529-1536.
Hirooka, T., R. C. Lundquist, P. Rogowsky, R. C. Tait, and C. I. Kado, unpublished results.
Hooykaas, P. J. J., M. Hofker, H. Den Dulk-Ras, and R. A. Schilperoort. 1984. Plasmid 11: 195-205.
Howell, S. T., L. L. Walker, and R. M. Walden. 1981. Nature 293: 483-486.
Howard, E. A., K. J. Danna, E. S. Dennis, and W. J. Peacock. 1985. In: Plant Genetics (M. Freeling, ed.) pp. 225-234. Alan R. Liss, Inc., New York
Iyer, V. N., H. J. Klee, and E. W. Nester. 1982. Mol. Gen. Genet. 188: 418-424.
Kado, C. I., and A. Kleinhofs. Int. Rev. Cytology, Supplement 11B: 47-80 (1980).
Klapwijk, P. M., P. van Beelen, and R. A. Schilperoort. 1979. Mol. Gen. Genet. 173: 171-175.
Klee, H. J., F. F. White, V. N. Iyer, M. P. Gordon, and E. W. Nester. 1983. J. Bacteriol. 153: 878-883.
Koukolikova-Nicola, Z., R. D. Shillito, B. Hohn, K. Wang, M. Van Montagu, and P. Zambryski. 1985. Nature 313: 191-196.
Krens, F. A., L. Molendijk, G. J. Wullems, and R. A. Schilperoort. 1982. Nature 296: 72-74.
Krens, F. A., R. M. W. Mans, T. M. S. Van Slogteren, J. H. C. Hoge, G. J. Wullems, and R. A. Schilperoort. 1985. Plant Mol. Biol. 5: 223-234.
Kwok, W. W., E. W. Nester, and M. P. Gordon. 1985. Nucleic Acids Res. 13: 459-470.
Landegent, J. E., I. De Wal Jansen, R. A. Baan, J. H. J. Hoeijmakers, and M. Van der Ploeg. 1984. Exptl. Cell. Res. 153: 61-72.
Landridge, W. H. R., B. J. Li, and A. A. Szalay. 1985. Plant

Cell Rept. 4: 355-359.
LeBon, J. M., C. I. Kado, L. J. Rosenthal and J. G. Chirikjian. 1978. Proc. Natl. Acad. Sci. USA 75: 4097-4101.
Lorz, H., B. Baker, and J. Schell. 1985. Mol. Gen. Genet. 199: 178-182.
Lundquist, R. C.,, T. J. Close, and C. I. Kado. 1984. Mol. Gen. Genet. 193: 1-7.
Lurquin, P. F. 1979. Nucleic Acids Res. 6: 3773-3784.
Lurquin, P. F., and C. I. Kado. 1977. Mol. Gen. Genet. 154: 113-121.
Lurquin, P. F., and R. E. Sheehy. 1982. Plant Sci. Lett. 25: 133-146.
Mertz, J. E., and J. B. Gurdon. 1977. Proc. Natl. Acad. Sci. USA 74: 1502-1506.
Nixon, B. T., C. W. Ronson, and F. M. Ausubel. 1986. Proc. Nat. Acad. Sci. 83: 7850-7854.
Okada, K., S. Hasezawa, K. Syono, and T. Nagata. 1985. Plant Cell Rept. 4: 133-136.
Otten, L., G. Piotrowiak, P. Hooykaas, M. Dubois, E. Szegedi, and J. Schell. 1985. Mol. Gen. Genet. 199: 189-193.
Owens, R. A., R. W. Hammond, R. C. Gardner, M. C. Kiefer, S. M. Thompson and D. E. Cress. 1986. Plant Mol. Biol. 6: 179-192.
Paszkowski, J., R. D. Shillito, M. Saul, V. Mankak, T. Hohn, and I. Potrykus. 1984. EMBO J. 3: 2717-2722.
Peerbolte, R., F. A. Krens, R. M. W. Mans, M. Floor, J. H. C. Hoge, G. J. Wullems, and R. A. Schilperoort. 1985. Plant Mol. Biol. 5: 235-246.
Potrykus, I., M. Saul., J. Petruska, J. Paszkowski, and R. D. Shillito. 1985. Mol. Gen. Genet. 199: 183-188.
Potrykus, I., J. Paszkowski, M. Saul, S. Kruger-Lebus, T. Muller, R. Schocher, I. Negrutiu, P. Kunzler, and R. D. Shillito. 1985. In: "Plant Genetics" M. Freeling, ed. Alan R. Liss, Inc., New York. pp. 181-199.
Powell, B. , G. K. Powell, R. O. Morris, P. M. Rogowsky, and C. I. Kado. 1987. manuscript submitted
Quayle, T. J. A., and C. I. Kado, submitted.
Reich, T. J., V. N. Iyer, and B. L. Miki. 1986. Bio/Technology 4: 1001-1004.
Riggs, C. D., and G. W. Bates. 1986. Proc. Natl. Acad. Sci. USA 83: 5602-5606.
Rochaix, J.-D., and J. Van Dillewijn. 1982. Nature 296: 70-72.
Rogers, S. G., D. M. Bisaro, R. B. Horsch, R. T. Fraley, N. L. Hoffmann, L. Brand, J. S. Elmer, and A. M. Lloyd. 1986. Cell 45: 593-600.
Rogowsky, P., T. J. Close, and C. I. Kado. 1987. In: Molecular Genetics of Plant-Microbe Interactions. (D. P. S. Verma and N. Brisson, eds.) Martinus Nijhoff, Dorrdrecht/Boston/Lancaster. pp. 14-19.
Rogowsky, P. M., T. J. Close, J. A. Chimera, J. J. Shaw, and C. I. Kado. 1987. J. Bacteriol., in press.
Rusconi, S., and W. Schaffner. 1981. Proc. Natl. Acad. Sci. USA 78: 5051-5055.
Scowcroft, W. R., P. Davies, S. A. Ryan, R. I. S. Brettell, M. A. Pallotta and P. J. Larkin. 1985. in "Plant Genetics" M. Freeling, ed. Alan R. Liss, Inc., New York. pp. 799-815.

Shaw, J. J., P. M. Rogowsky and C. I. Kado. 1987. Plant Mol. Biol. Rept. in press
Shillito, R. D., M. W. Saul, J. Paszkowski, M. Muller, and I. Potrykus. 1985. Bio/Technology 3: 1099-1103.
Spielmann, A., and R. B. Simpson. 1986. Mol. Gen. Genet. 205: 34-41.
Stachel, S. E., and E. W. Nester. 1986. EMBO J. 5: 1445-1454.
Stachel, S. E., E. Messens, M. Van Montagu, and P. Zambryski. 1985. Nature 318: 624-629.
Stachel, S., B. Timmerman, and P. Zambryski. 1986. Nature 322: 706-712.
Takamatsu, N., M. Ishikawa, T. Meshi, and Y. Okada. 1987. EMBO J. 6: 307-311.
Townsend, R., J. Watts, and J. Stanley. 1986. Nucl. Acids Res. 14: 1253-1265.
Wagner, T. E., P. C. Hoppe, J. D. Jollick, D. R. Scholl, R. L. Hodinka, and J. B. Gault. 1981. Proc. Natl. Acad. Sci. USA 78: 6376-6380.
Wigler, M., R. Sweet, G. K. Sim, B. Wold, A. Pellicer, E. Lacy, T. Maniatis, S. Silverstein, and R. Axel. 1979. Cell 16: 777-785.
Wigler, M., M. Perucho, D. Kurtz, S. Dana, A. Pellicer, R. Axel, and S. Silverstein. 1980. Proc. Natl. Acad. Sci. USA 77: 3567-3570.
Winans, S. C., P. R. Ebert, S. E. Stachel, M. P. Gordon, and E. W. Nester. 1986. Proc. Nat. Acad. Sci. USA 83: 8278-8282.
Yanofsky, M. F., S. G. Porter, C. Young, L. M. Albright, M. P. Gordon, and E. W. Nester. 1986. Cell 47: 471-477.

EFFECT OF N^6-BENZYL AMINOPURINE ON THE TRANSLATION AND TRANSCRIPTION ACTIVITY FROM APPLE CELLS

Ricardo J. Ordás, Roberto Rodriguez, Belen Fernández and Ricardo Sanchez

University of Oviedo, Laboratorio de Fisiologia Vegetal, Arias de Velasco s/n, Oviedo, Asturias, Spain

INTRODUCTION

In vitro culture studies of the effect and action mechanism of plant growth regulators are relevant in establishing, the basis of prediction and understanding the "in vitro" behavior.

Cytokinin, like other types of plant growth regulators, affects a number of physiological and metabolic processes (references in: Guern and Pèaud-Lenoël 1981; Amrhein 1983; Scott 1984). Although there is evidence that cytokinins regulate by a direct control of protein synthesis at the translation level (Klämbt 1977; Muren and Fosket 1977; Kulaeva 1981; Szweykowska et al. 1981) and possibly at the transcription level (Waliszewska-Wojtkowiak et al. 1985), their mechanism of action is still unknown. Thus in this work the attempt is made to obtain more information on the way it enhances cell division.

Abbreviations: BAP, N^6-benzyl aminopurine; 2,4-D, 2,4-dichlorophenoxyacetic; PAGE, polyacrylamide gel electrophoresis; RNA, ribonucleic acid; SDS, sodium dodecyl sulphate; TCA, trichloroacetic acid.

MATERIAL AND METHODS

Apple cell suspension culture: A cytokinin-requiring cell line (D2,14), cloned from leaf callus of apple (*Malus domestica* Borkh.) was used. Cell suspensions were routinely subcultured in an agitated M.S. liquid medium. (Murashige and Skoog 1962), containing 2,4-D (5 μM) and kinetin (0.5 μM). Cells were collected in the early stationary growth phase and washed with cytokinin-free control medium. The washed cells were transferred to fresh medium without cytokinin, 0.5 μM BAP was added to the medium after 0,3, 6, and 9 days of culture.

NATO ASI Series, Vol. H18
Plant Cell Biotechnology. Edited by M. S. S. Pais et al.

Cell suspensions were maintained with constant agitation at 25 °C and a 16-h photoperiod of white light (1.75 klux).

To measure the population density, aliquots of cell suspensions were added to chromic acid (100 g l^{-1} final concentration, 15 h at room temperature). Cells were mechanically dispersed and counted under the microscope using a Fuchs-Rosenthal counting cell.

In vivo synthesis of protein: In order to determine the speed of response, the cell suspension, grown for 5 days in an M.S. medium with 2,4-D (5 µM) (lacking cytokinin), was transferred at a density of 0.3 x 10^5 125 (98% viability) per ml to 5 ml of similar fresh media with or without BAP (0.5 µM). The suspensions were supplemented with 5 µCi ^{35}S-L-methionine (1000 Ci $mmol^{-1}$). The labeled amino acid was isotopically diluted with 10 µl unlabeled L-methionine (5 mM).

The labeled medium with BAP (0.5 µM) also was prepared with or without 25 mg actinomycin-D.

The cell suspensions were agitated at 25 °C under continuous white light in 25-ml flasks.

The reaction was terminated by cooling the samples (0.5 ml) on ice and adding 1.5 ml of 100 g l^{-1} TCA. After 12 h at 4 °C, the samples were heated to 90 °C for 15 min. After cooling, the precipitate was collected on Whatman GF/A glass fiber discs. The discs were washed first with 100 g^{-1} TCN and then dried at 60°C. The radioactivity of the discs was counted in a toluene-based scintillation mixture.

Analysis of the in vitro translation product: The translation product was analyzed using SDS-PAGE according to Laemmli (1970). Filtered suspensions were rinsed with 40 cell volumes of a mixture of 20 mM KCl-5 mM EDTA, then 10 cell volumes of 20 mM KCl. Acetone powders were prepared from intact tissue by homogenizing the cells with 5 vol of cold (-20 °C) acetone in a Potter-Elvejhem glass homogenizer for 15 s. The acetone-precipitated material was pelleted (5 min in a microfuge) and washed in cold acetone. Air-dried pellets were dissolved in 100 µl of Laemmli buffer (1970) at twice the concentration, according to Axelos

Pèaud-Lenoël (1980). The dissolved samples were diluted with 100 µl of water and were then heated in a boiling water bath for 2 min. After cooling to room temperature, samples (20-100 µl) of equal radioactivity (usually 100000 cpm) were fractionated and analyzed by SDS-PAGE (Laemmli 1970) in 1-mm-thick slab gels containing 60 to 200 g l^{-1} (w/v) acrylamide linear concentration gradients. Marker proteins of known molecular weight were included in parallel.

After electrophoresis, the gels were fixed with 200 g l^{-1} TCA, rinsed in water, and impregnated with sodium salicylate (Chamberlain 1979). The gels were dried on Whatman 3MM paper under vacuum. Radioactive peptides were located by fluorography at -70 ^{o}C using Kodak X-O mat X-ray film.

<u>5,6^{3}H-uridine incorporation into the transcribed RNA:</u> Culture conditions were the same as before but supplemented with 5 µCi 5,6^{3}H uridine (41 Ci mmol^{-1}) which was isotopically diluted with 10 µl unlabeled L-methionine (1 mM). The radioactivity estimation was also similar, but Whatman discs were washed first with 100 g l^{-1} TCA containing 10 mM uridine.

All experiments were repeated three times with similar results.

RESULTS AND DISCUSSION

The apple cell line used in these experiments is cytokinin-dependent as ilustrated in Fig. 1. The cell number progressively decreased as BAP (0.5 µM) application is delayed (3, 6, and 9 days). After BAP (0.5 µM) was added some cells began to divide.

These results are in agreement with the hypothesis that during a period in cytokinin-deficient medium cells undergo senescence processes which prevent them from responding to the exogenously applied BAP. There is a correlation between BAP application and cell number counted.

A significant increase in protein synthesis occurred (Fig. 2). The ^{35}S-methionine incorporation increased nearly 100% 240 min after BAP application which shows that the main effect is a translation, while transcription activity, as traced with 5-6^{3} H-uridine, remained without significant change between cells maintained with and without BAP.

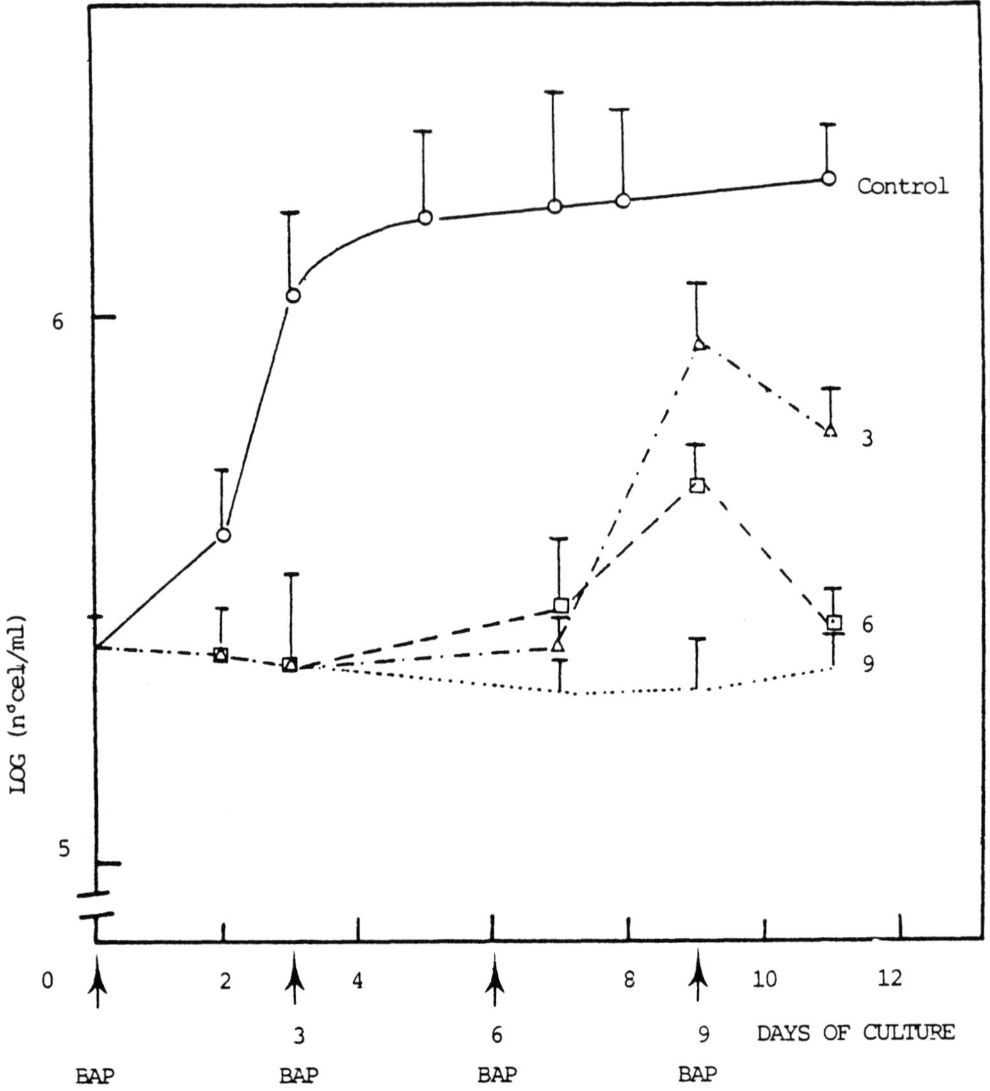

Fig. 1. Growth of suspension culture of apple. BAP was added to the medium at several times (0, control; 3; 6, and 9 days) during the culture period

The results of Fig. 2 support the hypothesis that the primary action of BAP is on translation. The application of actinomycin-D confirms this hypothesis since it did not change protein synthesis (Fig. 3). The processes studied are not dependent on de novo synthesis of RNA.

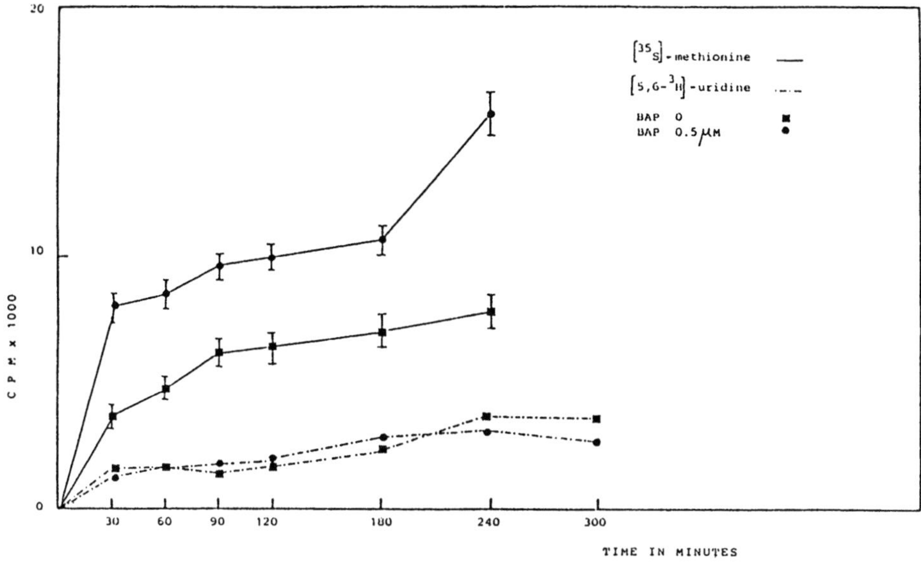

Fig. 2. Uptake of ^{35}S-methionine and 5-6^{3}H-uridine by apple cells in a culture medium lacking BAP

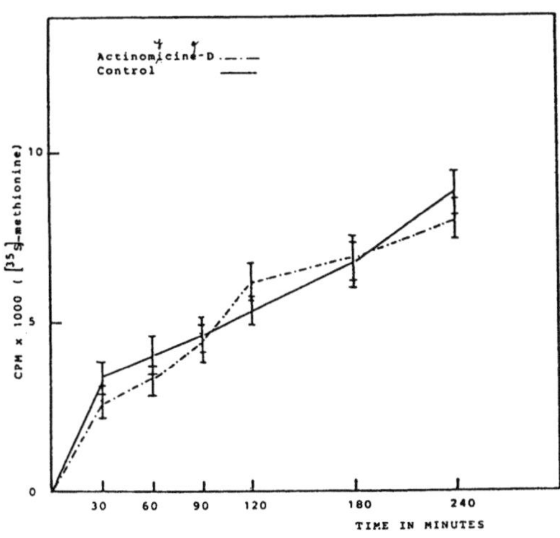

Fig. 3. Effect of actinomycin-D on the protein synthesis of apple cells cultivated with BAP (0.5 μM)

There were no qualitative differences on protein synthesis after BAP application (Fig. 4) which is consistent with the results of Gwòzdz and Wozny (1983), who showed that there were no qualitative differences in in vitro polypeptide synthesis. However,

Teyssedier de la Serve et al. (1984) found that qualitative changes occurred when cytostatic concentrations of BAP (10 μM) were applied.

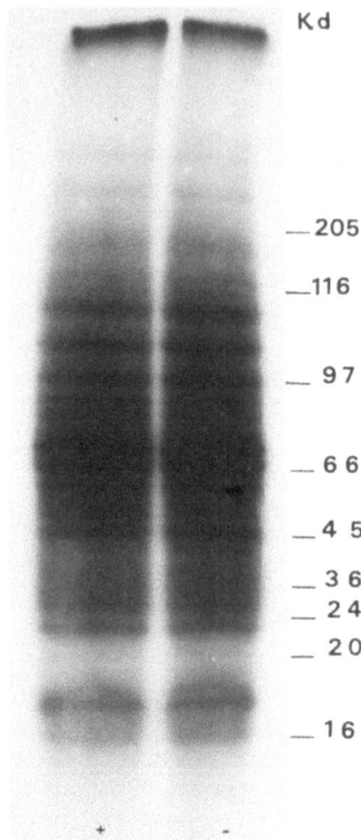

Fig. 4. A fluorogram of a gradient polyacrylamide gel after SDS electrophoretic separation of protein ^{35}S-labeled proteins from apple cells cultured for 20 h in medium containing or lacking a cytokinin. Stationary-phase cells were transferred to fresh medium lacking a cytokinin. Five days later the cells were collected on Miracloth, washed with medium, and transferred to fresh medium which either contained 0.5 μM BAP or lacked a cytokinin. - = extracts from cells cultured in medium lacking a cytokinin: + = extracts from cells cultured in medium containing 0.5 μM BAP

Taking account of earlier results obtained in our laboratory (Ordás et al. 1986), these results could also be explained by an increase in the rate of transformation of monosomes to polysomes. Similar results have also been published (Gwozdz and Wozny 1983).

It appears that BAP enhanced protein synthesis in apple cell culture in a specific way from the pre-existing RNA without modifying its synthesis.

REFERENCES

Amrhein N (1983) Growth. Progress in Botany 45:136-165

Axelos M and Pèaud-Lenoël (1980) The apoprotein of the light-harvesting chlorophyll a/b complex of tobacco cells as a molecular marker of cytokinin activity. Plant Sci Lett 19: 33-41

Chamberlain JP (1979) Fluorography detection of radioactivity in polyacrylamide gel with the water-soluble fluor, sodium salicylate. Anal biochem 98:132-135

Guern J and Pèaud-Lenoël C (1981) Metabolism and molecular activities of cytokinins. Springer-Verlag/Berlin-Heidelberg 1979

Gwozdz EA and Wozny A (1983) Cytokinin-controlled poliribosome formation and protein synthesis in cucumber cotyledons. Physiol Plant 59: 103-110

Klämbt D (1977) Cytokinin and cell metabolism. In: Pilet PE (ed). Plant growth regulation. Springer/Berlin-Heidelberg: 154-160

Kulaeva ON (1981) Cytokinin action on transcription and translation in plants. In: Guern J and Pèaud-Lenoël C (eds). Metabolism and molecular activities of cytokinins. Springer-Verlag/Berlin, Heidelberg:218-227

Laemmli VK (1970) Cleavage of structural proteins during the assembly of the head of bacteriophage T. Nature 227:680-685

Murashige T and Skoog F (1962) A revised medium for rapid growth and bioassays with tobacco tissue cultures. Physiol Plant 15:473-479

Muren RC and Fosket FE (1977) Cytokinin-mediated translational control of protein synthesis in cultured cells of Glycine max. J. Exp Bot 28:775-784

Ordás R, Rodriguez R, Fernández B and Sanchez R (1986) Effect of cytokinins on some nuclei parameters. 5th Congress of the FESPP (31 August-4 September 1986/Hamburg-FDG) 1:5

Scott TK (1984) Hormonal regulation of development II. The function of hormones from the level of the cell to the whole plant. In: Encyclopedia of plant physiology 10. Springer-Verlag/Berlin, Heidelberg, New York, Tokyo

Szweykowska A, Gwòzdz E and Spychala (1981) The cytokinin control of protein synthesis in plants. In: Guern J and Pèaud-Lenoël (eds). Metabolism and molecular activities of cytokinins. Springer-Verlag/Berlin:212-217

Teyssendier de la Serve B, Jouanneau JP and Pèaud-Lenoël C (1984) Incorporation of N^6-benzyladenine into messenger poly (A)-RNA of tobacco cytokinin concentration. Plant Physiol 74:669-674

Waliszewska-Wojtkowiak B, Schneider J and Szweykowska A (1985) Effect of kinetin on the transcription activity of chromatin from cucumber cotyledons. Biochem Physiol Pflanzen 180:413-422

CHARACTERISTICS OF NaCl-TOLERANT CALLI AND SOMACLONES OF TOMATO

Angel Luque, Rafael Robaina and Guillerma Garcia-Reina

Departamento de Biologîa, Universidad Politécnica de Canarias, Box 550, Las Palmas de Gran Canaria, Spain

INTRODUCTION

The genetic basis for NaCl tolerance in the tomato has been proved (Rush and Epstein 1976, 1981). Salt tolerance mechanisms are thought to be cell-based and do not depend on whole plant organization in *Lycopersicon* (Tal et al. 1978, Tal and Katz 1980, Taleinsk-Gertel et al. 1983). Different mechanisms for NaCl tolerance have been described: salt exclusion (Jia-Ping et al. 1981), osmotic regulation (Orton 1980) or organic synthesis (Rush and Epstein 1976). The objective of this work is to describe the characteristics of NaCl-tolerant morphogenetic calli and isolated somaclones of three tomato land races from the Canary Islands.

MATERIAL AND METHODS

Two *L. esculentum* land races from the island of Fuerteventura (*Rusa* and *Especial*), irrigated with brackish water and one wild genotype from volcanic fields of the island of Gran Canaria (*Salvaje*), were used as explant sources.

Murashige and Skoog medium (1962) supplemented with 0, 1.2, 2.4, 3.6, 5.8 g l^{-1} NaCl were used for seed germination. Seven weeks later, cotyledons, leaves and shoot apices were directly cultured in the same media containing 5 mg l^{-1} BA plus 0.5 mg l^{-1} IAA and the same NaCl concentrations.

After 210 days (six subcultures) the NaCl concentration of the cultures was increased proportionally (in g l^{-1}): control (0), 1.2 to 5.8, 2.4 to 10, 3.6 to 15 and 5.8 to 20. Trichomes were studied with light and polarized microscopy and transmission and scanning electron microscopy.

NATO ASI Series, Vol. H18
Plant Cell Biotechnology. Edited by M.S.S. Pais et al.

Three samples of each plant material, genotype and NaCl concentration were used for Cl^-, NO^-, Na^+ and K^+ analysis.

The inorganic component of the osmotic potential in plant material was theoretically calculated applying Vant Hoff's equation, and in the culture media was calculated according to Debergh et al. (1981).

RESULTS

NaCl-tolerant callus type. NaCl increments reduced the incidence of organogenetic cultures. Established calli (which were sometimes partially covered by less compact green callus) show a globular and microglobular appearance and did not lose their regeneration potential over six subcultures (210 days). The somaclones obtained from calli grown on 0 to 5.8 g l^{-1} NaCl grew quite normally, producing a basal morphogenetic callus, and rooted without difficulty irrespective of preculture, genotype or explant type.

After NaCl increments the predominant callus growth type was the "microglobular" one. The less compact callus areas became necrotic. Trichomes: Trichomes appeared only in calli and somaclones in NaCl concentrations over 3.6 g l^{-1}. No correlation exists between trichome density and NaCl concentration. Adult trichomes have eight cells, four in the stalk area and four in the globular area (Fig. 1). Inorganic contents: The endogenous Cl^- and Na^+ concentrations increase in all plant material and genotypes as the NaCl concentration of the media increases, showing the highest values in the somaclones (Fig. 2). Increased NaCl levels resulted in a concomitant decrease of NO^- and K^+ (Fig. 2). The increase (absolute values) of the osmotic potential (Fig. 3) of the culture medium rises from -6.03 in 0 g l^{-1} NaCl (0%) to -10.44 in 5.8 g l^{-1} NaCl (172%). Somaclones (408% increase) have a quite higher osmotic potential increase than seedlings (135% increase) at 5.8 g l^{-1} NaCl. The osmotic potential increased in seedlings and calli at a similar rate to the culture media (Fig. 3).

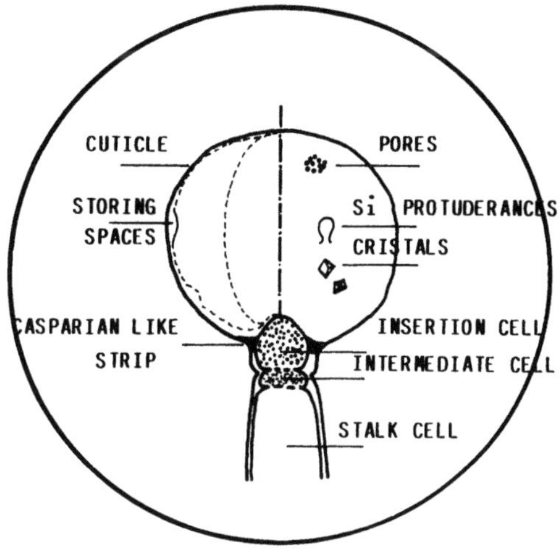

Fig. 1. Trichomes

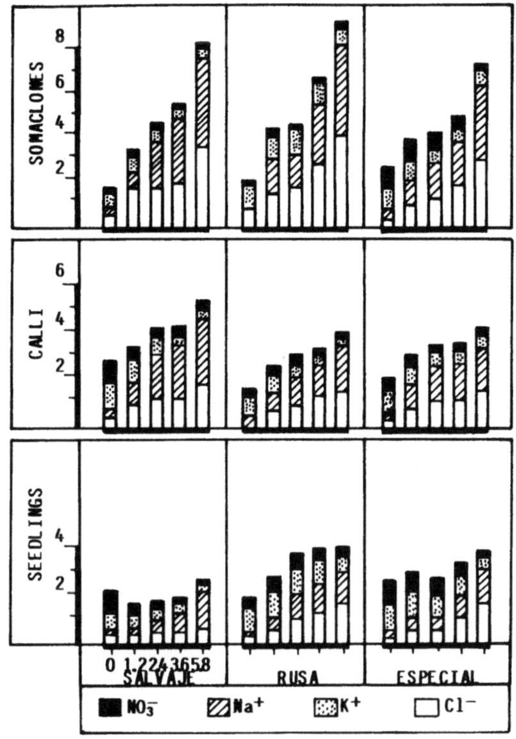

Fig. 2. Evolution of the endogenous concentration (µM. g^{-1} dry wt.)

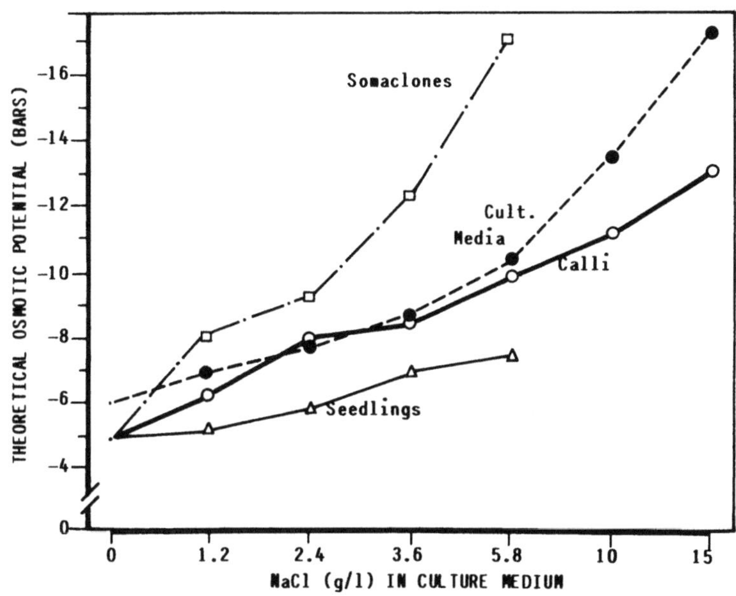

Fig. 3. Evolution of the theoretical osmotic potential

DISCUSSION

Direct callus formation in selective media appears to make selection more economical allowing us to obtain morphogenetic calli tolerant to 20 g l^{-1} NaCl and somaclones tolerant to 15 g l^{-1} NaCl. This procedure generates NaCl-tolerant cell strains during initial steps of the culture. Variability in the initial steps of the culture has been described by several authors (Larkin and Scowcroft 1981, Larkin et al. 1983, Sibi 1980). Higher NaCl tolerance is related to microglobular callus development, suggesting it to be a morphological marker for tolerance. NaCl tolerance in the tomato is related to the meristematic status of the cells (Taleinsk-Gertel et al. 1983). The hormonal balance employed (BA:IAA = 5:1) is of importance in the tomato as a generator of a callus system with a high meristematic/surface ratio. Reducing the globular diameter increases this ratio and the tolerance to NaCl.

The appearance of the trichomes in culture over 3.6 g l^{-1} NaCl suggest that their role is related to salinity. The trichomes are similar to those described in other plants (Fahn et al. 1977, Mozafar and Goodin 1970, Shymony et al. 1973) as saline glands.

The increase of Cl^- and Na^+ as a function of external NaCl concentration agrees with the results of other authors (Liu and Yeh 1984, Rush and Epstein 1981, Tal and Katz 1980, Taleink-Gertel et al. 1983). The Na^+ content of leaves from tomato cultivars becames toxic at levels over 5% (dry wt.) (Rush and Epstein 1981). At the 5.8 NaCl concentration, calli of all the genotypes contained around 5% Na^+ (dry wt.), and the somaclones 7.6% Na^+ in Especial, and around 10% Na^+ (dry wt.) in Salvaje and Rusa. The results suggest a better osmotic adjustment by inorganic ions (Cl^- and Na^+) and not by organic acids, in agreement with the results of Heyser and Nabors (1981a) in NaCl-tolerant cell lines of Nicotiana.

Considerably higher Cl^- and Na^+ concentrations (and Na/K ratio) were found in somaclones than in calli at all NaCl levels, and calli concentrations exceeded those found in seedlings. This response was observed in all genotypes. The somaclones retained the same mode of tolerance selected at the cellular level, which permits them, but not seedlings, to grow in saline media. The higher Cl^- and Na^+ levels in somaclones with respect to the calli (from which they were generated) suggest their origin in callus cells of a higher tolerance. The expression of a cryptic mechanism allowing the cells to tolerate the stress seems more probable than a mutation in the initial steps of the culture.

REFERENCES

Debergh P, Harbaoui Y and Lemeur R (1981) Mass propagation of globe artichoke (Cynara scolymus): evaluation of different hypotheses to overcome vitrification with special reference to water potential. Physiol Plant 53:181-187

Fahn A and Shymony C (1977) Development of the glandular and non-glandular leaves hairs of Avicennia marina Vierh. Bot J Linn Soc 74:37-46

Heyser JW and Nabors MW (1981a) Osmotic adjustment of cultured tobacco cells grown on sodium chloride. Plant Physiol 68: 1454-1459

Heyser JW and Nabors MW (1981b). Growth, water content and solute accumulation of two tobacco cell lines cultured on sodium chloride, dextran and polyethylene glycol. Plant Physiol 67:720-727

Jia-Ping Z, Roth EJ, Terzaghi W and Lark KG (1981) Isolation of sodium dependent variants from haploid soybean cell cultures. Plant Cell Reports 1:48-51

Larkin PJ and Scowcroft WR (1981) Somaclonal variation: a novel source of variability from cell cultures for plant improvement. Theor Appl Gen 60:197-214

Larkin PJ, Brettell R, Ryan S and Scowcroft WR (1983) Protoplast and variation from culture. Experientia Supplementum 46:51-56

Liu MC and Yeh HS (1984) Regeneration of NaCl-tolerant sugarcane plants from callus re-initiated from pre-selected differentiated shoots. Proc Nat Sci Council Taiwan 8:11-18

Murashige T and Skoog F (1962) A revised medium for rapid growth and bioassays with tobacco tissue cultures. Physiol Plant 15: 473-479

Mozafar A and Goodin JR (1970) Vesiculated hairs: a mechanism for salt tolerance in Atriplex halimus L. Plant Physiol 45: 62-65

Orton TJ (1980) Comparison of salt tolerance between Hordeum vulgare and H. jubatum in whole plants and callus cultures. Z Pflanzenphysiol 98:105-118

Rush DW and Epstein E (1976) Genotypic responses to salinity. Plant Physiol 57:162-166

Shymony C, Fahn A and Reinhold L (1973) Ultrastructure and ion gradients in the salt glands of Avicennia marina. New Phytol 72:38-45

Sibi M (1980) Variants epigenetiques et culture in vitro chez Lycopersicon esculentum L. Reunion EUCARPIA, Versailles, pp. 179-185

Tal M, Heinkin H and Dehan K (1978) Salt tolerance of the cultivated tomato: responses of callus tissue of Lycopersicon esculentum, L. peruvianum and Solanum penellii to high salinity. Z Pflanzenphysiol 86:231-240

Tal M and Katz A (1980) Salt tolerance in the wild relatives of the cultivated tomato: the effect of proline on the growth of callus tissue of Lycopersicon esculentum and L. peruvianum under salt and water stress. Z Pflanzenphysiol 98:283-288

Taleinsk-Gertel E, Tal M and Shannon MC (1983) The response to NaCl of excised fully differentiated and differentiating tissues of the cultivated tomato, Lycopersicon esculentum, and its wild relatives, L. peruvianum and Solanum penellii. Physiol Plant 59:659-663

INVESTIGATIONS ON THE MOLECULAR MECHANISMS OF RESISTANCE TO FUNGI IN BARLEY (*Hordeum vulgare* L.)

J.B. Andersen and V. Smedegaard-Petersen

Department of Plant Pathology, The Royal Veterinary and Agricultural University, Thorvaldsensvej 40, DK-1871 Frederiksberg C, Denmark

INTRODUCTION

In the breeding of barley for resistance against barley powdery mildew (*Erysiphe graminis* f.sp. *hordei*) the so-called specific resistance genes have been utilized for many years (Torp et al. 1977).

Almost all these resistance genes are dominant and display a 3:1 resistant to susceptible pattern when crossed to a recessive allele partner. Most of the genes have a dramatic effect on the development of the fungus and exert a limiting effect on the fungus at the stage of penetration or at one of the subsequent stages (Andersen and Torp 1986). These genes fit into the "gene for gene" hypothesis proposed by Flor and reviewed in 1971, stating that for every resistance gene in the plant there exists a corresponding avirulence gene in the fungus in order for resistance to be expressed (Flor 1971). If a virulence gene is present in the fungus, susceptibility will be the case. Unfortunately, virulence is frequently developed against the resistance genes, and the effect is eliminated.

In order to improve the effectivity of the resistance genes knowledge is needed on the mechanisms and molecular genetics involved. With sufficient knowledge of the function of genes in both barley and the fungus it will hopefully be possible to manipulate the genes in the plant to secure a longer-lasting effect.

MATERIAL AND METHODS

Two barley lines (*Hordeum vulgare* L.) near-isogenic with respect to the "Algerian" gene (Ml-a) for powdery mildew resistance were

NATO ASI Series, Vol. H18
Plant Cell Biotechnology. Edited by M.S.S. Pais et al.

used (Kølster et al. 1986). The lines are in theory 98.4% isogenic (1 crossing + 5 backcrossings, selecting the resistant offspring in each generation).

Two powdery mildew isolates (Erysiphe graminis DC. f.sp. hordei Em. Marshall) is used. Isolate A6 is virulent on the Ml-a line (i.e. the interaction between the fungus and the plant is compatible) and the isolate C15 is avirulent on the Ml-a line (i.e. incompatible interaction).

RNA is isolated according to the protocol of Manners and Scott (1985), and poly A^+ RNA is selected using Hybond mAP filters (Amersham International Ltd).

In vitro translation is performed using the rabbit reticulocyte system (Amersham International Ltd.). Separation of polypeptides is achieved with SDS-PAGE.

Differential hybridization of non-fractionated RNA populations was accomplished by immobilizing the first cDNA strand of one component on a filter, and hybridizing with the 3'-end labelled mRNA. After hybridization the non-hybridized mRNA is collected and divided into three fractions. The filter is overlaid with X-ray film and exposed. The non-hybridized mRNA is either separated in a denaturing agarose gel, transferred to filter and overlaid with X-ray film for autoradiography, used for in vitro translation, or used for cDNA preparation.

RESULTS AND DISCUSSION

Experimental design

For the investigation of resistance mechanisms against a biotrophic fungus (i.e. depending on a living host cell) like the barley powdery mildew fungus, certain factors must be taken into consideration.

First, the plant material must be able to display the characters to be investigated. Preferably isogenic or near-isogenic lines with and without the specific characters and with as an identical gene background as possible must be used (Daly 1972). Andersen and Torp (1986) found that differences in gene background could account for statistically significant effects on the level of compatibility.

Second, well-characterized fungal isolates that are either avirulent or virulent are needed.

Third, an experimental system that will permit the elimination of factors that are irrelevant to the project or amplify factors essential for the project are needed.

These prerequisites need to be fulfilled before using molecular gentics as a tool in investigating resistance mechanisms.

Current experiments

In our experimental system we utilize near-isogenic barley lines, prepared using the susceptible line "Pallas" (ml-a) as the mother variety (Kølster et al. 1986).

Table 1. The combinations of host lines and fungal isolates used in the experiments. At all sampling hours samples are taken in all combinations, i.e. six samples

Fungal isolate	Host genotype	
	Ml-a	ml-a
Non-inoculated	Control	Control
Avirulent isolate	Resistant	Resistant
Virulent isolate	Susceptible	Susceptible

Pallas is used as the susceptible control (ml-a) in the experiments, and the Pallas isoline P-01 with the "Algerian" gene (Ml-a) is used as the line with a resistance gene (Table 1).

Before the work commenced, the composition of the mRNA population in the two lines (non-inoculated) was investigated using in vitro translation. No differences between the lines could be found, and it was concluded that it should be possible to use the lines for studies of resistance mechanisms at the molecular level.

During this study, two isolates (C15 and A6) with a known virulence spectrum on the known resistance genes are used.

They differ with respect to virulence on the Algerian gene, as C15 is avirulent and A6 is virulent. The isolates are unfortunately not isogenic, but to our knowledge, no reports exist on gene-background effects in powdery mildew fungi. The isolates behave identically, when the susceptible interaction on the ml-a line is studied with respect to penetration efficiency. Attempts to prepare isogenic isolates will be made using the mutagenesis treatment described by Gabriel et al. (1982).

To distinguish between resistance-related reactions in the plant, and concomitant non-resistance related changes, the experimental system in Table 1 is used. The system is essentially the same as the system used by Davidson et al. (1987), Manners et al. (1985) and Manners and Scott (1985), and provides the possibility of separating the general responses to inoculation from the resistance-related responses.

ACKNOWLEDGEMENTS. The current project has received support from The Danisch Agricultural Research Council, The Carlsberg Foundation and The Tuborg Foundation.

REFERENCES

Andersen JB & Torp J (1986) Quantitative analysis of the early powdery mildew infection stages on resistant barley genotypes. J. Phytopathology 115:173-186

Daly JM (1972) The use of near-isogenic lines in biochemical studies of the resistance of wheat to stem rust. Phytopathol. 62:392-400

Davidson AD, Manners JM, Simpson RS & Scott KJ (1987) cDNA cloning of mRNAs induced in resistant barley during infection by _Erysiphe gramins_ f.sp. _hordei_. Plant Molec. Biol. 8:77-85

Flor HH (1971) Current status of the gene for gene hypothesis. Ann. Rev. Phytopathol. 9:275-296

Gabriel DW, Lisker N & Ellingboe AH (1982) The induction and analysis of two classes of mutations affecting pathogenicity in an obligate parasite. Phytopathol. 72:1026-1028

Kølster P, Munk L, Stølen O & Løhde J (1986) Near-isogenic barley lines with genes for resistance to powdery mildew. Crop Science 26:903-907

Manners JM, Davidson AD & Scott KJ (1985) Patterns of post-infectional protein synthesis in barley carrying different genes for resistance to the powdery mildew fungus. Plant Molec. Biol. 4:275-283

Manners JM & Scott KJ (1985) Reduced translatable messenger RNA activities in leaves of barley infected with _Erysiphe graminis_ f.sp. _hordei_. Physiol. Plant Pathol. 26:297-308

Torp J, Jensen HP & Jørgensen JH (1978) Powdery mildew resistance genes in 106 northwest European spring barley varieties. Kgl. Vet. - og Landbohøjsk. Årsskr. 75-102

THE STRUCTURE AND FUNCTION OF PLANT COATED VESICLES

Larry C. Fowke and Michael A. Tanchak
Department of Biology,
University of Saskatchewan,
Saskatoon, Saskatchewan,
Canada, S7N OWO

INTRODUCTION

Protoplasts have been used extensively for basic studies of plant cells and their organelles. Research has included studies of cell fusion (Gleba and Sytnik 1984), the plasma membrane (Leonard and Rayder 1985; Fowke 1986), the cytoskeleton (Willison and Klein 1982) and cell organelle isolation and characterization (Hall 1983). Recent research has shed light on the structure and function of a little known organelle in plants, the coated vesicle. Studies with plant protoplasts have clearly indicated that coated vesicles are involved in the endocytosis of externally supplied labels which are delivered sequentially to a number of membranous organelles and ultimately to the central vacuole of plants. This chapter summarizes some of the recent developments which have clarified our understanding of the function of plant coated vesicles.

ANIMAL COATED VESICLES

Animal coated vesicles are tiny organelles (approx. 80 - 120 nm in diam.) consisting of a membranous sac enclosed within a complex protein cage. The major constituents of the cage are a group of proteins known as clathrins, both high molecular weight or heavy chains of clathrin (180 KD) and lower molecular weight or light chains (33 and 36 KD). The clathrin molecules are arranged into triskelions and with additional proteins constitute the cage which in surface view resembles a soccer ball.

It is clear from a considerable body of evidence that coated vesicles have a number of important functions in animal cells including receptor mediated endocytosis, membrane recycling,

NATO ASI Series, Vol. H18
Plant Cell Biotechnology. Edited by M.S.S. Pais et al.

protein sorting and exocytosis (e.g. Fine and Ockleford 1984; Breitfield et al. 1985; Wileman et al. 1985). The most thoroughly studied of these is receptor mediated endocytosis. In this process ligands bind to specific receptors on clathrin coated regions of the plasma membrane (coated pits) and are internalized into coated vesicles which uncoat and deliver the ligand to different internal organelles including endosomes, Golgi bodies, multivesicular bodies and lysosomes. The ligand is thought to be separated from its receptor by the low pH within the endosome compartment of the cell. A number of important proteins (e.g. insulin, transferrin, low density lipoprotein) are internalized by receptor mediated endocytosis in animal cells.

PLANT COATED VESICLES

Coated vesicles remarkably similar to those in animal cells are widespread within the plant kingdom (Newcomb 1980). They are often found in cells actively involved in cell wall synthesis, for example, in dividing cells at the developing cell plate, rapidly expanding cells and in plant protoplasts which are regenerating a new cell wall (e.g. Fowke et al. 1985). Within these cells coated vesicles are generally localized near dictyosomes, the plasma membrane and a very interesting tubular membranous organelle bearing coated regions on its surface. This organelle has been identified in *Chara* cells, protoplasts and cells of a number of angiosperms and has been named the partially coated reticulum (Pesecreta and Lucas 1984,1985; Tanchak et al. 1984). Coated vesicles are particularly abundant in freshly isolated protoplasts derived from rapidly growing suspension cultured cells. It is possible to recognize two distict populations of coated vesicles in these protoplasts, larger ones associated with the plasma membrane and smaller ones located near dictyosomes (van der Valk and Fowke 1981). Coated pits are also very abundant in plant protoplasts. Comparisons between leaf protoplasts and cell suspension protoplasts indicate a direct correlation between the number of coated pits and the ability of protoplasts to regenerate a new cell wall (Fowke et al. 1983). The most frequently suggested function for plant coated vesicles, based primarily on their distribution, has been exocytosis either of cell wall protein or polysaccharide and the transport of new membrane to the plasma membrane (e.g. Robertson and Lyttleton 1982; Nakamura and Miki-Hiroshige 1982, Fowke et al. 1985). The possibility that endocytosis might be occurring has also been considered (e.g. Lucas and Franceschi 1981), but until recently only indirect evidence was available to support this suggestion.

Biochemical analyses of coated vesicles isolated from plant protoplasts and cultured cells have revealed a number of differences between plant and animal coated vesicles (Mersey et al. 1982, 1985; Depta and Robinson 1986). Plant clathrin heavy chain has a molecular mass of 190 KD compared to 180 KD for animal coated vesicles. Preparations of plant coated vesicles contain a large number of other proteins, many of which do not have counterparts in coated vesicle fractions from animal cells. The clathrin light chains of plants have not been clearly identified as yet. However, recent studies suggest that plant coated vesicles may have two clathrin light chains of 57 and 60 KD (Hawes, personal communication), again heavier than those in animal cells. Antibodies against animal clathrin light chains were found to cross react with proteins of these molecular masses.

ENDOCYTOSIS IN PLANTS

Within the last three years, major advances have been made in understanding the function of coated vesicles in plant cells. Protoplasts have played a key role in this research. One of the most interesting discoveries is the involvement of coated vesicles in an endocytotic pathway in plant cells. The initial work involved incubation of protoplasts in cationized ferritin with subsequent analysis of uptake by conventional thin sectioning and transmission electron microscopy (Joachim and Robinson 1984 - bean leaf protoplasts; Tanchak et al. 1984 - soybean suspension culture protoplasts). A summary of the pathway established for ferritin uptake by protoplasts of soybean is presented below and is illustrated diagrammatically in Figure 1.

Ferritin particles bind uniformly to the surface of freshly isolated soybean protoplasts. The binding is probably due to electrostatic attraction between the negatively charged plasma

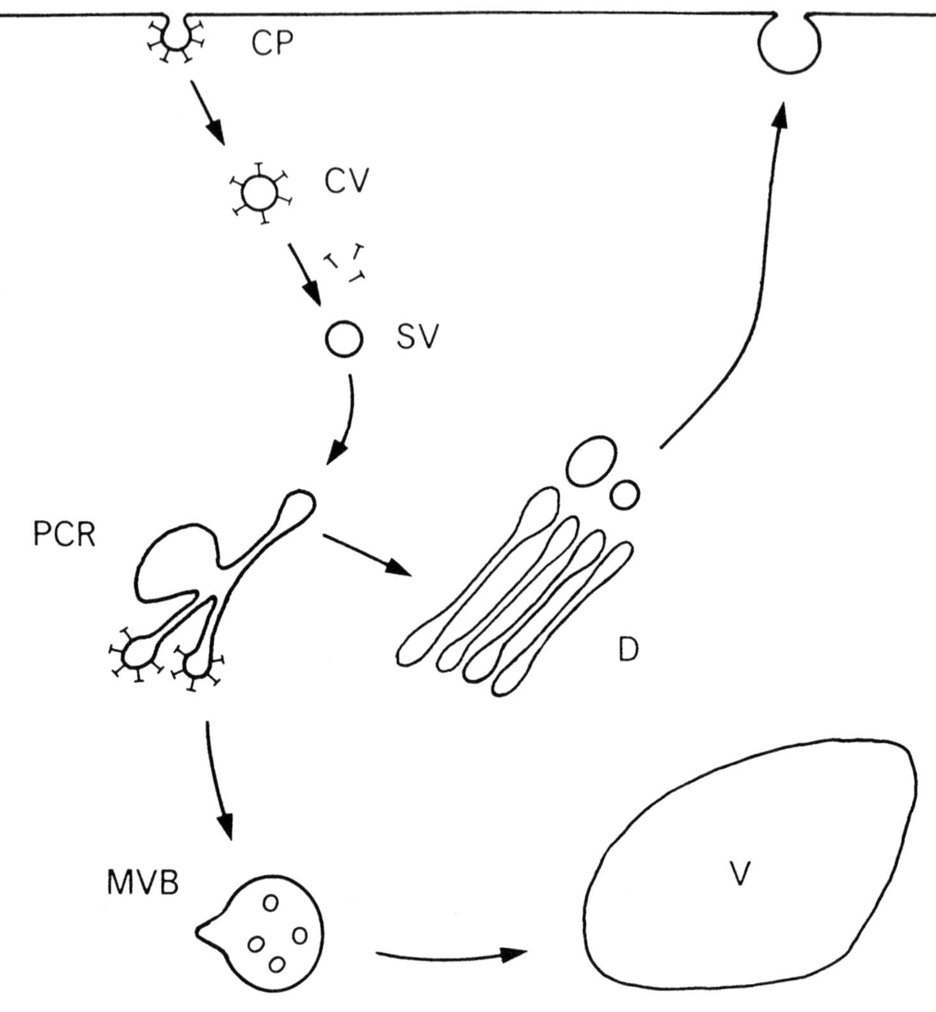

Fig. 1. Diagrammatic representation of the endocytotic pathway in soybean protoplasts. External label and plasma membrane are endocytosed via coated pits (CP), to coated vesicles (CV) which uncoat to form smooth vesicles (SV). They are then delivered to the partially coated reticulum (PCR) and dictyosomes (D). This transfer of plasma membrane to dictyosomes may compensate for the movement of dictyosome vesicles to the cell surface during matrix secretion. Label may be transported to the lytic compartments of the cell, the multivesicular bodies (MVB) and vacuoles (V) to be degraded. Based on Tanchak et al. 1984, Tanchak and Fowke 1987 and unpublished results.

membrane and the cationized ferritin. Within seconds the ferritin is internalized via coated pits to coated vesicles in the cytoplasm. The presence of partially coated vesicles shortly thereafter suggests that coated vesicles rapidly loose their coats as they do in animal cells.

After approximately 5 minutes the ferritin appears within the tubules of the partially coated reticulum. Three dimensional reconstructions based on serial sectioning indicate that despite its close association with dictyosomes, the partially coated reticulum can exist as a separate organelle in the cytoplasm (Pesecreta and Lucas 1985; Tanchak and Fowke unpublished). The partially coated reticulum is morphologically similar to some animal endosomes. This coupled with the fact that it labels with externally supplied ferritin suggests that it may be functionally equivalent to endosomes in animal cells. Further studies are necessary to test this hypothesis. These might involve an investigation of the pH of the partially coated reticulum using fluorescent pH indicators or the isolation of this organelle and an examination of its chemistry.

Shortly after the partially coated reticulum of soybean protoplasts becomes labelled, ferritin appears in several cisternae of the dictyosomes, almost exclusively in the periphery of individual cisternae. Similar results have been obtained with ferritin uptake by bean leaf protoplasts (Joachim and Robinson 1984) and lectin-gold conjugate uptake by both bean leaf and carrot suspension culture protoplasts (Hillmer et al. 1986). The appearance of the label in dictyosomes in these different protoplasts is significant and clearly demonstrates movement of membrane from the surface of the cell to dictyosomes. It is well established that dictyosomes package and secrete matrix materials during cell wall formation (e.g. Willison 1981; Robinson and Kristen 1982). Freshly isolated cell suspension protoplasts initiate wall regeneration rapidly (Fowke and Gamborg 1980; Willison and Klein 1982) and secrete large quantities of matrix polysaccharides, thus increasing dramatically the flow of membrane to the cell surface. Dictyosomes have been directly implicated in this process in soybean protoplasts (Griffing et al. 1986). Coated vesicle traffic may thus represent an important mechanism for membrane retrieval to compensate for dictyosome activity.

Multivesicular bodies in soybean protoplasts label with ferritin later than the partially coated reticulum and dictyosomes and accumulate ferritin with time (Tanchak, unpublished results). This time course of labelling suggests that the multivesicular bodies receive their label from the partially coated reticulum or dictyosomes. It is interesting to note that the multivesicular bodies in plant protoplasts are quite similar to their counterparts in animal cells (Willingham and Pastan 1984; Harding et al. 1985; Hornick et al. 1985). They are approximately the same size, are bounded by a single membrane and contain a number of small vesicles which appear to form by invagination of the bounding membrane. They typically possess a lamellar plaque on their cytoplasmic surface and often bear small tubular protrusions into the cytoplasm. Occasionally coated regions are clearly recognizable on their cytoplasmic surface as well. Multivesicular bodies in animal cells are generally considered to be a prelysosomal compartment but in some cases can contain acid hydrolases and thus are technically lysosomes (e.g. Morales et al. 1985). Some of the multivesicular bodies of soybean protoplasts contain peroxidase (Griffing and Fowke 1985) and acid phosphatase (Griffing unpublished results) and thus also constitute part of the lytic compartment in these protoplasts. The morphological observations and the results from the uptake experiments with soybean protoplasts certainly suggest that multivesicular bodies accumulate externally supplied labels and are likely involved in degradative processes. Recent uptake studies with zoospores of the fungus *Aphanomyces euteiches* are consistent with the soybean protoplast results (Cerenius et al. unpublished results). Cationized ferritin binds to the surface of zoospores and is internalized by coated membranes during collapse of the contractile vacuole. During subsequent encystment of the zoospore the ferritin is delivered to multivesicular bodies. It is not known whether these organelles contain degradative enzymes, however, structures resembling multivesicular bodies in zoospores of *Thraustochytrium* do stain histochemically for acid phosphatase (Kazama 1973).

Finally, after long exposures to cationized ferritin (1 hr. or longer) the label is delivered to the central vacuole and small peripheral vacuoles in soybean protoplasts (Tanchak and Fowke 1987). Aggregates of small vesicles with associated ferritin particles resembling the contents of multivesicular bodies are observed in a number of vacuoles, suggesting that multivesicular bodies have fused with the vacuoles thus releasing their contents. The ultimate fate of much of the endocytosed ferritin is therefore the vacuoles of the cell. Since vacuoles are generally considered to be the lysosomal compartment of plant cells (Matile 1975) the ferritin is likely eventually degraded. A parallel situation exists in animal cells. Multivesicular bodies apparently fuse with existing lysosomes where their contents are degraded by lysosomal enzymes (Wileman et al. 1985).

In summary, cationized ferritin is endocytosed by coated pits and passed sequentially to coated vesicles, smooth vesicles, the partially coated reticulum and dictyosomes, mutivesicular bodies and finally to the cell vacuoles. The significance of this endocytotic pathway is unknown, however, a number of possibilities exist. First, the movement of plasma membrane to dictyosomes during endocytosis suggests that the pathway may have an important role in membrane recycling. Membrane moved to the surface of the cell by dictyosome vesicles during matrix secretion could be retrieved by this mechanism. Second, the delivery of ferritin to the vacuoles of plant cells suggests that the pathway may be involved in uptake and degradation of externally supplied molecules. This pathway might, for example, be responsible for detoxification reactions in cells of roots which are exposed to a variety of substances in solution in the soil. The recent report of uptake of metal salts by coated pits and coated vesicles of maize root cap cells (Hubner et al. 1985) is consistent with this idea although the metal salts only reached the partially coated reticulum, dictyosomes and small vacuoles and did not end up in the multivesicular bodies or central vacuole. Third, the endocytotic pathway may also have a role in the uptake of other molecules which have not been identified. Because the pathway bears such a striking resemblance to receptor mediated endocytosis in animal cells it is tempting to speculate that it may also be involved in the uptake of very specific molecules in plant cells. It is possible, for example, that this pathway might be involved in host pathogen interactions. Elicitor molecules from fungi would be likely candidates for such an uptake mechanism. Further research is necessary to explore this and other possibilities.

OTHER COATED VESICLE FUNCTIONS

Indirect evidence suggests that coated vesicles may also be involved in the sorting and transport of proteins in plant cells. Examination of isolated membranes from soybean protoplasts reveals the enrichment of glucan synthase I in both dictyosome and coated vesicle fractions (Griffing et al. 1986). This enzyme produces a β-1-4 linked glucan. Ray (1980) suggests that glucan synthase I is involved in forming the backbone of the hemicellulose xyloglucan. However, it has also been suggested that it may represent the cellulose synthase proenzyme which becomes cellulose synthase when incorporated into the plasma membrane (Hayashi and Matsuda 1981). Perhaps the enzyme is packaged in the dictyosomes and is shipped to the plasma membrane via coated vesicles. Glucan synthase I is known to increase in activity at the time cultured carrot protoplasts begin to incorporate substantial amounts of glucose into the newly forming cell wall (Langebartels et al. 1981).

Cytochemical studies provide further evidence which suggests that coated vesicles are involved in the transport of proteins in plant cells. Peroxidase, for example, is localized in components of the secretory system of soybean protoplasts (Griffing and Fowke 1985). These include dictyosomes, multivesicular bodies, some vacuoles and dictyosome-associated coated vesicles. It is tempting to speculate that peroxidase is packaged by dictyosomes and then transported to the vacuoles via certain coated vesicles. In animal cells, Golgi-associated coated vesicles have been implicated in the transport of lysosome destined enzymes (Brown and Farquhar 1984).

Many proteins manufactured on ribosomes of the endoplasmic reticulum are subsequently modified in the dictyosomes before being transported within the cell. The above examples suggest that plant coated vesicles may play a role in the movement of these proteins. It is also possible that coated vesicles could be involved in actual sorting of proteins within dictyosomes prior to transport.

CONCLUSIONS

Recent research with plant protoplasts has contributed significantly to our understanding of the structure, chemistry and function of plant coated vesicles. The most important discovery is the existance in plant cells of an endocytotic pathway which resembles its counterpart in animal cells. The role of this pathway in plants has not been determined but it likely is invlolved in membrane recycling from the plasma membrane to dictyosomes, detoxification reactions and perhaps also in endocytosis of specific macromolecules. Details of these different coated vesicle functions in plants should be forthcoming in the near future.

ACKNOWLEDGEMENTS

A special thanks to those in our laboratory who have contributed to this work, in particular Larry Griffing, Brent Mersey and Pat Rennie. Thanks also to Dennis Dyck for preparing the figure. The support of the Natural Sciences and Engineering Research Council of Canada is gratefully acknowledged.

REFERENCES

Breitfield PP, Simmons CF, Strous GJAM, Geuze HJ, Schwartz AL (1985) Cell biology of the asialoglycoprotein receptor system: a model of receptor-mediated endocytosis. Int Rev Cytol 97: 47-95

Brown WJ, Farquhar MG (1984) The mannose-6-phosphate receptor for lysosomal enzymes is concentrated in *cis* Golgi cisternae. Cell 36: 295-307

Depta H, Robinson DG (1986) The isolation and enrichment of coated vesicles from suspension-cultured carrot cells. Protoplasma 130: 162-170

Fine RE, Ockleford CD (1984) Supramolecular cytology of coated vesicles. Int Rev Cytol 91: 1-43

Fowke LC (1986) The plasma membrane of higher plant protoplasts. In: Chadwick CM, Garrod, DR (eds) Hormones, receptors and cellular interactions in plants. Cambridge University Press, Cambridge pp 217-239

Fowke LC, Gamborg OL (1980) Applications of protoplasts to the study of plant cells. Int Rev Cytol 68: 9-51

Fowke LC, Griffing LR, Mersey BG, Tanchak MA (1985) Protoplasts for studies of cell organelles. In: Fowke LC, Constabel F

(eds) Plant protoplasts. CRC Press, Boca Raton pp 39-52
Fowke LC, Rennie PJ, Constabel F (1983) Organelles associated with the plasma membrane of tobacco leaf protoplasts. Plant Cell Reports 2: 292-295
Gleba YY, Sytnik KM (1984) Protoplast fusion. Springer Verlag, Berlin
Griffing LR, Fowke LC (1985) Cytochemical localization of peroxidase in soybean suspension culture cells and protoplasts: intracellular vacuole differentiation and presence of peroxidase in coated vesicles and multivesicular bodies. Protoplasma 128: 22-30
Griffing LR, Mersey BG, Fowke LC (1986) Cell-fractionation analysis of glucan synthase I and II distribution and polysaccharide secretion in soybean protoplasts. Planta 167: 175-182
Hall JL (1983) Plasma membranes. In: Hall JL, Moore AL (eds) Isolation of membranes and organelles from plant cells. Academic Press, London, pp 55-81
Harding C, Levy MA, Stahl P (1985) Morphological analysis of ligand uptake and processing: the role of multivesicular endosomes and CURL in receptor-ligand processing. Europ J Cell Biol 36: 230-238
Hayashi T, Matsuda K (1981) Biosynthesis of xyloglucan in suspension-cultured soybean cells. Evidence that the enzyme system of xyloglucan synthesis does not contain β-1-4 glucan 4-β-glucosyltransferase activity (EC 2.4.1.12). Plant Cell Physiol 22: 1571-1584
Hillmer S, Depta H, Robinson DG (1986) Confirmation of endocytosis in higher plant protoplasts using lectin-gold conjugates. Europ J Cell Biol 41: 142-149
Hornick CA, Hamilton RL, Spaziani E, Enders GH, Havel RJ (1985) Isolation and characterization of multivesicular bodies from rat hepatocytes: an organelle distinct from secretory vesicles of the Golgi apparatus. J Cell Biol 100: 1558-1569
Hubner R, Depta H, Robinson DG (1985) Endocytosis in maize root cap cells. Evidence obtained using heavy metal salt solutions. Protoplasma 129: 214-222
Joachim S, Robinson DG (1984) Endocytosis of cationic ferritin by bean leaf protoplasts. Europ J Cell Biol 34: 212-216
Kazama F (1973) Ultrastructure of Thraustochytrium sp. zoospores. III. Cytolysosomes and acid phosphatase distribution. Arch Microbiol 89: 95-104
Langebartels C, Seitz U, Seitz HV (1981) β-glucan synthetase activities in regenerating protoplasts from carrot suspension cultures. Plant Sci Lett 22: 327-335
Leonard RT, Rayder L (1985) The use of protoplasts for studies on membrane transport in plants. In: Fowke LC, Constabel F (eds) Plant protoplasts. CRC Press, Boca Raton, pp 105-118
Lucas WJ, Franceshi VR (1981) Characean charasome-complex and plasmalemma vesicle development. Protoplasma 107: 255-267
Matile P (1975) The lytic compartment of plant cells. Springer-Verlag, New York
Mersey BG, Fowke LC, Constabel F, Newcomb EH (1982) Preparation of a coated vesicle-enriched fraction from plant cells. Exptl Cell Res 141: 459-464
Mersey BG, Griffing LR, Rennie PJ, Fowke LC (1985) The isolation of coated vesicles from protoplasts of soybean. Planta 163: 317-327
Morales C, Clermont Y, Hermo L (1985) Nature and function of

endocytosis in Sertoli cells of the rat. Am J Anat 173: 203-217

Nakamura S, Miki-Hiroshige H (1982) Coated vesicles and cell plate formation in the microspore mother cell. J Ultrastruct Res 80: 302-311

Newcomb EH (1980) Coated vesicles: their occurrence in different plant cell types. In: Ockleford CJ, Whyte A (eds) Coated vesicles. Cambridge University Press, Cambridge, pp 55-68

Pesecreta TC, Lucas WJ (1984) Plasma membrane coat and a coated vesicle-associated reticulum of membranes: their structure and possible interrelationships in Chara corallina. J Cell Biol 98: 1537-1545

Pesecreta TC, Lucas WJ (1985) Presence of a partially-coated reticulum in angiosperms. Protoplasma 125: 173-184

Ray PM (1980) Cooperative action of β-glucan synthetase and UDP-xylose transferase of Golgi membranes in the synthesis of xyloglucan-like polysaccharides. Biochim Biophys Acta 629: 431-444

Robertson JG, Lyttleton P (1982) Coated and smooth vesicles in the biogenesis of cell walls, plasma membranes, infection threads, and peribacteroid membranes in root hairs and nodules of white clover. J Cell Sci 58: 63-78

Robinson DG, Kristen U (1982) Membrane flow via the Golgi apparatus of higher plant cells. Int Rev Cytol 77: 89-127

Tanchak MA, Fowke LC (1987) The morphology of multivesicular bodies in soybean protoplasts and their role in endocytosis. Protoplasma 138: 173-182

Tanchak MA, Griffing LR, Mersey BG, Fowke LC (1984) Endocytosis of cationized ferritin by coated vesicles of soybean protoplasts. Planta 162: 481-486

Wileman T, Harding C, Stahl P (1985) Receptor-mediated endocytosis. Biochem J 232: 1-14

Willingham MC, Pastan I (1984) Endocytosis and exocytosis: current concepts of vesicle traffic in animal cells. Int Rev Cytol 92: 51-92

Willison JHM (1981) Secretion of cell wall material in higher plants. In: Tanner W, Loewus FA (eds) Encyclopedia of Plant Physiology, New Series, Vol 13B. Springer-Verlag, Berlin, pp 513-541

Willison JHM, Klein AS (1982) Cell wall regeneration by protoplasts isolated from higher plants. In: Brown RM (ed) Cellulose and other natural polymer systems. Plenum, New York pp 61-85

van der Valk P, Fowke LC (1981) Ultrastructural aspects of coated vesicles in tobacco protoplasts. Can J Bot 59: 1307-1313

NATURAL PRODUCTS FROM HIGHER PLANTS AND PLANT CELL CULTURE

Michael W. Fowler and Alan H. Scragg.
The Wolfson Institute of Biotechnology,
University of Sheffield,
Sheffield S10 2TN,
United Kingdom

INTRODUCTION

The economic importance of plants is immense with plants being used as a source of food, construction materials, fabrics and paper and fuel. In addition, plants contribute to a wide array of chemicals used in industry including pharmaceuticals, foods, flavours, dyes, pigments and agrochemicals. The potential for chemical synthesis within the plant kingdom is vast, something of the order of 2 x 10^4 structures are known from plants. New chemical structures reported from plants each year are currently running in excess of 1500.

In general plant-derived chemicals come under the heading of secondary metabolites. "Secondary metabolite"is a relatively loose definition for some constituents of cells not essential for their survival but it does not include such substances as simple sugars and cellulose. Specific secondary products are not widely distributed in the plant kingdom, each is typically restricted to a limited number of species. These plant chemical products range from highly purified compounds, such as many pharmaceuticals, to complex mixtures as for example food additives and perfumes (Table 1).

MEDICINALS

Among the most important secondary plant products are those used as medicinals. At the turn of the nineteenth century plant extracts and potions contributed the major part

NATO ASI Series, Vol. H18
Plant Cell Biotechnology. Edited by M.S.S. Pais et al.

TABLE 1. Commonly Used Plant Products

Pharmaceuticals
Food
Flavours and Fragrances
Enzymes
Agrochemicals
Dyes and Pigments
Oils

of the medicinals used. The discovery and development in the period 1920-1950 of synthetic drugs and microbially based antibiotics lead to a marked reduction in the use of plant products. Table 2 shows the wide range of plant-derived drugs used in orthodox medicine. Table 3 gives the top 12 pure compounds from higher plants as prepared by Farnsworth and Morris (1976), which also gave the most commonly encountered higher plant extracts.

Poisons: The possible first uses of plant products as medicinals were as poisons used in warfare, euthanasia, hunting and fishing. Hemlock was used by the Greeks as a poison and the Middle Ages in Europe saw the sophisticated use of plant poisons for asassination, for example by the Borgias. The poisons can be grouped biochemically according to their toxic principles, and include a wide range of chemical types. Cardiac glycosides and alkaloids are the prime poisons which often act as neurotoxins, for example, curare.

Alkaloids: The main plant-derived pharmaceuticals belong to the group known as alkaloids (Table 4). The alkaloids are a heterogeneous class of natural products difficult to define but which consist of basic nitrogen-containing hetercyclic compounds. The important alkaloid groups are as follows.

Pyrrolidine alkaloids. This is a small class of alkaloids sub-divided into pyrrolidines or tropanes. It is this latter

TABLE 2. Plant Derived Pharmaceuticals

Drug	Chemical	Plant	Use
aloes	anthraquinone	Aloe sp.	cathartic
caffeine	purine	Camellia sinesis	stimulant
castor oil	lipid	Ricinus communis	cathartic
cocaine	alkaloid	Erythroxylum coca	anaesthetic
codeine	alkaloid	Papaver sp.	analgesic
colchicine	alkaloid	Liliaceae	gout treatment
digitoxin	cardiac glycoside	Digitalis purpurea	heart disease
digoxin	cardiac glycoside	Digitalis lanata	heart disease
emetine	alkaloid	Cephaelis ipecacuanha	emetic
emodin	anthraquinone	several plants	cathartic
ephedrine	alkaloid	Ephedra	bronchodilator
hyoscine	alkaloid	Solanaceae	sedative
hyoscyamine	alkaloid	Solanaceae	antispasmodic
lobeline	alkaloid	Lobelia sp.	respiratory stimulant
morphine	alkaloid	Papaver somniferum	narcotic
ouabain	cardiac glycoside	Apocynaceae	heart disease
papaverine	alkaloid	Papaver sp.	heart disease
physostigmine	alkaloid	Physostigma venenosum	glaucoma
quinine	alkaloid	Cinchona sp.	malaria
quinidine	alkaloid	Cinchona sp.	heart disease
reserpine	alkaloid	Rauwolfia sp.	tranquilliser
senna	anthraquinone	Cassia sp.	cathartic
strychnine	alkaloid	Strychnos sp.	stimulant
tubocurarine	alkaloid	Strychnos sp.	muscle relaxant
theophylline	purine	Camellia sinensis	stimulant
theobromine	purine	Theobroma cacao	stimulant
vinblastine	alkaloid	Catharanthus roseus	leukaemia
vincristine	alkaloid	Catharanthus roseus	leukaemia
valepotriates	terpene	Valeriana wallichii	sedative

From Vickery and Vickery (1981)

group which contains alkaloids of commercial value such as cocaine, a local anaesthetic which is isolated from Erythroxylon coca. Other examples are hyocyamine and atropine isolated from Datura stramonium which are used as anti-spasmodics, and scopolamine, an anti-hypertensive.

Pyridine and Piperidine alkaloids. The most extensively studied alkaloid in this group is nicotine from Nicotiana tabaccum. Other members are coniine, the principle alkaloid of hemlock (Corium maculatum), and ricinine from the castor bean (Ricinus communis).

Indole alkaloids. The simple indole alkaloids are widely

TABLE 3. Most Commonly Encountered Pharmaceuticals from Higher Plants

Steroids (from diosgenin)

Codeine

Atropine

Reserpine

Pseudoephedrine

Ephedrine

Hyoscyamine

Digoxin

Scopolamine

Digitoxin

Pilocarpine

Quinidine

distributed through the plant kingdom and many are physiologically active. Examples are the anti-leukaemics vincristine, and vinblastine, the anti-hypertensive and tranquilizer reserpine, the stimulant strychnine, and cardiac drugs, ajmalicine and serpentine.

Quinoline alkaloids. This is a small class of alkaloids of which the most important are the antimalarial and bittering agent, quinine and cardiac depressant quinidine both isolates from Cinchona species.

Isoquinoline alkaloids. The simple isoquinolines have as their best known member mesccline, a hallucinogen from the cactus Lophophora williamsii. The more complex have as members the alkaloids from Papaver (the opium poppy), morphine, codeine and thebaine which have potent analgesic properties. Another opium derivative, papaverine, is used as a muscle relaxant. Tubocurarine from Strychnos species is used as skeletal muscle relaxant in surgery, and in tetanus and strychine poisoning. Another alkaloid of this group is berberine, used as a medicinal agent in Japan. A member of this group in the sub-group Ipecac is emetine, which is used as an emetic and expectorant, and for treating amoebic dysentery.

Colchicine. The most important member of a group of alkaloids found in the seeds and corn of autumn crocus and other Colchicum species. It has been used for years as a treatment for gout but its ability to cause metaphasic arrest has seen its wide use in biology.

TABLE 4 Medicinal Alkaloids

Pyrrolidine alkaloids	atropine	antispasmodic	Datura stramonium
	lyocyamine		
	scopolamine	anticholinergic	Datura metel
	cocaine	anesthetic	Erythroxylum coca
Pyridine and piperidine	nicotine		Nicotiana tabacum
			Conium maculatum
	coniine	poison	hemlock
	ricinine	poison	castor bean Ricinus communis
Indole	vinchristine	antileukemic	C. roseus
	vinblastine	antileukemic	C. roseus
	serpentine/ ajmalicine	cardiac	C. roseus
	reserpine	antihypertensive	Rauwolfia serpentina
Quinoline	quinine	bittering agent antimalarial	Cinchona
	quinidine	cardiac depressant	Cinchona
Isoquinoline	mescaline	hallucinagem	Lophophora williansii
	papaverine	smooth muscle relaxant	Papaver
	tubocurarine	skeletal muscle relaxant	Strychros
	berberine		Berberis
	emetine	antiamoebic	Cephaelis ipecacuntia
	morphine	analgesic	Papaver somniferum
	codeine	analgesic	Papaver somniferum
	thebaine	analgesic	Papaver somniferum
	colchicine	mitotic inhibitor	Colchicum

Stimulants: Widely used but not normally harmful are a wide range of stimulants most of which contain caffeine or a related purine. The leading caffeine beverage is tea prepared from the leaves of Camellia sinensis, followed by Coffea arabica containing caffeine and the theophylline. The third species the cocoa bean from Theobroma cacao contains up to 1%

of the purine theobromine. The *Cola nitida*, cola 'nuts' contain caffeine and is now commercially important as a basis of soft drinks.

Steroids. Virtually all members of the hormone group of drugs are prepared by conversion of plant derived steroidal sapogenins and particularly diosgenin from *Dioscorea*.

Terpenoids. The anti-tumour compounds, tripdiolide and triptolide have been isolated from *Tripterqium wilfordii*. The sporins of *Panax ginseng* are regarded as the active principles of ginseng which has an increasing market. Among valuable terpenoid molecules are those making up natural rubber from *Hevea brasiliensis*. A degraded triterpenoid, quassin, is a bittering agent extracted from the wood of *Picrasma excelsa* or *Quassia amara*. The roots of *Valeriana wallichii* are used as a sedative in Germany with the active constitutents, monoterpene valepotriates.

Quinones. The quinone ubiquinone-I0 has a large market as a heart drug in Japan. Another compound, shikonin, is used as a dye and pharmaceutical in Japan and is extracted from the roots of *Lithospermum erythrorhizon*. The anthraquinones from *Cassia* and other plants have also been used as cathartics.

Cardenolides. Various *Digitalis* species are capable of synthesising the cardiac glycoside digitoxin. Digitoxin is used to treat certain heart diseases and is extracted from the leaves of the foxglove *Digitalis purpurea* although digoxin is preferred by many doctors.

ESSENTIAL OILS

Mainly monoterpenes, the essential oils are isolated from a vast array of plant species each providing its characteristic complex mixture. Although the amounts of material produced are small their value may be high, with the most expensive used in perfumery. The essential oils are mixtures normally extracted by steam distillation, solvent extraction or absorption into purified fats. Components such as geraniol, citronellol and linalool form the base of many perfumes.

Spices and flavouring oils are generally more diverse than perfumes, containing other components such as phenols, aromatic aldehydes and sulphur compounds.

AGROCHEMICALS

Apart from the discovery of plant growth regulators the most important discovery relates to the pyrethroids. These are extracted from the flowers of *Chrysanthemum cinerariafolium* and are extremely potent insecticides. Although under strong competition from synthetic pyrethroids they still have a place in the domestic market. Originally nicotine was used to destroy insect pests.

DYES AND PIGMENTS

The use of plants as a source of dyes and pigments has a very long history with examples such as indigo (*Indigofera tinctoria*), woad (*Isatis tinctoria*) and madder (*Rubia tinctorum*).

PLANT CELL CULTURE

It is against the background of a wide range of well established higher plant products that plant cell culture as an alternative source has developed. Many of the plant products which cannot be chemically synthesized are produced from plantation grown crops, for example *Digitalis* and *Cinchona*. The concept of using plant cell cultures to produce various products gained from in 1956 with Routier and Nickell (1956), and subsequently Tulecke and Nickell (1959). A wide range of plant products have been detected in plant cell cultures (Fowler, 1983; Curtin, 1983; Sahai and Knuth, 1985) and this has led to the possibility of the production of the chemicals in a factory setting. This alternative system has certain advantages over the plantation such as stability of supply, freedom from disease and climatic changes and a closer relationship between supply and demand.

The development of new technology is, at least in its initial stages, expensive and this must also be true of the bioreactor growth of plant cell suspensions. Therefore there are only a limited number of possible candidates for this type of factory production, those of high value required in relatively low volumes. Some examples are shown in Table 5.

Table 5

Possible Costs and Requirements of Plant Cell Culture Products

Compound	Use	Price $ Kg	World Annual Requirement Kg
ajmalicine (serpentine)	circulatory diseases	1,500	3,000
codeine	sedative	650	7,700 (U.S)
digoxin	heart stimulant	3,000	6,000 (U.S)
diosgenin	steroids	674	200,000
quinine	antimalarial, bittering agent.	100	300-500,000
shikonin	dye, antibacterial	4,500	150 (JAPAN)
vincristine, vinblastine	anti-leukemic	5,000,000	5

Data taken from Curtin (1983) and Sahai & Knuth (1985) & McHale (1986).

The most valuable appear to be the anti-leukaemics, vinblastine and vincristine, which only occur in very small quantities in the plant Catharanthus roseus, and the cheapest quinine at 100$ Kg. The annual requirements for these products varies from 5-500,000 Kg but if one assumes that 10% of the market share would be required at first, a typical volume would be 300 Kg. The natural plant products which consist of mixtures such as jasmine or rose oil, have not been included because at this stage mixtures are too difficult to reproduce. Current estimates of the cost of a product produced by plant cell culture range from $500 Kg (Goldstein et al., 1980) to $1500 Kg (Fowler, 1983).

To be economic the plant cell process must have as high a productivity as possible, productivity being defined as gram product/l culture/day. The productivities and scale of microbial and plant cell cultures are compared in Table 6.

Table 6

Batch Economics Microbial Products and Plant Cell Culture Products

	Selling Price ($ Kg.)	Batch Cycle (days)	Product $g.l^{-1}$	Biomass $g.l^{-1}$	Productivity $g.l^{-1}.d^{-1}$
Penicillin G[a]	41.6	7	10-15	-	1.4-2.1
Monosodium glutamate[a]	2.0	1.5	80-100	-	53.3-66.6
Citric acid[a]	1.4	4.5	120-150	-	30-37.5
Rosmarinic acid	-	14	5.5	25.7	0.91
Diosgenin[a]	674	15-16	0.43	11.3	0.75

[a] Values from Sahai and Knuth (1985)

Productivity on an annual basis of a bioreactor or process is affected by a number of factors; the rate of growth, the yield of biomass, the yield of product and the number of runs which can be carried out per year. The latter is a combination of downtime, for example, the time required to clean and resterilise the vessel; to fill with medium if sterilized empty and a percentage of runs lost due to contamination, holidays or maintenance. Superimposed on these features is the kinetics of secondary product formation by plant cell cultures. In general the secondary products are formed after growth has ceased, extending the bioreactor time and requiring a change to production medium forming a two-stage process with an extended run time.

Plant cells grow slowly with doubling times measured in days, such that improvements in doubling time to give values of below 2 days will not affect productivity as much as increases in yield (Scragg, 1986). Therefore, a high yield of secondary product is essential if a process is to be developed. As has been stated plant cell cultures have been shown to be capable of synthesising a wide range of secondary products and a number of these at high levels (Table 7).

TABLE 7 Plant cell cultures giving high yields of secondary products

Product	Species	Yield % dry wt.
rosmarinic acid	Coleus blumei	21.4
anthraquinone	Morinda citrifolia	18.0
benzylisoquinolines	Coptis japonica	15.0
acetoside	Syringa vulgaris	15.0
shikonin	Lithospermum erythrorhizon	12.4
berberine alkaloids	Berberis wilsonae	10.0
shikimic acid	Galium mollugo	10.0
diosgenin	Dioscorea deltoidea	7.8
nicotine	Nicotiana tabacum	5.0
serpentine	Cathatanthus roseus	2.2
ajmalicine	Catharanthus roseus	1.8

Unfortunately many of the products formed at high levels are not those which are economically attractive if one refers to Table 5. However, if we examine some of these cases we can perhaps gain an insight into the possible production.

production of alkaloids by cultures of Catharanthus roseus. Interest in this species is high because of the two dimeric alkaloids vinblastine and vincristine. However, despite considerable effort these have not been detected in the culture. However, two other alkaloids were found to accumulate, serpentine and ajmalicine. Here again a production medium was formulated where 2,4-D was replaced by NAA (Zenk et al., 1977). Later a cell line capable of growth and product formation in the same medium was developed (Morris, 1986). Using autofluorescence or RIA assays higher producing cell lines have been isolated (Deus and Zenk, 1982) but have proved to be unstable. A similar case is found with nicotine levels and Deus-Neumann and Zenk (1984) list a number of examples of stable and unstable plant culture and product systems.

Low Yielding Cultures: In contrast to the above many of the commercially interesting compounds are only produced in trace amounts or not at all (compare Table 5 with Table 6). Examples of this are codeine and morphine from cultures of Papaver somniferum, hyoscyamine and scopolamine from Hyoscyamus, quinine and quinidine from Cinchona and digitalis from D. lanata. This failure to form products is not a reflection of the lack of experimental effort applied but rather a lack of knowledge of the pathways and controls involved in secondary product synthesis.

In addition to the points raised above there are a number of other features must be considered in the development of a plant cell culture process. One key consideration is that of downstream processing. Little research has been carried out in this area but no problems were encountered with products such as rosmarinic acid and shikonin. A second concern relates to the way in which the public and regulatory bodies will regard a product produced by plant cell culture. Will it be regarded as 'nature derived' as it has been extracted from plant cells or will it be 'nature identical'? If the latter, will it require the extensive testing that any new product receives? This will perhaps only be resolved when a product becomes available. In the case of shikonin it is only sold in Japan, but it is perhaps not just for convenience that it has gone on sale in

HIGH PRODUCING CELL LINES

Rosmarinic acid is one of the caffeic acid derivatives which was detected at high levels (15% dry weight) without selection (Zenk et al., 1977). A simple production medium (sucrose only) was developed which gave values of up to 23% rosmarinic acid and combined with a fed-batch system raised the biomass level to 25.7g.l^{-1}. This in turn gave production of 5.6g.l^{-1} and 0.91g.l^{-1} day^{-1}. Another caffeoyl derivative is acetoside (verbascoside) which is produced by cultures of Syringa vulgaris at levels of up to 15% (Ellis, 1983). In both cases the cultures have spontaneously accumulated the product once growth had begun to decline.

The naphthroquinones and anthraquinones represent another group of compounds which accumulate at high levels in plant cell suspensions. One naphthoquinone, shikonin, is extracted from the roots of Lithospermum with a yield of 1-2%. Callus cultures of L. erythrorhizon were found to contain the red pigment and by selection and medium optimization, a two-stage production process was developed. A yield of 12% was obtained and in 1983 Mitsui announced the first commercial process using plant cell culture for the production of shikonin (Curtin, 1983). Anthraquinones have been produced at very high levels by Morinda citrifolia by replacing 2,4-D with NAA and increasing the sucrose level to 7% (Zenk et al., 1975). The treatment of cultures of Galium mollugo with the herbicide glyphosate increase the shikimate level to 10%. Cell lines of Coptis and Berberis were found to produce protoberberine alkaloids (Breuling et al., 1985; Fukui et al., 1982) and the Coptis line increased its yield to 15% as growth improved. The steroid diosgenin was found regularly at levels of 0.1% in various suspension cultures of Dioscorea deltoidea (Kaul and Staba, 1968). However, by transfering the cells to a production medium a yield of 7.8% diosgenin was obtained.

Another case where increase in sucrose levels, reduction in phosphate and the replacement of 2,4-D with NAA increases product formation is with nicotine from Nicotiana tabacum (Mantell et al., 1983). Another important example is the

the U.S.A. as a lipstick colourant which avoids FDA approval.

Plant cell culture still remains a possible source of plant synthesised chemicals but it may take considerably more research and time before being credible as an alternative technology. However when one considers the advances made over the last few years in plant cell culture, one is encouraged.

REFERENCES

Breuling M, Alfermann W, Reinhard E (1985) Plant Cell Report, 4. 220

Curtin ME (1983) Biotechnology, 649

Deus B, Zenk MH (1982) Biotec Bioeng, 24. 1965

Deus-Neumann B, Zenk MH (1984) Planta Medica, 50. 427

Ellis BE (1983) Phytochemistry 22. 1941

Farnsworth NR, Morris RW (1976) Amer J Pharm, 148. 46

Fowler MW (1983) "In" Mantell SH, Smith H (eds) Plant Biotechnology. Cambridge University Press, London, 3

Fukui H, Nakagawa K, Tsuda S, Tabata M (1982) "In" Fujiwara A (ed) Plant Cell Culture 1982. Maruzen Press, Tokyo, 313

Goldstein WE, Ingle MB, Lasure L (1980) "In" Staba EJ (ed) Plant Tissue Culture as a Source of Biochemicals. CRC Press, Boca Raton, Florida, 191

Kaul B, Staba EJ (1968) Lloydia, 31. 171

Mantell SH, Pearson DW, Hazell LP, Smith H (1983) Plant Cell Reports, 2. 73

Morris P (1986) Planta Medica, 1982. 121

Routier JB, Nickell LG (1956) US Patent 2, 747. 334

Sahai O, Knuth M (1985) Biotechnology Progress, 1. 1

Scragg AH (1986) "In" Morris P, Scragg AH, Stafford A, Fowler MW (eds) Secondary Metabolism in Plant Cell Culture. Cambridge University Press, Cambridge, 202

Tulecke W, Nickell, LG (1959) Science, 130. 863

Vickery ML, Vickery B (1981) Secondary Plant Metabolism. The Macmillan Press Ltd, London.

Zenk MH, El-Shagi H, Ulbrich B (1977) Naturwisn , 64. 585

Zenk MH, El-Shagi H, Arens H, Stockigt J, Weiler EW, Deus B (1977) "In" Barz W, Reinhard E, Zenk MH (eds) Plant Tissue Cultures and its Biochemical Application. Springer-Verlag, Berlin. 77

WIB06A

PHYSIOLOGICAL AND BIOCHEMICAL CHARACTERISTICS OF PHOTOAUTOTROPHIC PLANT CELL CULTURES

Wolfgang Hüsemann

Lehrstuhl für Biochemie der Pflanzen
Universität Münster
D-4400 Münster, Federal Republic of Germany

INTRODUCTION

During the past 25 years plant cell and tissue cultures have successfully been used for basic research and biotechnological application, including the selection of agronomically and industrially important traits. As photosynthesis is a characteristic feature of plant cells, the development of photosynthetically active chloroplasts in plant cell cultures should greatly increase their applicability for diverse studies.

Cultured plant cells from different plant species are able to form chlorophyll and develop fully functional chloroplasts under the appropriate light conditions. These green cell cultures grown in the presence of an exogenous sugar in the nutrient medium (photomixotrophic mode of nutrition), are used in several laboratories for studies on the nutritional and cultural conditions for chloroplast differentiation, as well as for the induction of high photosynthetic CO_2 assimilation rates, as recently reviewed (Dalton and Peel 1983).

In 1967 Bergmann succeeded for the first time in growing chlorophyllous cell cultures from higher plants (*Nicotiana tabacum*) photoautotrophically in a sugar-free medium but in the presence of CO_2-enriched atmosphere (1% CO_2; V/V). About 10 years later, two groups, Hüseman and Barz (1977) and Yamada and Sato (1978) independently started again the research on photoautotrophism in cultured plant cells. Sustained photoautotrophic growth of in vitro cultured plant cells in the light with CO_2 as the sole carbon source has been achieved exclusively with about 18 different plant species of dicotyledonous angiosperms with exception of the moss *Marchantia polymorpha* and the monocot *Asparagus officinalis* (Table 1).

NATO ASI Series, Vol. H18
Plant Cell Biotechnology. Edited by M. S. S. Pais et al.

Amaranthus retroflexus	Glycine max *
Arachis hypogaea *	Gossypium hirsutum **
Asparagus officinalis	Hyoscyamus niger
Catharanthus roseus	Marchantia polymorpha *
Chenopodium rubrum *	Morinda lucida
Cytisus scoparius	Nicotiana tabacum
Datura stramonium	Peganum harmala
Daucus carota	Ruta graveolens
Digitalis latana	Solanum tuberosum
Euphorbia characias	Spinacea oleracea *

Table 1. Photoautotrophic plant cell cultures

*) Cells exhibitung complete photoautotrophic growth in a simple mineral salt culture medium without any organic constituents and with CO_2 (1%-2%;v/v) as the sole carbon source.

**) Cells that can be grown photoautotrophically under normal air (After Husemann 1985)

Recent progress and still existing problems in selecting cell cultures with high photosynthetic potential capable of sustained photoautotrophic growth has been reviewed by Yamada and Sato (1984), Horn and Widholm (1984), Hüsemann (1984, 1985), Yamada (1985).

In this context, some still unresolved problems should be mentioned. To date, we do not understand exactly the complexity of interactions of nutritional, physiological, biochemical as well as molecular-biological factors in controlling chloroplast differentiation in cultured cells, especially from the monocotyledons. Another problem concerns the demand of cultured cells for high CO_2 concentrations far above ambient air level for photoautotrophic growth.

This paper now contributes to some essential experimental results to characterize the physiological and biochemical specifications involved with the expression of photoautotrophism in cultured plant cells. Some preference will be contributed to the authors studies on photoautotrophic cell cultures from Chenopodium rubrum.

INDUCTION AND MAINTENANCE OF PHOTOAUTOTROHPIC CELL CULTURES

Major prerequisites for establishing photoautrophic cell cultures are a high chlorophyll content and photosynthetic competence of the cells. As cell cultures have a special demand for phytohormones and sugars, and are differentially sensitive to light (intensity and quality of the photosynthetic active radiation) as well as to the composition of the gaseous atmosphere in the culture vessels (CO_2; O_2; ethylene), they are more or less heterogenous in their specific characters. Therefore, the establishment of photoautotrophism in cultured plant cells will be a process of serial selection. Special considerations such as:

- lowering the sugar content in the culture medium and simoultaneously increasing the CO_2-partial pressure above ambient air level,
- preventing oxygen and ethylene accumulation in the culture atmosphere,
- establishing high light intensities (6,000-8,000 lux) and
- maintaining a well balanced ratio of auxin/cytokinin in the culture medium

have proved advantageous to trigger photosynthetic development.

Why it is difficult to induce chloroplast differentiation in cultured cells remains still to date unclear. This applies to the mode of repression of sugars as well as to the benefit of low oxygen concentrations on chlorphyll formation and development of the photosynthetic potential of the cells. Besides, little is known about the influence of auxins - necessary for cell growth - on the morphological as well as biochemical differentiation of the chloroplasts (Dalton and Peel 1983, Dalton 1984, Horn and Widholm 1984, Richter 1984).

Due to the extensive studies in the past 10 years, it is possible now to propagate cells from about 20 different plant species (Table 1) under photoautotrophic conditions in small vials (Yamada and Sato 1978, Kato 1979), in two-tier culture flasks (Hüsemann and Barz 1977, Hüsemann 1984), as well as in fermenter and continuous culture systems (Dalton 1980, Yamada et al. 1981, Peel 1982, Hüsemann 1983, 1984).

Summerizing the experimental results from several research groups, photoautotrophically cultured cells pass through successive stages of cell growth and metabolic activities. Cytoplasmic and mitochondrial activities dominate during the phase of rapid division, while chloroplast-associated activities increase after the transition to stationary growth phase, including unbalanced growth. This has been shown for the accumulation of cell components (chlorophyll, protein, carbohydrates), for the in vitro activities of enzymes related to different metabolic pathways as well as to the pattern for photosynthetic product formation and fine structure of the cells (Hüsemann et al 1984, Nato et al 1985, Herzbeck and Hüsemann 1985).

On the other hand, continuous growth under photoautotrophic conditions has been reported for *Spinacea oleracea* (Dalton 1980), *Asparagus officinalis* (Peel 1982) and *Chenopodium rubrum* (Hüseman 1983). Steady-states have been achieved for several cultural and cell specific parameters such as nutrient uptake, cell division acitivity, cellular content of chlorophyll and protein, photosynthetic CO_2 assimilation rates and enzyme activities as shown in Table 2.

	Batch-culture day 5	Batch-culture day 14	Continuous culture (steady-state)
Cell doubling time (h)	50	120	100
Protein ($\mu g/10^6$ cells)	380	420	240
Chlorophyll ($\mu g/10^6$ cells)	17	28	12
Starch ($\mu g/10^6$ cells)	25	80	5
Photosynthetic activity: (μmol CO_2/mg chlorophyll x h)	80	65	100
Enzyme acitivty: (nmol CO_2/mg protein x min)			
RuBP-carboxylase	105	190	72
PEP-corboxylase	85	65	82

Table 2. Unbalanced and steady-state growth in photoautotrophic cell cultures from *Chonopodium rubrum* (After Hüsemann 1987)

SIMILARITIES AND DIFFERENCES BETWEEN PHOTOAUTOTROPHIC CELL CULTURES AND LEAF MESOPHYLL CELLS

The question as to whether the physiological and biochemical properties of photoautotrophically cultured plant cells resemble those of mesophyll cells can now be affirmed in many respects.

Barz and Hüsemann (1982) found a great similarity in the chlorophyll and carotenoid composition between photoautotrophic cell cultures and leaf cells from *Peganum harmala*.

Investigations with cell cultures from *Chenopodium rubrum* (Hüsemann et al 1980) and *Glycine max* (Martin et al 1984) document a great similarity in the pattern and composition of the lipids of photoautotrophic cell cultures and green leaves. The more photoautotrophic the cells, the more leaf-like the lipids are. This is shown by a drastic increase in the total amount of glycerollipids and increased proportions of monogalactosyldiacylglycerols and digalactosyldiacylglycerols, both associated with the chloroplasts. Furthermore, photosynthetic cells always contain higher proportions of the more unsaturated linolenic acid (18:3) compared to the linoleic acid (18:2) and oleic acid (18:1).

Membrane transport studies indicate that at least photoautotrophically cultured cells from *Chenopodium rubrum* have similar membrane properties as the leaf mesophyll cells. This applies to the electrical membrane potential of about -200 mV (Ohkawa et al 1981), the water transport properties and hydraulic conductivity (Buchner et al 1981), the protonmotive force dependent and carrier mediated transport of hexoses as well as for a cell wall bound acid invertase, possibly also involved in hexose transport (Gogarten and Bentrup 1984, Gogarten 1986).

Finally, photoautotrophically cultured cells resemble a mesophyll cell in their photosynthetic capacities as has been demonstrated by several research groups (Hüsemann et al 1979, 1984, Yamada et al 1982, Yamada 1985, Bender et al 1985, Neumann 1987). This topic including some remarkable special features of the green cell cultures concerning the involvement of the PEP-carboxylase in CO_2 assimilation and the blue light dependent chloroplast differentiation will be discussed more in detail later.

To date, photoautotrophic cell suspension cultures have only been grown at CO_2 concentrations (1-2%; v/v) far above ambient air levels. A 2% CO_2 content in the gaseous atmosphere of the culture vessels is in equilibrium with a total amount of 780 nmol

(880 nmol) inorganic carbon (CO_2 + HCO^-_3)/ml culture medium at pH 4.5 (5.5), this is approximately 70-times higher as compared with normal air.

Are the lack of stomata, high diffusion resistance to CO_2 and/or insufficient or missing CO_2-uptake and concentration mechanisms the underlying reasons for the high CO_2 concentrations needed by in vitro cultured plant cells for photoautotrophic growth ?

We should urgently try to select cell lines capable of photoautotrophic growth under normal air. Comparison between the normal and variant cell lines can provide clues to answer this question.

Recently, Blair et al (1986) reported the adaptation of photoautotrophic cell suspensions of <u>Gossypium</u> <u>hirsutum</u> to growth in normal air.

PHOTOSYNTHETIC CO_2 ASSIMILATION AND CARBON METABOLISM

The early photosynthetic products after short-term $^{14}CO_2$-photosynthesis (10 sec.) in photoautotrophic cell cultures from C_3-plants confirmed the C_3-pathway of CO_2 assimilation mediated by a predominant $C_1 \rightarrow C_5$-carboxylation using the RubisCO. But 14-C-labeling of malate and aspartate during photosynthesis is rather high and accounts for 10% to 40% of total radioactivity are incorporated. Decarboxylation of 14-C-labeled malate using the NAD(P)-dependent malic enzyme showed that about 80% of total 14-C-radioactivity was located in the C-4 (β-carboxyl) position of the malate molecule, thus confirming $C_1 \rightarrow C_3$-carboxylation mediated by the PEP-carobxylase (Hüsemann et al 1979, Nishida et al 1980, Yamada et al 1982, Yamada 1985, Bender et al 1985, Herzbeck 1985).

The metabolic function of malate in photoautotrophic cell cultures from <u>Nicotiana</u> <u>tabacum</u> (Yamada et al 1982) and <u>Chenopodium</u> <u>rubrum</u> (Herzbeck and Hüsemann 1985, Herzbeck 1985) has been studied. It was found that malate, photosynthetically formed or exogenously supplied, is rapidly metabolized and carbon is preferentially channelled into amino acids (58%), intermediates of the citric acid cycle (11%) and protein (10%). The build up of sugars and starch (10%-20%) from exogenously supplied (U-14-C)-malate may be accomplished by the RuBP-carboxylase dependent refixation and recycling of 14-CO_2 evolved from respiratory 14-CO_2 or from malic enzyme dependent decarboxylation of 14-C-malic acid.

High PEP-carboxylase and correspondingly high proportions of CO_2 incorporation into C_4-acids are supposed to be involved in anaplerotic reactions in providing additional carbon skeletons for biosynthetic reactions of the tricarboxylic acid cycle and possibly contribute to an intracellular carbon transport in photosynthetically active cell cultures from C_3-plants. Thus, high PEP-carboxylase activity seems to be a characteristic feature of photosynthesizing cell cultures from C_3-plants such as Nicotiana tabacum, Daucus carota, Cataranthus roseus, Cytisus scoparius and Chenopodium rubrum. On a chlorophyll basis, the PEP-carboxylase activity of cultured green cells may be 4- to 8-fold higher than that of leaves of the corresponding C_3-plant (Yamada 1985). The most stimulating conditions to reach maximum in vitro activities of the PEP-carboxylase are high cell division rates during exponential culture growth, an increased (NH_4^+): (NO_3^-) ratio in the nitrogen supply of the cells as well as a photomixotrophic mode of nutrition (Hüsemann et al 1979, Barz and Hüsemann 1982, Nato et al 1985, Yamada et al 1982). The PEP-carboxylase from photoautotrophic cell cultures of Nicotiana tabacum was characterized as a C_3-type enzyme due to its kinetic properties (Sato et al 1986).

Meanwhile, NAD/NADP-dependent malic enzyme as well as pyruvate-orthophosphate-dikinase, supposed to be key enzymes of C_4-photosynthesis and crassulacean acid metabolism (CAM), have also been found in photoautotrophic cell cultures from C_3 plants (Herzbeck 1985, Hüsemann 1987, Sato et al 1986).

In this context, recent findings of Aoyagi and Bassham (1986) on the presence of the malic enzyme and pyruvate-orthophosphate-dikinase in mesophyll cells from wheat, a C_3-plant, are very remarkable and fit well into the hypothesis (Sato et al 1986), that the major difference in the C_3- and C_4-mode of CO_2 fixation might be a quantitative rather than a qualitative difference in enzyme activity. Summarizing these results, it is discussed that the enzymes (PEP-carboxylase, malic enzyme, pyruvate-Pi-dikinase) might be involved in intracellular carbon transport in C_3-plant cells. The question arises, whether the typical C_4-photosynthesis or the CAM can be expressed in photoautrotrophic cell cultures from the corresponding plants. To date, no such findings are available.

LIGHT-DEPENDENT CHLOROPLAST DIFFERENTIATION

Light plays a crucial role in chloroplast development. In contrast to the intact plant, in photomixotrophic plant cell cultures chloroplast development is strictly blue light dependent (Richter 1984), as has been shown for callus cultures from *Nicotiana tabacum* (Bergmann and Berger 1966), *Crepis capillaris* (Hüsemann 1970), *Chenopodium rubrum* (Richter et al 1986, 1987). The availability of photoautotrophic cell cultures now offers new perspectives in analyzing the possible "blue-light syndrome" in chloroplast development, excluding any unspecific changes in the pattern of chloroplast development caused by the sugars, phytohormones and vitamins in the culture medium under photomixotrophic growth conditions. In a detailed study using cell suspension cultures from *Chenopodium rubrum* (Richter et al 1987) clearly proved that chloroplast differentiation in cultured cells is predominantly under blue light control, irrespective of whether the cells are grown photomixotrophically or photoautotrophically. Chlorophyll formation and steady-state concentrations of the m-RNAs coding for plastid proteins (large subunit of the RubisCO; light-harvesting-chlorophyll-protein, LHCP) are promoted by blue light. Red light (660 nm) suppresses these developmental processes. Besides, DNA-protein-complexes with active RNA-polymerase (transcriptional active complex, TCA) have been isolated from blue and red light grown cells. The activity of TCA from "blue light cells" was extremely higher than that of cells grown under red light.

Essentially new in this context is the observation that red light (660 nm) does not completely abolish chlorophyll formation in photoautotrophically cultured cells but reduced the chlorophyll content to about 1/3 of that in blue light grown cells. This means, photoautotrophic cell cultures from *Chenopodium rubrum* can maintain differentiation, multiplication and function of the chloroplasts even under continuous red light illumination at rates sufficient to allow sustained photoautotrophic growth, though at reduced rates compared to blue light (Hüsemann, unpublished). This is corroborated by the fine structure of the chloroplasts, possessing a less well developed system of grana thylakoids in red light compared to blue light. It seems that chloroplasts from photoautotropically cultured cells represent rather stable and functionally independent entities, because they are indispensible for the survival of the cells in contrast to those in photomixotrophically cultured cells.

We now have to consider a blue light photoreceptor of yet unknown properties which seems to adopt at least partially the function of the phytochrome system, generally involved in the photocontrol of chloroplast development in the intact plant (Mohr 1984).

BIOTECHNOLOGICAL APPROACHES

1. Screening for xenobiotics: Resistance and metabolism

The current state of cell cultures as a screening system for xenobiotics has been reviewed by Mumma and Davidonis (1983) and Hughes (1984). Green cell cultures obviously reveal the effect of toxins on plant seedlings better than chlorophyll-free cell cultures do. For example, simazine, atrazine and metribuzine discriminate between photosynthesizing and nonphotosynthesizing cells (Gressel 1979). Photosyntetic cell cultures are already used as a tool to select for photosynthetic herbicide resistance (Horn and Widholm 1984, Wang et al 1986). If secondary modes of action of herbicides not affecting photosynthesis are involved, they can now be assayed with photoautotrophic and heterotrophic cell lines of the same plant species. Even without plant regeneration, biocide resistant photoautotrophic cell cultures will provide a clue as to the basics for biocide toxicity, biocide metabolism and detoxification strategies of the cells as well as for the mode and the site of biocide action on the level of the single photoautotrophic cell. Such strategies will enable us to screen for biocide action under quite different metabolic states of the plant cells.

2. Studies on Secondary Metabolites

The usefulness of photosynthesizing plant cell cultures in analyzing chloroplast associated pathways for secondary metabolism has become evident, but studies on this topic are still scarce.

Wink and Hartmann (1980) described the enhanced formation of quinolizidine alkaloids lupanine and sparteine in green cell cultures from *Lupinus polyphylus* because two enzymes of this biosynthetic pathway, lysine decarboxylase and 17-oxosparteine synthase, are localized in the chloroplasts.

Yamada and Watanabe (1980) selected a vitamin B high-producing cell strain from photosynthetic cell cultures of *Cytisus scoparius*.

Recently, Drawert et al (1984) and Jordan et al (1986) compared the different potential of heterotrophic and photosynthetic cell cultures from *Ruta graveolens* for the formation of the volatile compounds of photomixotrophic and heterotrophic cell cultures. It is postulated that the characteristic constituents of the photomixotrophic cells - methylketons and aliphatic esters - are at least partially the result of the chloroplast metabolism.

Igbavboa et al (1985) studied the switch from the formation of the chloroplast-bound lipoquinones to the synthesis of anthraquinones in cell cultures from *Morinda lucida* by reversion from the photoautotrophic to the heterotrophic mode of nutrition. They found that photoautotrophy is strictly correlated with lipoquinone synthesis, whereas heterotrophy correlates exclusively with anthraquinone formation. This pattern reflects the situation in the intact plant, where lipoquinones are chloroplast-associated, whereas anthraquinones occur in roots. These experimental findings are remarkable because both metabolites share the same pathway up to the chorisminic/isochorisminic acid.

In contrast to these "positive results" on chloroplast associated secondary product formation, there are some examples where these metabolic links are obviously missing.

Rower (1986) and Tyler et al (1986) reported that photoautotrophic cell cultures from *Catharanthus roseus* did not accumulate vindoline or dimeric alkaloids.

Hagimori et al (1984) only observed digitoxin formation in regenerating shoots from photoautotrophic cell cultures from *Digi-*

talis purpurea, but not in the unorganized photoautotrophic callus cells.

To date we cannot reliably decide whether the presence of photosynthetically active chloroplasts in cultured plant cells will increase their capacity for secondary product formation. But the availability of photoautotrophic plant cell cultures offers the unique possibility for studying secondary metabolism in relation to the mode of nutrition of the plant cell.

ACKNOWLEDGEMENT

The author's studies on photoautotrophic cell cultures have been supported by the Deutsche Forschungsgemeinschaft.

REFERENCES

Aoyagi K and Bassham JA (1986) Appearance and accumulation of C_4 carbon pathway enzymes in developing wheat leaves. Plant Physiol 80:334-340

Barz W, Herzbeck H, H semann W, Schneiders G and Mangold H (1980) Alkaloids and lipids of heterotrophic, photomixotrophic and photoautotrophic cell suspension cultures of Peganum harmala. Planta Med 40:127-136

Barz W and Hüsemann W,(1982) Aspects of photoautotrophic cell suspension cultures. In: Fujiwara (ed). Plant Tissue Culture 1982. Maruzen 1982 Tokyo:245-248

Bender L, Kumar A and Neumann KH (1985) On the photosynthetic system and assimilate metabolism of Daucus and Arachis cell cultures. In: Neumann KH, Barz W and Reinhard E (eds). Primary and secondary metabolism of plant cell cultures. Springer Verlag 1985 Berlin-Heidelberg:24-42

Bergmann L (1967) Wachstum grüner Zellsuspensionskulturen von Nicotiana tabacum var. "Samsun" mit CO_2 als einziger Kohlenstoffquelle. Planta 74:243-249

Bergmann L and Berger C (1966) Farblicht und Plastidendifferenzierung in Zellkulturen von Nicotiana tabacum var "Samsun". Planta 69:58-69

Blair LC, Widholm JM and Chastain C (1986) The initiation and characterization of a cotton photoautotrophic suspension culture and adaptation of the culture to grow in atmospheric air. In: Domers DA, Gegenbach BG, Biesboer DD, Hackett WP and Green CE (eds). VI International Congress of Plant Tissue and Cell Culture. Book of abstracts. University of Minnesota 1986 Minneapolis:410

Buchner KH, Zimmermann U and Bentrup FW (1981) Turgor pressure and water transport properties of suspension-cultured cells of Chenopodium rubrum. Planta 151:95-102

Dalton CC (1980) Photoautotrophy of spinach cells in continuous culture: Photosynthetic development and sustained photoautotrophic growth. J Exp Bot 31:291-391

Dalton CC (1983) Photosynthetic development of Ocimum basilicum cells on transition from phosphate to fructose limitation. Physiol Plant 59:623-626

Dalton CC and Peel E (1983) Product formation and plant cell specialization: A case study of photosynthetic development in plant cell cultures. Prog Ind Microbiol 17:109-166

Dalton CC and Peel E (1983) Product formation and plant cell specialization: A case study of photosynthetic development in plant cell cultures. Prog Ind Microbiol 17:109-166

Drawert F, Berger R, Godelmann R, Collin S and Barz W (1984) Characterization of volatile constituents from photomixotrophic cell suspension cultures of Ruta graveolens. Z. Naturforsch 39c:515-530

Gogarten JP (1986) Untersuchungen zum Zuckertransport an photoautotrophen Suspensionszellen von Chenopodium rubrum. Doctoral Thesis, University of Giessen FRG

Gogarten JP and Bentrup FW (1984) Properties of a hexose carrier at the plasmalemma of green cell suspension cells from Chenopodium rubrum. In: Cram WJ, Janacek K, Rybova R and Sigler K (eds). Membrane transport in plants. Academia Publishing House of the Czechoslovak Academy of Sciences 1984 Praha:183-188

Gressel J (1979) A review of the place of in vitro cell culture systems in studies of action, metabolism and resistance of biocides affecting photosynthesis. Z Naturforsch 34c:905-913

Hagimori M, Matsumoto T and Mikami Y (1984) Photoautotrophic culture of undifferentiated cells and shoot-forming cultures of Digitalis purpurea L. Plant Cell Physiol 25:1099-1102

Herzbeck H (1985) Untersuchungen zum Kohlenstoffmetabolismus in photoautotrophen Zellsuspensionkulturen von Chenopodium rubrum. Doctoral Thesis, University of Münster FRG

Herzbeck H and Hüsemann W (1985) Photosynthetic carbon metabolism in photoautotrophic cell suspension cultures of Chenopodium rubrum. In Neumann KH, Barz W and Reinhard E (eds). Primary and secondary metabolism of plant cell cultures. Springer Verlag 1985 Berlin-Heidelberg:15-23

Horn ME and Widholm JM (1984) Aspects of photosynthetic plant tissue cultures. In: Collins GB and Petolino JF (eds). Application of genetic engineering to crop improvement. M Nijhoff/ Dr. Junk Publ 1984 Dordrecht:113-161

Hüsemann W (1970) Der Einfluss verschiedener Lichtqualitäten auf Chlorophyllgehalt und Wachstum von Gewebekulturen aus Crepis capillaris. Plant Cell Physiol 11:315-322

Hüsemann W (1981) Growth characteristics of hormone and vitamin independent photoautotrophic cell suspension cultures from Chenopodium rubrum. Protoplasma 109:415-431

Hüsemann W (1984) Photoautotrophic cell cultures. In: Vasil IK (ed). Cell culture and somatic cell genetics of plants. vol 1, Laboratory techniques. Academic Press 1984 New York:182-191

Hüsemann W (1985) Photoautotrophic growth of cells in culture. In: Vasil IK (ed). Cell culture and somatic cell genetics of plants. Vol. 2 Cell, growth, nutrition, cytodifferentiation and cryopreservation. Academic Press 1985 New York:213-251

Hüsemann W and Barz W (1977) Photoautotrophic growth and photosynthesis in cell suspension cultures of Chenopodium rubrum. Physiol Plant 40:77-81

Hüsemann W, Plohr A and Barz W (1979) Photosynthetic characteristics of photomixotrophic and photoautotrophic cell suspension cultures of Chenopodium rubrum. Protoplasma 100:101-112

Hüsemann W, Radwan SS, Mangold HK and Barz W (1980) The lipids in photoautotrophic and heterotrophic cell suspension cultures of Chenopodium rubrum. Planta 147:379-383

Hüsemann W, Herzbeck H and Robeneck H (1984) Photosynthesis and carbon metabolism in photoautotrophic cell suspension cultures of Chenopodium rubrum at different phases of batch growth. Physiol Plant 62:349-355

Hüsemann W (1987) The photosynthetic potential and growth of photoautotrophic cell suspension cultures from Chenopodium rubrum - a case study. In: Marin B (ed). Plant vacuoles. Their importance in plant cell compartmentation and their applications in biotechnology. Nato ASI Series, Plenum Publishing Corporation 1987 New York:64-76

Hughes K (1983) Selection for herbicide resistance. In: Evans DA, Sharp WR, Ammirato PV and Yamada Y (eds). Handbook of plant cell culture, vol. 1, Techniques for propagation and breeding. McMillan Publishing Co 1983 New York:442-460

Igbavboa U, Sieweke HJ, Leistner E, Rower I, Hüsemann W and Barz W (1985) Alternative formation of anthraquinones and lipoquinones in heterotrophic and photoautogrophic cell suspension cultures of Morinda lucida. Planta 166:537-544

Jordan M, Rolfs CH, Barz W, Berger R, Kollmannsberger H and Drawert F (1986) Characterization of volatile constituents from heterotrophic cell suspension cultures of Ruta graveolens. Z Naturforsch 41c:809-812

Katoh K (1983) Kinetics of photoautotrophic growth of Marchantia polymorpha L cells in suspension culture. Physiol Plant 59: 242-248

Martin AA, Horn ME, Widholm JM and Rinne RW (1984) Synthesis, composition and location of glycerollipids in photoautotrophic soybean cell cultures. Biochem Biophys Acta 796:146-154

Mohr H (1984) Phytochrome and chloroplast development. In: Baker NR and Barber J (eds). Chloroplast biogenesis. Elsevier 1984 Amsterdam:305-347

Mumma RO and Davidonis GH (1983) Plant tissue culture and pesticide metabolism. In: Hutson DH and Roberts TR (eds). Progress in pesticide biochemistry and toxicology, Vol. 3. John Wiley & Sons Ltd 1983 Chichester:255-278

Nato A, Hoarau J, Brangeon J, Hirel B and Suzuki A (1985) Regulation of carbon and nitrogen assimilation pathways in tobacco cell suspension cultures in relation with ultrastructural and biochemical development of the photosynthetic apparatus. In: Neumann KH, Barz W and Reinhard E (eds). Primary and secondary metabolism of plant cell cultures. Springer Verlag 1985 Berlin-Heidelberg:43-57

Neumann KH and Bender L (1987) Photosynthesis in cell and tissue culture systems. In: Green CE, Somers DA, Hackett WP and Biesboer DD (eds). Plant tissue and cell culture. Alan R. Liss, Inc 1987 New York:151-165

Onkawa T, Kohler K and Bentrup FW (1981) Electrical membrane potential and resistance in photoautotrophic suspension cells of Chenopodium rubrum. Planta 151:88-94

Pean M, Chagvardieff P, Carrier P, Dimon B and Tapie P (1986) Oxygen exchanges in autotrophic and mixotrophic plant cell suspension measured by mass spectrometry. In: Somers DA, Gengenbach BG, Biesboer DD, Hackett WP and Green CE (eds). VI International congress of plant cell culture, Book of abstracts. University of Minnesota 1986 Minneapolis:409

Peel E (1982) Photoautotrophic growth of suspension cultures of Asparagus officinalis cells in turbidostats. Plant Sci Lett 24:147-155

Radwan SS, Mangold KH, Hüsemann W and Barz W (1979a) Lipids in plant tissue cultures. VII Heterotrophic and mixotrophic cell cultures of Chenopodium rubrum. Chem Phys Lipids 24:79-84

Richter G, Hundrieser J, Gross M, Schultz S, Bottlander K and Schneider CH (1984) Blue light effects in cell cultures. In: Senger H (ed) Blue light effects in biological systems. Springer Verlag 1984 Berlin Heidelberg:387-396

Richter G, Einspanier R, Hüsemann W, Dudel A and Wessel K (1986) Gene expression in blue-light dependent chloroplast differentiation of cultured plant cells. In: Akoyunoglou G and Senger H (eds). Regulation of chloroplast differentiation. Alan R Liss, Inc 1986 New York:549-558

Richter G, Dudel A, Einspanier R, Dannhauer I and Hüsemann W (1987) Blue light control of mRNA and transcription rate in chloroplast differentiation of photomixotrophic and photoautotrophic cell cultures of Chenopodium rubrum. Planta in press

Röwer I (1986) Etablierung, Wachstumseigenschaften und Sekundarstoffwechsel photosynthetisch aktiver Zellen höheren Pflanzen. Doctoral Thesis, University of Münster FRG

Sato F, Koizumi N, Takeda S and Yamada Y (1986) Photoautotrophic culture of tobacco cells and their character: Carboxylation enzyme and herbicide resistance of photoautotrophic cells. In: Somers DA, Gengenbach BG, Biesboer DD, Hackett WP and Green CE (eds). VI International congress of plant tissue and cell culture, Book of abstracts University of Minnesota 1986 Minneapolis:408

Tyler RT, Kurz WG and Panchuk BD (1986) Characteristics of photoautotrophic cultures or periwinkle (Catharantus roseus). In: Somers DA, Gegenbach BC, Biesboer DD, Hackett WP and Green CE (eds). VI International congress of plant tissue and cell culture, Book of abstracts. University of Minnesota 1986 Minneapolis:409

Wang WC, Lazzeri PA, Myers JR, Hildebrand DF and Collins GB (1986) Variant selection for atrazine tolerance in tobacco. In Somers DA, Gegenbach BC, Biesboer DD, Hackett WP and Green CE (eds). VI International congress of plant tissue and cell culture, Book of abstracts: University of Minnesota 1986 Minneapolis: 408

Wink M and Hartmann T (1980) Production of quinolizidine alkaloids by photomixotrophic cell suspension cultures: Biochemical and biogenetic aspects. Planta Med 40:149-155

Yamada Y (1985) Photosynthetic potential of plant cell cultures. In: Fiechter A (ed). Advances in biochemical engineering/biotechnology. Plant cell culture. Springer engineering/ biotechnology. Plant cell culture. Springer Verlag 1985 Berlin Heidelberg:89-98

Yamada Y and Sato F (1978) The photoautotrophic culture of chlorophyllous cells. Plant Cell Physiol 19:691-699

Yamada Y, Sato F and Watanabe K (1982) Photosynthetic carbon metabolism in cultured photoautotrophic cells. In: Fujiwara A (ed). Plant Tissue culture 1982. Maruzen 1982 Tokyo:249-250

STRESS-INDUCED SECONDARY METABOLISM IN PLANT CELL CULTURES

Peter Brodelius

Institute of Biotechnology, Swiss Federal Institute of Technology, Hönggerberg, CH-8093 Zürich, Switzerland

INTRODUCTION

Stress on whole plants is generally defined as a condition which has a negative effect on the increase of the dry matter of the plant. Here, stress is defined as an external constraint influencing the secondary metabolism of cultivated plant cells. Of particular interest is the induction of enzymes of secondary metabolism. An increased yield of a target secondary metabolite is therefore considered as a positive effect of the stress even though the growth of the cells may be more or less restricted. From a biotechnological point of view stress may be an important alternative to other methods (e.g. strain selection) to increase the productivity of plant cell cultures. Stress on cultivated plant cells may be caused by a number of different cultivation conditions or modifications of the growth medium as summarized in Table 1. In addition, immobilization of cells is often considered as a stress factor. Some of the stress-inducing agents will be discussed in the following and elucidated by a few examples. Emphasis will be placed on the use of fungal elicitors to induce or enhance the production of secondary metabolites.

OXYGEN

Oxygen limitations may be an important factor in the large-scale cultivation of plant cells. In particular immobilized cells may experience an oxygen limitation which may alter the metabolism of the cells. We have used in vivo 31-phosphorus NMR to study the effects of immobilization on cell metabolism (Vogel and Brodelius 1984). A major advantage of this technique is the fact that cell metabolism may be studied under non-invasive condi-

NATO ASI Series, Vol. H18
Plant Cell Biotechnology. Edited by M. S. S. Pais et al.

Table 1. Various stress factors which may be employed on cultivated plant cells

Nutritional stress
sugar, nitrogen, phosphate
Hormonal stress
Light stress
pH stress
Aeration stress
Osmotic stress
Chemical stress
antibiotics, heavy metal ions, abiotic elicitors
Infectional stress
virus, microorganisms, biotic elicitors

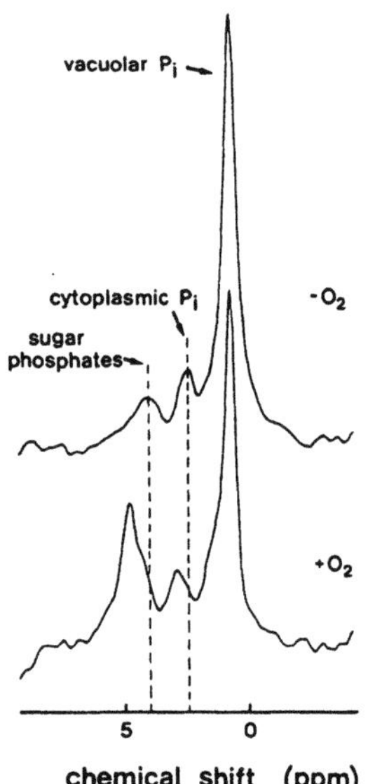

Fig. 1. A comparison of the downfield region of the 31-phosphorus NMR spectrum of C. roseus cells under aerobic and anaerobic conditions. Note the different chemical shift positions of the sugar phosphate and cytoplasmic inorganic phosphate resonances. The shift corresponds to a pH change of 0.5 (Vogel and Brodelius 1984)

tions. The chemical shift of certain signals (i.e. inorganic phosphate and sugar phosphates) are pH-dependent and therefore they may be used to estimate the oxygenation of the cells. Limited oxygen transfer leads to cell respiration resulting in an acidification of the cytoplasm as illustrated in Fig. 1 due to lactate and carbon dioxide formation. No such shift has been observed for cells entrapped in agarose or alginate at cell concentrations of up to 50% (w/v) (Vogel and Brodelius 1984). Furthermore, the ADP/ATP ratio which reflects the energy level of the cells may be estimated from 31-P NMR spectra. No change in this ratio could be observed upon immobilization, indicating that the metabolism of the immobilized cells is similar to that of freely suspended cells (Vogel and Brodelius 1984).

The critical bead diameter for unlimited oxygen transfer has been estimated by respiration measurements for various cell concentrations entrapped in alginate (Hulst et al. 1985). Based on experimental values and computer simulations, the critical bead diameters summarized in Table 2 could be calculated. The model was based on Michael-Menten kinetics and a diffusivity of oxygen within the gel matrix equal to 80% of that in water (Hulst et al. 1985).

Table 2. Critical bead diameters for various concentrations of *Daucus carota* cells entrapped in alginate at various oxygen surface concentrations (Hulst et al. 1985)

Cell concentration (%)	Bead diameter (mm) Oxygen concentration at bead surface (% saturation)				
	2	20	50	75	95
5.6	0.55	1.74	2.76	3.38	3.80
14.1	0.39	1.23	1.95	2.39	2.69
50.6	0.20	0.64	1.02	1.25	1.40

PHOSPHATE

In a number of studies it has been established that phosphate limitations may lead to an enhanced formation of secondary metabolites. An increased formation of indole alkaloids in cell suspension cultures of *Catharanthus roseus* has been observed at low phosphate concentrations as illustrated in Fig. 2 (Knobloch and Berlin 1983). For immobilized cells such a phosphate limitation may actually be created by the cells themselves. For instance, cells of *C. roseus* take up phosphate from the medium within a few hours and store it within the vacuoles, while cells of *Daucus carota* take up the phosphate slowly from the medium as it is needed for biosynthetic purposes (Brodelius and Vogel 1985). Thus, transfer of immobilized cells of *C. roseus* to fresh medium results in the creation of a phosphate gradient

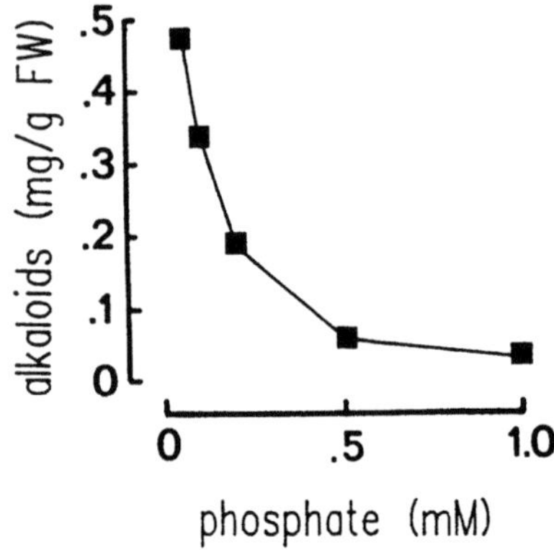

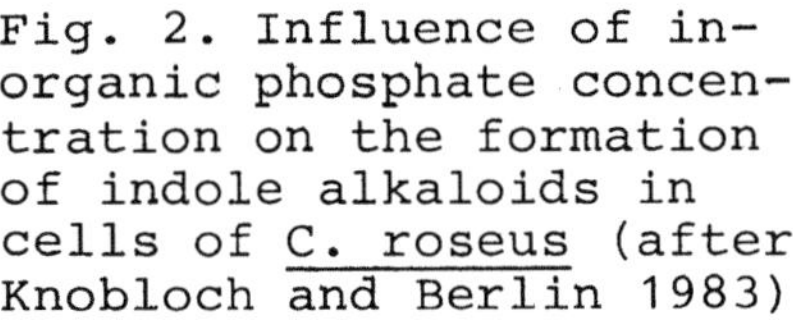
Fig. 2. Influence of inorganic phosphate concentration on the formation of indole alkaloids in cells of C. roseus (after Knobloch and Berlin 1983)

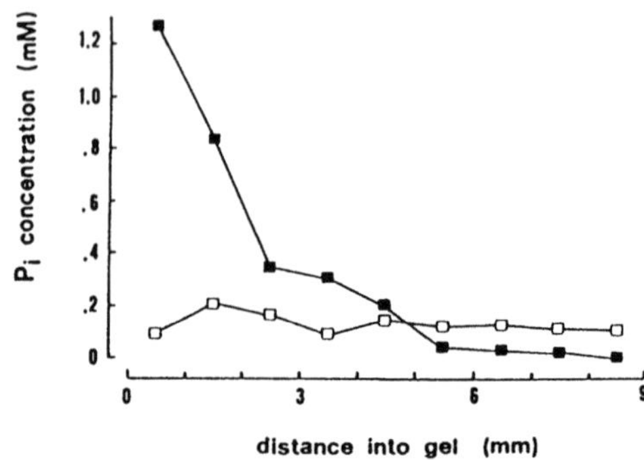

Fig. 3. Inorganic phosphate concentration profile in an agarose gel containing cells of C. roseus. (□) gel without cells; (■) gel with cells

within the gel as illustrated in Fig. 3. The increased productivity observed for alginate-entrapped cells of C. roseus (Brodelius 1986) may be due to such phosphate gradients. This example clearly illustrates the formation of a nutrient gradient within an immobilized plant cell preparation leading to stress (phosphate limitation) within at least part of the cell population.

LIGHT

Extensive studies on the induction of flavonoid biosynthesis in suspension cultures of Petersilium hortense by UV-light have been carried out by Hahlbrock and co-workers (1982). These excellent investigations probably represent the most thoroughly studied pathway of secondary metabolism in plant cell cultures. The induction patterns of both mRNAs and enzymes involved in the biosynthesis of flavonoids have been established. It has been concluded that groups of enzymes stand under the same gene regulation and the enzymes involved in the flavonoid biosynthesis may be divided into two groups. While the first group (group I, i.e. the initial three enzymes of the general phenylpropanoid pathway) can be induced by other treatments (e.g. elicitors; see below), the second group of enzymes (group II, i.e.

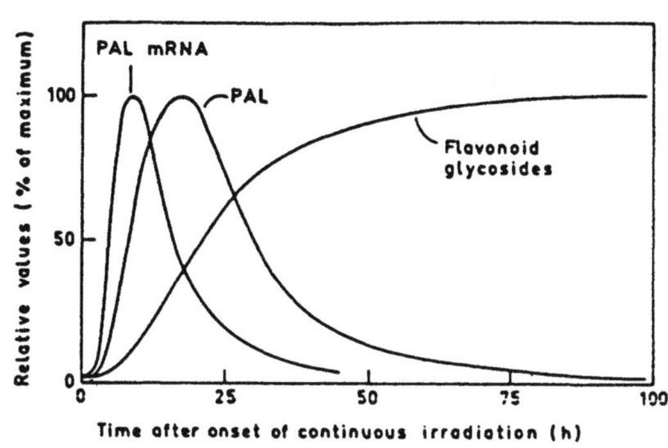

Fig. 4. Time courses of the light-induced changes in phenylalanine ammonia lyase mRNA and enzyme activity and in the amount of flavonoids accumulated in the cells (Hahlbrock et al. 1982)

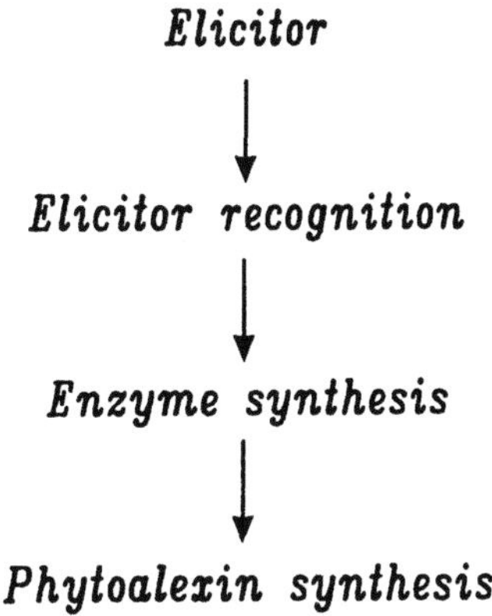

Fig. 5. Schematic representation of the steps involved in phytoalexin biosynthesis after elicitation

calchone synthase and subsequent enzymes) can only be induced by light. Furthermore, it has been shown that the mRNA coding for a particular enzyme can be isolated from the cultured cells before the enzyme activity can be shown, as demonstrated for phenylalanine ammonia lyase (PAL) in Fig. 4. The final product is isolated at a later stage after induction.

MICROBIAL ELICITORS

Microbial elicitors have received increasing attention during recent years. They may be applied to plant cell cultures for the following purposes:

1. Increase the yield of a target substance;
2. Studies on enzymology of secondary metabolism;
3. Studies on plant defense mechanisms.

Here, the first two applications will be discussed.

Enzymes of secondary metabolism may be induced in cell cultures in various ways. The use of fungal elicitors for this purpose has proven most valuable. When plants are infected by pathogenic organisms (e.g. fungi) they respond by producing phytoalexins which are toxic to the invading organism (Fig. 5). During the infection process elicitors (oligosaccharins) are liberated from the pathogen or the plant itself which trigger phytoalexin biosynthesis. Similarly, phytoalexin production may be induced in cell cultures by administration of fungal or other elicitors.

Elicitors can be of biotic (natural) or abiotic character. Table 3 lists some compounds which have been used as elicitors to induce product formation in plant cell cultures. In most studies on the effects of elicitors on the biosynthetic capacity of plant cell cultures, the elicitor has been obtained from a pathogenic microorganism (most often a fungus). This can be achieved simply by autoclaving a culture of the organism and then using the supernatant as elicitor. In other studies more or less purified elicitor (glucan) preparations have been employed.

Table 3. Some examples of elicitors used to induce secondary metabolism in plant cell cultures

Biotic elicitors:	Abiotic elicitors:
chitosan	heavy metal ions
glucans	organic solvents
glycoproteins	detergents
conidia	pesticides

In one case a biotic elicitor (oligosaccharin) has been extensively characterized (Sharp et al. 1984). It has been shown that the smallest active glucan from Phytophthora megasperma, which induces phytoalexin production in plants of Glycine max, is a hepta glucoside. The structure of this elicitor is shown in Fig. 6. The slightest change in structure results in an inactive oligosaccharin. It has been suggested that oligosaccharins may serve as a new class of hormones regulating such functions as growth, development, reproduction and defense against disease in plants (Albersheim and Darvill 1985).

The effects of fungal elicitors on the parsley culture producing flavonoids after induction with UV-light (see above) have been studied (Hahlbrock et al. 1982). An elicitor prepared from P. megasperma induces the formation of furanocoumarins in the cultures. Also in this case the first three enzymes of the general phenylpropanoid pathway are induced (group I). However,

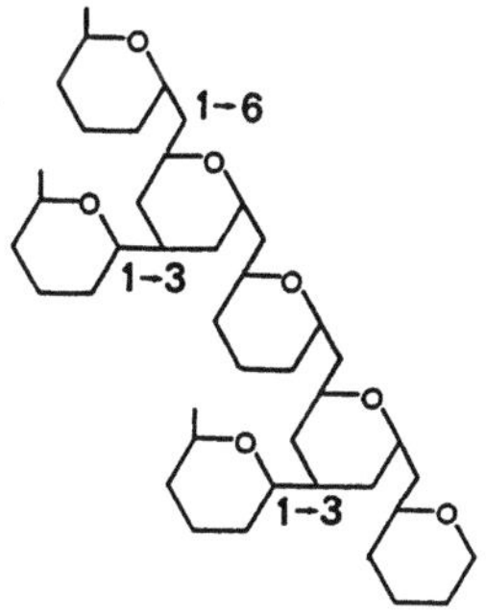

Fig. 6. Structure of the smallest active glucan (oligosaccharin) inducing phytoalexin formation in soybean. The glucan is liberated from the cell wall of P. megasperma by an enzyme from the plant

1-6
1-3
1-3

OLIGIOSACCHARIN

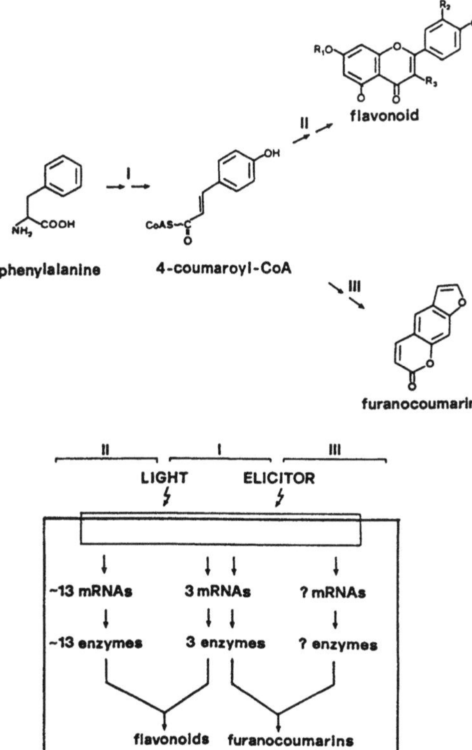

Fig. 7. Schematic diagramm summarizing the interrelationship between different groups of inducible enzymes involved in the biosynthesis of secondary metabolites (after Hahlbrock et al. 1982)

the subsequent enzymes of the biosynthetic pathway have been grouped into a third category (group III). This group of enzymes is only induced by the elicitor. UV-light cannot induce enzymes of group III. These studies are an excellent example of how the action of different stress agents can lead to the induction of different secondary products. As summarized in Fig. 7,

Table 4. Some examples of secondary products that have been induced in plant cell cultures by treatments with microbiol elicitors

Plant species	Microbial species[a]	Product induced	Concentration[b] Before	After	Incubation time(h)	Reference
Bidens pilosa	Phythium aphanidermatum	Phenylhepta-triyne	0	3.2(a)	48	Dicosmo et al., 1982
Canavalia ensiformis	Pithomyces chartarm	Medicarpin	0	0.43(a)	36	Gustine et al., 1978
Cephalotaxus barringtonia	Vericillium dahliae	Harringtonine alkaloids	0.01	0.51(b)	120	Heinstein, 1985
Cinchona ledgriana	Aspergillus niger	Anthraquinones	3	15 (b)	600	Wijnsma et al., 1985
Daucus carota	Chaetomium globosum	6-Methoxy-mellein	0	1 (c)	48	Kurosaki and Nishi, 1983
Dioscorea deltoides	Rhizopus arrhizus	Diosgenin	25	72 (b)	72	Rokem et al., 1985
Glycine max	Phytophthora megasperma	Glyceollin	0	0.05(b)	100	Ebel et al., 1984
Glycine max	Saccharomyces cerevisiae	Glyceollin	0	0.2 (b)	10	Funk et al., 1987
Gossypium arboreum	Vertcillium dahliae	Sesquiterpene aldehyde	Trace	96 (b)	120	Heinstein, 1985
Papaver somniferum	Botrytis sp.	Sanguinarine	Trace	6.6 (b)	24	Eilert et al., 1985
Papaver somniferum	Fusarium moniliforme	Morphine	0.07	1.40(b)	n.s.	Heinstein, 1985
		Codeine	0.08	1.44(b)	n.s.	
Petroselium hortense	Alternaris carthami	Bergapten	0	1.6 (c)	48	Tietjen et al., 1983
Phaseolus vulgaris	Colletotrichum lindemuthianum	Phaseollin	0	170 (a)	48	Robbins et al., 1985
Ruta graveolens	Rhodotorula rubra	Rutacridon--epoxides	0	0.23(b)	72	Eilert et al., 1982
Thalictrum rugosum	Saccharomyces cerevisae	Berberine	20	50 (b)	96	Funk et al., 1987

[a]Source of elicitor

[b](a) $\mu g\ g^{-1}$ Fresh weight; (b) $mg\ g^{-1}$ dry weight; (c) $\mu g\ ml^{-1}$ culture; n.s.= not stated.

UV-light induces enzymes of groups I and II leading to the synthesis of flavonoids, while elicitors induce enzymes of groups I and III resulting in furanocoumarin biosynthesis.

Some examples of the induction of secondary metabolites in plant cell suspension cultures are listed in Table 4. There are two fundamentally different types of elicitations observed. In some cultures the product is synthesized only after elicitor treatment, while in other cultures an enhanced product formation is seen after elicitation (e.g. diosgenin, morphine and berberine). Most of the former products may be classified as phytoalexins with antimicrobial activity (e.g. medicarpin, glyceollin and phaseollin). These are formed in the parent plant as a response to a species-specific pathogenic infection. However, in culture this specificity is <u>not</u> expressed. Elicitors prepared from non-pathogenic microorganisms may also be used to induce phytoalexins in cell cultures. Most other compounds found in Table 4 have antimicrobial activity but they should not be considered as phytoalexins until it has been shown that they are induced in the intact plant as a response to microbial attack.

In our studies we have used a glucan preparation from <u>Saccharomyces cerevisiae</u> as elicitor. The elicitor is prepared from yeast extract by ethanolic precipitation (Hahn and Albersheim 1978). This yeast glucan is effective in inducing phytoalexin (glyceollin) production in soybean cells as illustrated in Fig. 8 (Funk et al. 1987). Furthermore, it has been used to increase the productivity of berberine in cell cultures of <u>Thalictrum rugosum</u> (Funk et al. 1987).

It is of great importance to establish an optimum amount of elicitor to be added to the culture. This amount should be based on the amount of biomass in the culture and not on the volume of the culture. The berberine production as a function of incubation time after treatment with various concentrations of elicitor is shown in Fig. 9.

The growth stage of the cells is another important factor for an optimal effect of elicitor treatments. Figure 10 shows the berberine productivity of <u>T. rugosum</u> cells after treatment with the yeast elicitor at different growth stages. The maximum yield was observed when the cells were treated on day 6 after

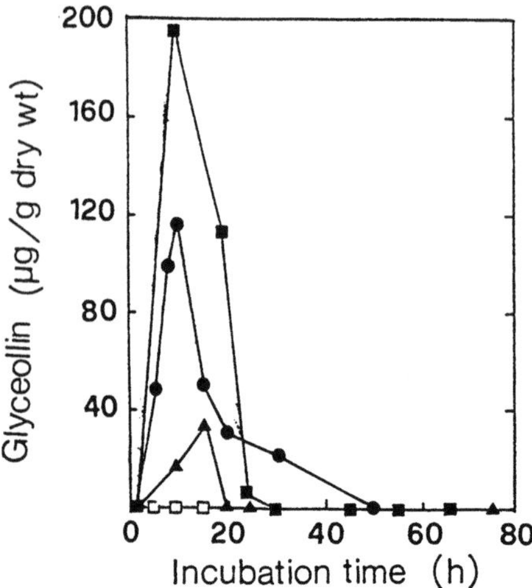

Fig. 8. Glyceollin production in cell suspension cultures of G. max after addition of various concentrations of elicitor. (□) 0; (▲) 2; (●) 10; (■) 20 µg elicitor per mg dry weight (after Funk et al. 1987)

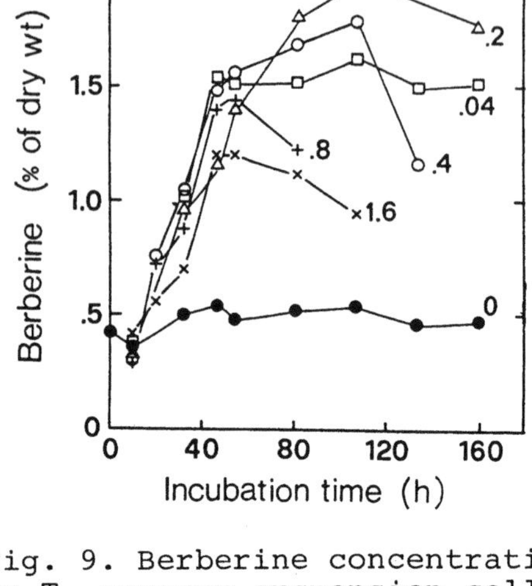

Fig. 9. Berberine concentration in T. rugosum suspension cells as a function of incubation time after additions of various concentrations of elicitor as indicated (mg glucan g^{-1} wet wt. cells) (Funk et al. 1987)

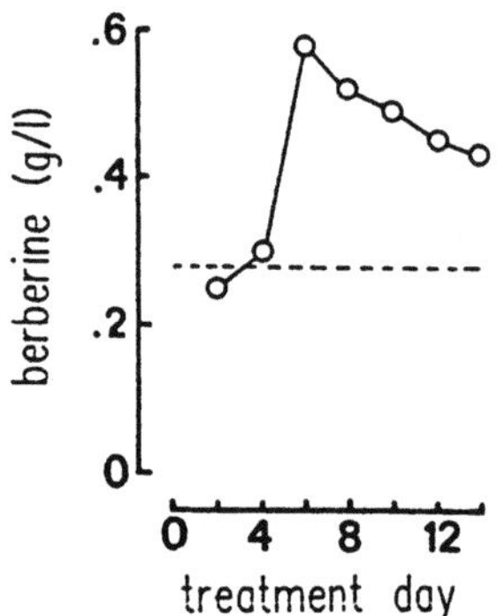

Fig. 10. Final berberine concentration in cultures of T. rugosum after elicitor treatment at various stages throughout an incubation. The broken line represents the concentration in untreated cells

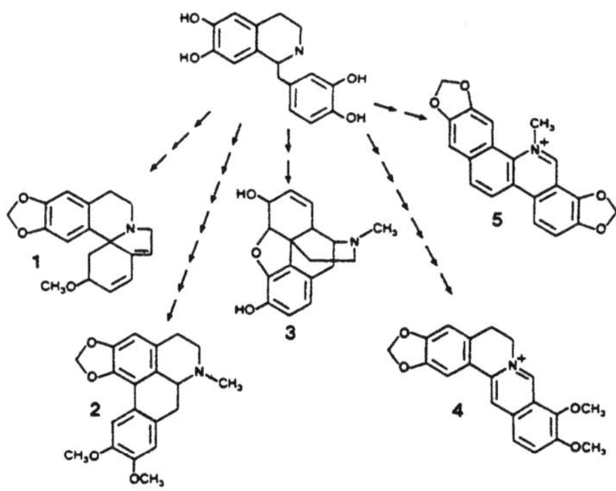

Fig. 11. Some examples of isoquinoline alkaloids derived from the common precursor norlaudanosoline. 1 Erythraline; 2 dicentrine; 3 morphine; 4 berberine; 5 sanguinarine

inoculation into fresh medium which corresponds to a mid-exponential growth stage. The yield of berberine was increased to around twice the amount in non-treated cells without increasing the fermentation time.

The biosynthesis of berberine from the precursor norlaudanosoline has been extensively studied (Zenk et al. 1985). All the enzymes involved have been isolated and partly characterized. Norlaudanosoline is a common precursor of a wide variety of alkaloids. Figure 11 gives some examples. This common precursor is synthesizted from two molecules of tyrosine. The immediate precursors to norlaudanosoline are dopamine and 3,4-dihydroxyphenacetaldehyde. It has not been established how these two substances are synthesized from tyrosine in the plant cell. The use of elicitors may be a valuable tool for investigating early steps of isoquinoline alkaloid biosynthesis, if it is assumed that the same pathway is operative in elicitor-treated cells as in non-treated cells.

Feeding experiments have shown that an enhanced tyrosine decarboxylase (TDC) activity is present in cells of T. rugosum after elicitor treatment as illustrated in Fig. 12 (Gügler et al. 1988). In non-treated cells (Fig. 12A) a high intermittant concentration of tyrosine is observed, while in elicitor-treated cells (Fig. 12B) a high tyramine concentration is seen. In the latter case tyrosine is rapidly decarboxylated by the induced TDC as it is taken up by the cells. There is no change in the uptake rate of the amino acid after elicitor treatment (Gügler et al. 1988). Furthermore, a good correlation between the enhanced berberine production and TDC activity within the cells has been demonstrated (Gügler et al. 1988). TDC may also be induced in other plant cell suspension cultures producing isoquinoline alkaloids. For example, TDC may be induced in cells of Eschscholzia californica producing benzophenanthridine alkaloids (Marques and Brodelius 1988). The amount of induced TDC activity in these cell cultures is highly dependent on elicitor concentration and incubation time after addition of elicitor as illustrated in Fig. 13 and 14 respectively. Cells of E. californica react more rapidly than cells of T. rugosum (Fig. 14). The optimal amount of elicitor for TDC induction is approxi-

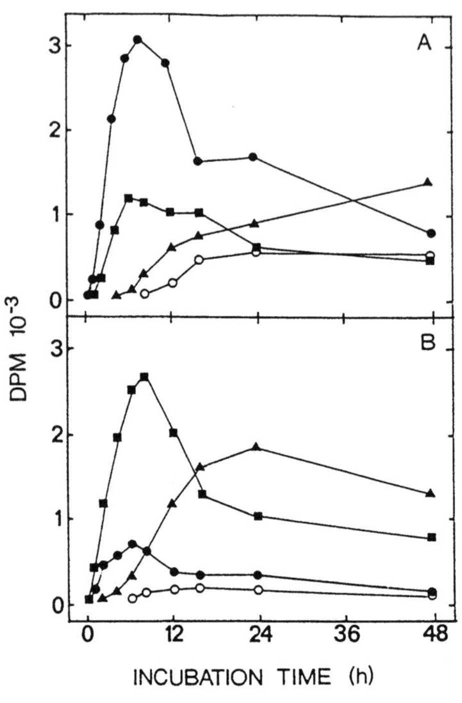

Fig. 12. Conversion of C-14 tyrosine (1 mM) added to cell cultures of T. rugosum. A Non-treated cells; B elicitor-treated cells (0.2 mg glucan g^{-1} wet wt. cells). (●) Tyrosine; (■) tyramine; (▲) dopamine; (o) DOPA (after Gügler et al. 1988)

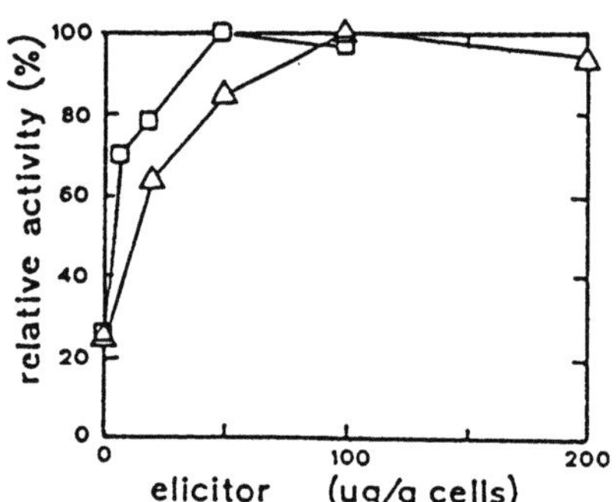

Fig. 13. Relative TCD activity as a function of yeast elicitor concentration for suspension cultures of T. rugosum (Δ) and E. californica (□)

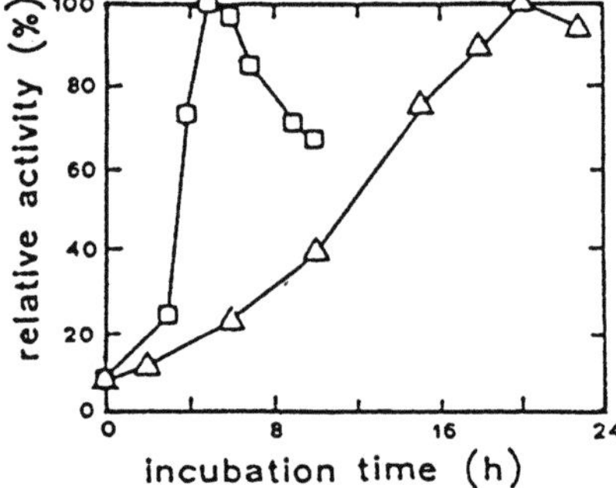

Fig. 14. Relative TCD activity as a function of incubation time after addition of yeast elicitor to suspension cultures of T. rugosum (Δ) and E. californica (□)

mately the same for the two cultures. Furthermore, this elicitor concentration is also optimal for berberine synthesis in the latter cell culture, indicating a correlation between TDC activity and berberine biosynthesis.

The results obtained indicate that TDC may play an important role in the regulation of isoquinoline alkaloid biosynthesis.

However, further studies are required before any final conclusions may be drawn. Nevertheless, the use of fungal elicitors to induce enzymes of secondary metabolism may be a valuable tool for studying the key enzymatic steps of a particular pathway. Other enzymes involved in norlaudanosoline biosynthesis should be investigated in elicitor-treated cells. Preliminary studies have shown that norlaudanosoline synthase condensing the two precursors does not appear to be induced by elicitor treatment (Hähner and Brodelius, unpublished). It is interesting to note that other intermediates (e.g. 3,4-dihydroxyphenylacetaldehyde) cannot be detected after feeding of radiolabelled tyrosine to the cells (cf. Fig. 12). Obviously, the turnover of this intermediate is high and therefore its formation may be rate-limiting in the biosynthesis of isoquinoline alkaloids in plant cell cultures.

CONCLUSIONS

Stress on plant cells in culture may be used to increase the productivity of secondary metabolites, as briefly reviewed in this paper. It is generally believed that the production of such compounds may be influenced by a number of stress factors. Different stress factors may result in the formation of different secondary products within the same cell culture. Fungal elicitors have received great attention in recent years. In fact, an induced or enhanced product formation has been shown for a large number of plant cell cultures after elicitor treatment (cf. Table 4). In addition to this increased productivity such substances may be employed to gain a better understanding of the genetic regulation and the enzymology of secondary metabolism in plant cells.

REFERENCES

Albersheim P and Darvill AG (1985) Oligosaccharins. Sci. Amer. 253(3):44-50

Brodelius P (1986) Immobilization of cultivated plant cells and protoplasts. Proceedings BIO FAIR TOKYO '86, Tokyo, Japan, October 15-19, 1986, pp. 173-180

Brodelius P and Vogel HJ (1985) A phosphorus-31 nuclear magnetic resonance study of phosphate uptake and storage in cultured Catharanthus roseus and Daucus carota plant cells. J. Biol. Chem. 260:3556-3560

Dicosmo F, Norton RA and Towers GHN (1982) Fungal culture-filtrate elicits aromatic polyacetylenes in plant tissue culture. Naturwissenschaften 69:550-551

Ebel J, Schmidt WE and Loyal R (1984) Photoalexin synthesis in soybean cells: elicitor induction of phenylalanine ammonia-lyase and chalcone synthase mRNAs and correlation with phytoalexin accumulation. Arch. Biochem. Biophys. 232:240-248

Eilert U, Engel B, Reinhard E and Wolters B (1982) Acridone epoxides in cell cultures of Ruta species. Phytochemistry 22:14-15

Eilert U, Kurz WGW and Constabel F (1985) Stimulation of sanguinarine accumulation in Papaver somniferum cell cultures by fungal elicitors. J. Plant Physiol. 119:65-76

Funk C, Guegler K and Brodelius P (1987) Increased secondary product formation in plant cell suspension cultures after treatment with a yeast carbohydrate (elicitor). Phytochemistry 26:401-405

Gustine DL, Sherwood RT and Vance CP (1978) Regulation of phytoalexin synthesis in jackbean callus cultures. Stimulation of phenylalanine ammonia-lyase and O-methyltransferase. Plant Physiol. 61:226-231

Gügler K, Funk C and Brodelius P (1988) Elicitor-induced tyrosine decarboxylase in berberine synthesizing suspension cultures of Thalictrum rugosum. Eur. J. Biochem., 170:661-666

Hahlbrock K, Kreuzaler F, Ragg H, Faulz E and Kuhn DN (1982) In: Biochemistry of differentiation and morphogenesis (L. Jaenicke, ed.) pp. 34-43, Springer Verlag, Berlin, Heidelberg

Hahn MG and Albersheim P (1978) Host-pathogen interactions. XIV. Isolation and partial characterization of an elicitor from yeast extract. Plant Physiol. 62:107-111

Heinstein PF (1985) Future approaches to the formation of secondary natural products in plant cell suspension cultures. J. Nat. Prod. 48:1-9

Hulst AC, Tramper J, Brodelius P, Eijkenboom LJC and Luyben KChAM (1985) Immobilized plant cells: respiration and oxygen transfer. J. Chem. Tech. Biotech. 35B:198-204

Knobloch K-H and Berlin J (1983) Influence of phosphate on the formation of the indole alkaloids and phenolic compounds in cell suspension cultures of Catharanthus roseus. Plant Cell Tissue Organ. Cult. 2:333-340

Kurosaki F and Nishi A (1983) Isolation and antimicrobial activity of the phytoalexin 6-methoxymellein from cultured carrot cells. Phytochemistry 22:669-672

Marques I and Brodelius P (1987) Elicitor-induced L-tyrosine decarboxylase from plant cell suspension cultures. I. Induction and purification. Plant Physiol., in press

Robbins MP, Bolwell GP and Dixon RA (1985) Metabolic changes in elicitor-treated bean cells. Selectivity of enzyme induction in relation to phytoalexin accumulation. Eur. J. Biochem. 148:563-569

Rokem JS, Tal B and Goldberg I (1985) Methods for increasing diosgenin production by Dioscorea cells in suspension cultures. J. Nat. Prod. 48:210-222

Sharp JE, Valent B and Albersheim P (1984) Purification and partial characterization of a glucan fragment that elicits phytoalexin accumulation in soybean. J. Biol. Chem. 259:11312-11320

Tietjen KG, Hunkler D and Martern U (1983) Differential response of cultured parsley cells to elicitors from two non-pathogenic strains of fungi. 1. Identification of induced products as coumarin derivatives. Eur. J. Biochem. 131:401-407

Vogel HJ and Brodelius P (1984) An in vivo ^{31}P NMR comparison of freely suspended and immobilized Catharanthus roseus plant cells. J. Biotechnol. 1:159-170

Wijnsma R, Go JTKA, van Weerden IN, Harkes PAA, Werpoorte R and Baerheim-Svendsen A (1985) Anthraquinones as phytoalexins in cell and tissue cultures of Cinchona sp. Plant Cell Rep. 4:241-244

Zenk MH, Rueffer M, Amann M and Deus-Neumann B (1985) Benzylisoquinoline biosynthesis by cultivated plant cells and isolated enzymes. J. Nat. Prod. 48:725-738

ELICITATION AND METABOLISM OF PHYTOALEXINS IN PLANT CELL CULTURES

W Barz, S Daniel, W Hinderer, U Jaques, H Kessmann, J Köster and K Tiemann

Westfälische Wilhelms-Universität
Lehrstuhl für Biochemie der Pflanzen
Hindenburgplatz 55
D-4400 Münster, FRG

INTRODUCTION

In the last decade cell suspension cultures of higher plants have convincingly been established as very suitable experimental systems to study the biosynthesis, the enzymology and the metabolic regulation of the accumulation of secondary plant constituents. Numerous compounds of very different chemical structure have been isolated from such cultures and in some cases cell cultures have turned out to be high accumulating systems of secondary plant products (Barz et al 1977, Barz and Ellis 1981, Yamada and Fujita 1983, Berlin 1984, Whitaker and Hashimoto 1986). Though quite successfully employed in biochemical investigations of different plant products there are also repeated reports in the literature that plant cell cultures failed to accumulate the characteristics secondary constituents found in differentiated tissues. Other observations also point to the interesting fact that cell suspension cultures synthesize and accumulate compounds which had hitherto not been found in the parent plant (Barz and Ellis 1981, Berlin 1984). Substantial evidence proves that cell cultures in relation to the differentiated tissue show a shift in the production of secondary products and that they synthesize new compounds (Barz and Ellis 1981; Yamada and Fujita 1983, Berlin 1984). Therefore, cell cultures can be considered as a potential source of new products. Furthermore, there is increasing evidence in the literature that higher plants also possess the ability to synthesize compounds only under very specific environmental and physiological conditions such as stress or infection by microbial pathogenes. Such compounds which normally are not found in the healthy plant tissue are a very interesting field of research because such compounds quite often show interesting

NATO ASI Series, Vol. H18
Plant Cell Biotechnology. Edited by M. S. S. Pais et al.

physiological properties for example a pronounced antimicrobial activity (i.e. phytoalexins). If higher plants possess the genetic potential to synthesize new products under stress conditions it should also be possible to induce the synthesis and the accumulation of such products in plant cell suspension cultures (Dixon 1986, Funk et al 1987). This concept to use cell suspension cultures as a source for new secondary constituents should be pursued with great intensity because several cases of the induced accumulation of hitherto unknown compounds in cell cultures have been published (Barz and Ellis 1981, Berlin 1984).

This report will first deal with a general description of the elicitation of phytoalexins and other plant products in cell suspension cultures and in a second part experiments will be discussed dealing with the elicitation and the metabolism of pterocarpan phytoalexins in cell suspension cultures of chickpea (<u>Cicer arietinum</u> L.).

ELICITATION OF PHYTOALEXINS IN PLANT CELL CULTURES

Phytoalexins are low molecular weight, antimicrobial compounds, synthesized by plants after microbial infection. These "postinfectional inhibitors" are part of the defense reaction of higher plants against phytopathogenes. Numerous phytoalexins of very different chemical structures have been isolated from a great number of plants (Bailey and Mansfield 1982, Brooks and Watson 1985). Inspection of the chemical structures of phytoalexins demonstrate a great diversity in chemical structure. The structure of phytoalexins seem to be characteristic for the genus or family of plant taxonomy. Solanaceae mainly synthesize sesquiterpenoids and Papillionaceae are known to synthesize isoflavones and pterocarpans as antimicrobial defense compunds (Bailey and Mansfield 1982, Brooks and Watson 1985). Though very different in chemical structure phytoalexins are by definition only formed under stress or infection conditions which documents the great potential for the formation of new secondary compounds. More than 200 phytoalexins are now known and due to continued efforts in many laboratories the number of new phytoalexins is steadily increasing. Apart from their biological function as defense compounds phytoalexins are of interest because they represent compounds with pronounced antimicrobial and possibly other pharmacological properties. In addition to the isoflavones, pterocarpans and sesquiterpenoids various alkaloids, diterpenoids, polyacetylenic compounds and derivatives of other basic skeletons have been described (Bailey and Mansfield 1982, Brooks and Watson 1985). The production of phytoalexins by plants can be stimulated not only by infection, but also by elicitors of fungal origin or other natural organic compounds (biotic elicitors) and finally by a variety of abiotic elicitors such as heavy metal ions (Bailey and Mansfield 1982, Darvill and Albersheim 1984, Brooks and Watson 1985, Dixon 1986, Funk et al 1987).

Since a biochemical analysis of the molecular events of a host-parasite-interaction in differentiated plant tissue represents a very complicated task it is very advantageous that elicitation of phytoalexin formation can be performed in plant cell cultures. Several reports dealing with such different compounds as isoflavones and pterocarpans (Ebel et al 1976, Ebel 1986, Funk et al 1987), stilbenes (Fritzemeier et al 1983), furanocoumarins (Tietjen et al 1983), chalcones (Ayabe et al 1987), alkaloids (Yamada and Fujita 1983, Funk et al 1987), and various other classes of natural products, provide ample evidence that phytoalexin biochemistry and enzymology can be studied in cell suspension cultures. It has been shown that the mechanisms of elicitation of phytoalexin biosynthesis are highly specific and only involve that part of the biosynthetic machinery which is required for phytoalexin formation (Robbins et al 1985, Daniel et al 1986). As far as presently known the elicitor signalled to de novo synthesis of enzymes of phytoalexin biosynthesis and corresponding mRNAs (Bailey and Mansfield 1982, Darvill and Albersheim 1984, Brooks and Watson 1985, Dixon 1986, Ebel 1986). The term "elicitor" describes any substance that stimulates the biosynthesis of phytoalexins. Biotic elicitors are derived from microorganisms or they are endogenous compounds of the infected plant. The latter group of elicitors which are often released by mechanical wounding or enzymatic hydrolysis of polymeric compounds in plants are in most cases pectic fragments from the plant cell walls (Darvill and Albersheim 1984, Brooks and Watson 1985, Dixon 1986). Two types of cell wall polysaccharides from fungi have now been shown to serve as elicitors, namely certain β-linked glucans and chitosan. In addition, glycoproteins and proteins have been isolated from various fungi and used as elicitors (Darvill and Albersheim 1984, Dixon 1986, Ebel 1986). Another group of elicitors from microorganisms are enzymes with polygalacturonase activity (Dixon 1986) which release pectic fragments from plant cell walls. The various aspects of elicitor structure and their mode of action have been summarized (Darvill and Albersheim 1984, Brooks and Watson 1985, Dixon 1986, Ebel 1986). Availability of fungal elicitors has greatly stimulated research on phytoalexins in plant cell suspension cultures because small amounts of such water soluble oligomers lead to a very rapid induction of phytoalexin synthesis and a pronounced accumulation of these compounds in the cell cultures (Bailey and Mansfield 1982, Darvill and Albersheim 1984, Brooks and Watson 1985, Dixon 1986, Funk et al 1987). Such well established experimental systems will and have already greatly facilitated the analysis of the sequence of reaction steps from the first perception of an elicitor signal by the plant cell via the mechanism of gene activation followed by protein synthesis and phytoalexin formation (Darvill and Albersheim 1984, Dixon 1986, Ebel 1986). In addition to these aspects of biochemistry and molecular phytopathology such elicitation systems of plant cell cultures represent a very valuable approach to recognize the still not fully known potential of higher plants for the production of new and biologically active secondary compounds.

CONSTITUTIVE AND INDUCIBLE SECONDARY METABOLISM IN *CICER ARIETINUM* L.

The chickpea plant, *Cicer arietinum* L., is an important crop plant in semiarid areas of the world and it is cultivated because of its high protein content of the seeds. The most widespread fungal disease of chickpea is caused by the deuteromycete *Ascochyta rabiei* (Pass.) Lab. which leads to the characteristic symptoms of a "leaf spot disease" also known as "Ascochyta blight" (Nene 1982, Saxena and Singh 1984).

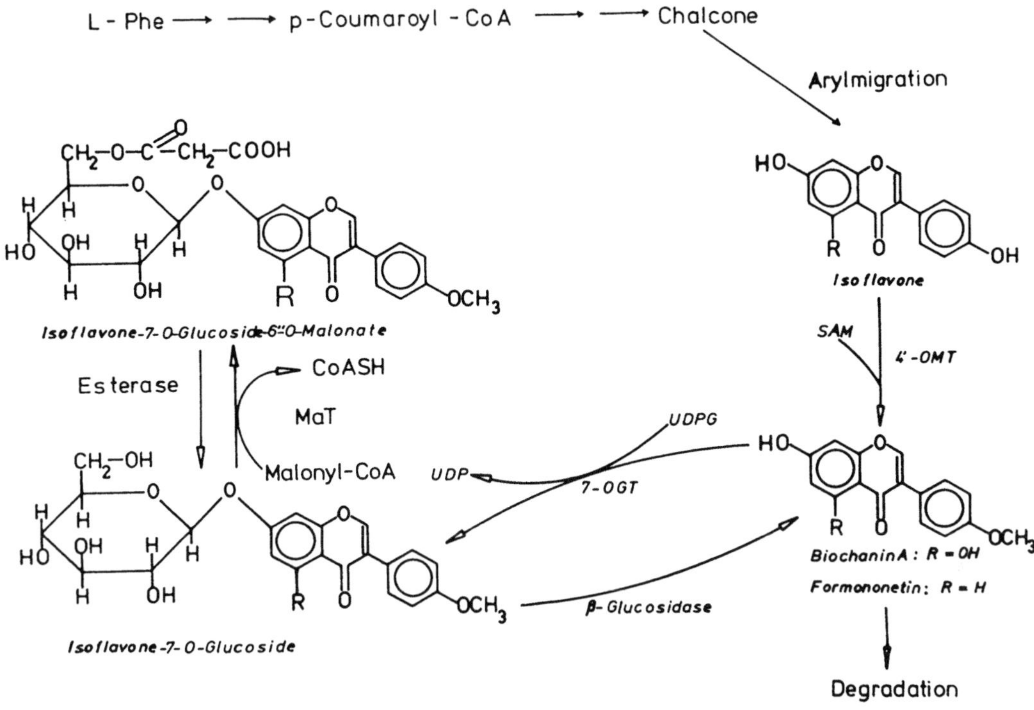

Fig. 1: Biosynthesis, conjugation reactions and proposed metabolism of the isoflavones formononetin and biochanin A in <u>Cicer arietinum</u> L.
4'OMT: S-Adenosylmethionine: isoflavone 4'-O-methyltransferase;
7-OGT: UDPG: isoflavone 7-O-glucosyltransferase;
MaT: Malonyl-CoA: isoflavone 7-O-glucoside-6"-O-malonyltransferase

<u>Ascochyta</u> <u>rabiei</u>-resistant cultivars have been produced by several breeding programs (Saxena and Singh 1984). Various investigations have led to the assumption (Nene 1982) that resistance of chickpea towards <u>A. rabiei</u> is explained by a high content of phenolic constituents. Investigations in the authors' laboratory have been devoted towards an analysis of the constitutive phenolic compounds of this plant. The question was whether different cultivars of chickpea differ in their level of constitutive phenolic compounds.

The main phenolic compounds of chickpea are the isoflavones biochanin A and formononetin which occur as the 7-O-glucoside-6"-O-malonates (Köster et al 1983). An increasing number of malonyl glucosides has been found in recent years in higher plants (Barz et al 1985), where they are considered to be constituents of vacuoles (Matern et al 1983). Due to their labile chemical character such malonyl conjugates have quite often been overlooked as a result of inproper isolation techniques (Köster et al 1983).

The enzymology of the formation and the subsequent metabolism of the isoflavone malonylglucosides have extensively been studied in _Cicer arietinum_ (Fig. 1). Starting with the isoflavone aglyca the following enzymes have been characterized substantially 1) a UDP-glucose depending O-glucosyltransferase (Köster and Barz 1981) which showed a high substrate specificity for the isoflavone skeleton and a strict position specificity for the hydroxyl group in position 7; 2) a malonyl-CoA-depending malonyl transferase (Köster et al 1984) which transfers the malonyl moiety to the carbon atom 6 of the glucose in an isoflavone 7-O-glucoside; 3) a highly substrate specific malonyl-esterase (Hinderer et al 1986) which removes the malonic acid and yields the isoflavone 7-O-glucoside. This enzyme is membrane-bound and has been found to be completely different from the numerous esterases known to occur in higher plants (Hinderer et al 1986); 4) isoflavone glucoside-specific glucohydrolases (Hösel et al 1975) which occur in chickpea in organ-specific patterns of isoenzymes. The metabolic grid shown in Fig. 1 readily explains the observation that the formononetin-malonylglucoside represents a metabolically very active plant product which shows permanent turnover (Jaques et al 1985). This steady state in the concentration of the formononetin malonylglucoside results from equal rates of synthesis and degradation (Barz and Köster 1981).

The metabolic activity of the formononetin conjugate in chickpea tissue possibly provides for the plant a system from which formononetin can be drawn for pterocarpan phytoalexin biosynthesis under conditions where the general phenylpropanoid pathway is not acitve or blocked. Such a metabolic alternative for phytoalexin formation can be regarded as a protective measure for the very rapid accumulation of defense compounds; this hypothesis is presently under experimental investigation.

Comparative analysis of the level and accumulation of isoflavones and isoflavoneglucoside conjugates in different cultivars of chickpea which are either highly resistant (ILC 3279) or very susceptible (ILC 1929) to Ascochyta rabiei have unequivocally shown that there is no difference in isoflavone conjugate concentration (Weigand et al 1986). These and other investigations seem to exclude the constitutive secondary compounds in chickpea as a source of differential protection towards this fungal parasite. On the other hand it could be demonstrated that chickpea cultivars ILC 3279 and ILC 1929 significantly differ in their ability to accumulate the pterocarpan phytoalexins medicarpin and maackiain (Weigand et al 1986). The resistant cultivar readily accumulates high amounts of the phytoalexins whereas the susceptible cultivar only accumulates low amounts of medicarpin. This result supports evidence from other plant systems (Hahn et al 1985) where phytoalexins must be assumed as an important system of chemical defense.

Maackiain and medicarpin have quite often been found as phytoalexins in numerous leguminous plants (Ingham 1982) and especially the latter compound appears to be a widespread phytoalexin. The chickpea plant accumulates the (6aR:11aR)-enantiomers (Kessmann and Barz unpublished) though the (6aS:11aS)-enantiomers have also been found in nature (VanEtten et al 1983). The biosynthesis of medicarpin and maackiain has been studied by Dewick and associates (Dewick 1975, Dewick 1977, Dewick and Ward 1977). The results of these feeding experiments with various phenylpropanoid precursors and suitably substituted intermediates showed that the phenylpropanoid pathway from L-phenylalanine through the stages of cinnamic acids and a chalcone leads to the isoflavone formononetin as an intermediate. The pathway suggested by Dewick and associates from formononetin to the pterocarpan phytoalexins is shown in Fig. 2 where a sequence of hydroxylation reactions in the side chain phenylring, reduction of an isoflavone to an isoflavanone with the final closure of the dihydrofuranring has been postulated. This pathway finds analogies in the postulated sequences leading to other pterocarpan phytoalexins such as phaseollin, pisatin and glyceollin (Brooks and Watson 1985, Smith and Banks 1986). Enzymatic evidence for these reactions as shown in Fig. 2 has been missing until quite recently (see below).

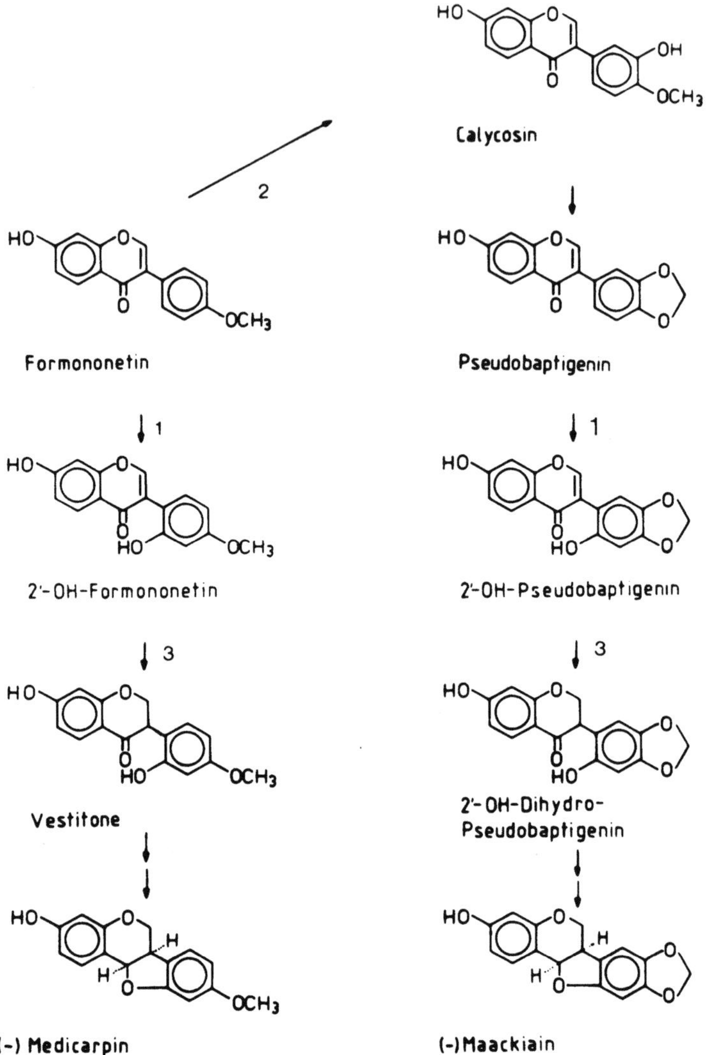

Fig. 2: Biosynthesis of pterocarpan phytoalexins in chickpea (Cicer arietinum L.).
The enzymes marked by numbers have been isolated from elicitor induced chickpea cell cultures

1) Formononetin 2'-hydroxylase (Hinderer et al 1987)
2) Formononetin 3'-hydroxylase (Hinderer et al 1987)
3) NADPH: 2'-hydroxyisoflavone oxidoreductase (Tiemann et al 1987)

ELICITOR-INDUCED ISOFLAVONE AND PHYTOALEXIN METABOLISM IN CHICKPEA CELL SUSPENSION CULTURES

Cell suspension cultures of chickpea cultivars ILC 1929 and ILC 3279 were routinely grown in PRL-4c medium (Kessmann and Barz

1987). Both cultures showed identical growth rates and they both accumulated formononetin and biochanin A aglyca and the glucosyl conjugates essentially as in the intact plants (Kessmann and Barz 1987). On the other hand the cell cultures of the two cultivars dramatically differed in their ability to synthesize and accumulate the phytoalexins medicarpin and maackiain (Kessmann and Barz 1986, Kessmann and Barz 1987). Analysis of isoflavone and phytoalexin accumulation after transfer of cells into new medium revealed that medicarpin and maackiain accumulated both in the cells and in the growth medium. Maximum of phytoalexin accumulation was consistently found 24 hours after inoculation of cells into new medium (Kessmann and Barz 1987). In agreement with data from chickpea plant tissue (Weigand et al 1986) medicarpin represented the most prominent phytoalexin. Figure 3 also shows that the cells of cultivar ILC 3279 are about three to four times more efficient in accumulating the phytoalexins. This difference between the two cell culture lines had also been observed at the level of the intact plant (Weigand et al 1986). The experiments also demonstrated that yeast extract, a component of the PRL-4c medium, was the inducing agent for phytoalexin accumulation in the cell cultures (Kessmann and Barz 1987). In the absence of yeast extract the cell cultures were practically devoid of phytoalexins under otherwise identical conditions (Kessmann and Barz 1987). After treatment of the cell cultures with yeast extract the phytoalexins predominantly accumulate in the growth medium. In general, it seems to be a common property of phytoalexin metabolism in plant cell cultures that substantial amounts of these newly synthesized compounds are found in the growth medium. The adherent mechanisms of excretion remain to be elucidated.

The yeast extract did not effect the accumulation of isoflavones and isoflavone conjugates (Kessmann and Barz 1987). This may support the assumption that constitutively formed formononetin did not function as a precursor of induced phytoalexin biosynthesis (Kessmann and Barz 1987). Subsequent studies have shown that yeast extract represents a very potent biotic elicitor for pterocarpan formation and relevant biosynthetic enzymes in chickpea cell cultures (Daniel et al 1986, Tiemann et al 1987,

Hinderer et al 1987). Therefore, yeast extract may be a suitable elicitor preparation also for biotechnological application (Funk et al 1987, Kessmann and Barz 1987).

Chickpea cell suspension cultures derived from cultivar ILC 3279 have been used to measure the specific activities of enzymes involved in the biosynthesis and degradation of isoflavones and isoflavone conjugates. Parallel to the main phase of the isoflavone conjugate accumulation the enzyme activities of phenylalanine ammonialyase, chalcone synthase, chalcone isomerase and isoflavone glucosyltransferase (Fig. 1) showed a pronounced maximum. In contrast, the malonyl esterase activity exhibited a minimum of specific activity at the time of maximum of isoflavone conjugate accumulation (Daniel et al 1986, Daniel et al in prep.).

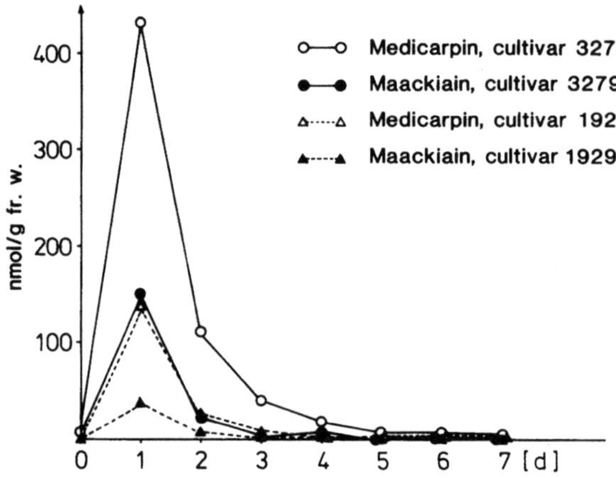

Fig. 3. Accumulation of medicarpin and maackiain in the growth medium of cell culture ILC 3279 and ILC 1929 after transfer of 2.5g (fresh weight) cell suspension into PRL-4c medium supplemented with yeast extract (2.5 g/l). n=3

The chickpea cultures not only respond to yeast extract as an elicitor material but also to an elicitor preparation derived from mycelium or culture fluid of A. rabiei (Daniel et al 1986, Kessmann and Barz 1986). Comparative studies revealed that cell cultures of chickpea cultivar ILC 3279 accumulated much greater amounts of medicarpin and maackiain than cell culture ILC 1929 after elicitor treatment (Kessmann and Barz 1987). This observation will form the basis for future molecular biological investigations to identify this remarkable difference of genotypes. The investigation with the A. rabiei-elicitor have again shown that constitutive isoflavone metabolism is practically not effected.

In order to determine whether the elicitor specifically effects the activities of enzymes involved in phytoalexin biosynthesis 14 enzymes of primary and secondary metabolism were measured after elicitor treatment (Daniel et al 1986). In case of cell culture ILC 3279 the elicitor application led to an increase in the specific activity of the following enzymes: glucose-6-phosphate dehydrogenase, phenylalanine ammonia lyase, chalcone synthase and isoflavone specific 7-O-glucosyltransferase. Maximum enzyme activity was reached some 4-8 hours after elicitor application which correspondes quite well with the observation that maximum phytoalexin levels were reached some 8-10 hours after elicitor application to the cell cultures. Various other enzymes were not effected by elicitor treatment. Other elicitor induced changes in metabolic pathways which seem to be independent from phytoalexin biosynthesis have been reviewed (Darvill and Albersheim 1984, Dixon 1986, Ebel 1986).

As shown in Fig. 3 the phytoalexins formed after induction with either yeast extract or A. rabiei-elicitor are not permanently accumulated in the medium but rather disappear within short periods of time. Intensive studies using cell cultures and cell free aliquots of the nutrient medium have demonstrated that the rapid disappearance of pterocarpan phytoalexins is due to peroxidatic polymerization (Barz et al 1986, Daniel et al 1986). The activity of peroxidases in the growth medium steadily increased during one growth cycle and it is therefore not possible to determine the optimum time for phytoalexin induction during the growth cycle of chickpea cell cultures (Daniel et al in prep.). Subsequent investigations have also shown that the differences in elicitor-induced phytoalexin accumulation between the two chickpea cell cultures cannot be explained by a different activity or level of peroxidases but rather by a different rate of biosynthesis.

3-Hydroxypterocarpans are readily oxidized by peroxidases essentially as shown for various other compounds of similar chemical structure (Barz and Köster 1981). The observation that peroxidatic destruction of phenolic compounds occurs in the growth medium of plant cell cultures is of general importance both for studies on other phytoalexins (Brooks and Watson 1985) and for the biotechnological application of cell cultures in biotransformation reactions (Barz and Ellis 1981, Berlin 1984). Biotransformation reactions with cell cultures normally require that both substrate and product will not be altered by enzymes of the growth medium. Biotransformation studies with peroxidase-sensitive compounds in plant cell cultures will therefore require the selection of peroxidase-free cell culture lines.

When grown in yeast-extract containing medium chickpea cell cultures regularly accumulate the isoflavones biochanin A and formononetin together with their glucosyl conjugates and, furthermore, upon each inoculation step phytoalexin formation is induced (Fig. 3). The yeast extract of the PRL-4c medium was substituted by small amounts of the amino acid glycine (2 mg/l). This substitution did not effect the excellent growth rates of the cultures (Kessmann and Barz 1987). However, after longer periods of growth of the cell culture ILC 3279 on medium without yeast extract remarkable changes in the expression of secondary metabolism occurred. The data in Table 1 document that the cells of cell culture ILC 3279 now accumulated relatively small amounts of isoflavones and isoflavone conjugates. Phytoalexins were not detected in these cell cultures regardless of the growth stage. Though practically devoid of isoflavonoids the cell cultures ILC 3279 can still be induced by either yeast extract (Hinderer et al 1987, Tiemann et al 1987) or A. rabiei-elicitor for phytoalexin biosynthesis.

These cell cultures were used for isolation and characterization of new enzymes of pterocarpan phytoalexin biosynthesis after induction with yeast extract or with an elicitor from A. rabiei.

As outlined above elicitor induced phytoalexin formation is best explained by assuming both transcription and translation processes involving the genes which code for the enzymes involved in phytoalexin biosynthesis (Dixon 1986, Ebel 1986). Evidence for the involvement of induced transcription and translation processes has been compiled in various laboratories (for review see Bailey and Mansfield 1982, Darvill and Albersheim 1984, Dixon 1986, Ebel 1986, Smith and Banks 1986).

Compound	ILC 3279+YE	ILC 3279-YE
Biochanin A	55	39
Formononetin	41	6
Biochanin A-Glucoside	101	--
Formononetin-Glucoside	50	32
Biochanin A-Glucoside-6"-Malonate	510	8
Formononetin-Glucoside-6"-Malonate	230	4
Medicarpin	50	--
Maackiain	19	--
	1056 (100%)	89 (8.4%)

Tab. 1: Comparison of phenolic constituents in chickpea cell suspension cultures ILC 3279 after growth with (+YE) or without (-YE) yeast extract on day 3 of growth cycle; (data in nmol/g fr.w.) n=3

In case of elicitor induced phytoalexin biosynthesis in chickpea cell cultures it was shown, that the transcription inhibitor actinomycin D and the translation inhibitor cycloheximid both substantially inhibited medicarpin and maackiain accumulation (Fig. 4). These data support the assumption that the elicitor leads to de-novo synthesis of enzymes required for phytoalexin biosynthesis and corresponding mRNAs.

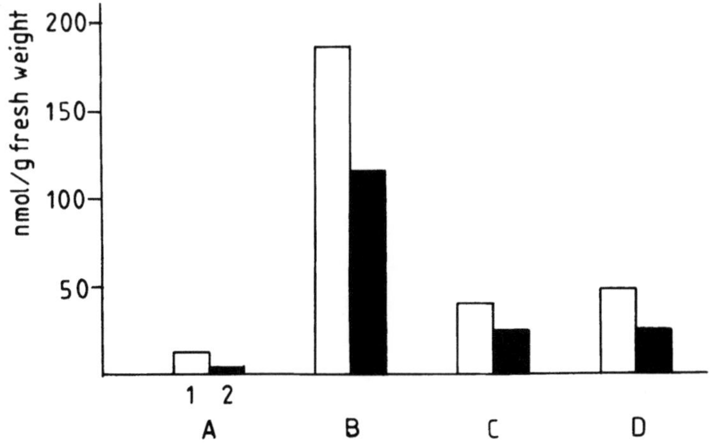

Fig. 4: Accumulation of medicarpin (1) and maackiain (2) in chickpea cell culture ILC 3279 after elicitor treatment and effect of actinomycin D and cycloheximid on elicitor-induced phytoalexin biosynthesis.

Cells (40 ml culture) were treated on day 3 of growth cycle. Cells and medium were extracted 8 h after elicitor application as described (Kessmann and Barz 1987).

A) control; B) 5 mg *A. rabiei* elicitor; c) 5 mg elicitor + 1 mg cycloheximid; D) 5 mg elicitor + 1,8 mg actinomycin D

The biosynthesis of medicarpin and maackiain (Fig. 2) has been previously studied by feeding experiments (Dewick 1975, Dewick 1977, Dewick and Ward 1977, Smith and Banks 1986). The sequence involves enzyme reactions which are specifically involved in phytoalexin accumulation and which are not part of general phenylpropanoid biosynthesis (Fig. 2). Knowledge of such specific enzymes of phytoalexin biosynthesis is very valuable with regard for the chemical structure of the intermediates involved and the metabolic regulation of such inductive pathways. There is, however, only limited knowledge of such specific enzymes of phytoalexin biosynthesis because a comparatively small number of such enzymes has been characterized (Tab. 2).

Using elicitor induced chickpea cell cultures it has recently been possible to identify and characterize a series of 2'- and 3'-hydroxylases which form the oxygen substitution pattern in the phenyl ring of medicarpin and maackiain (Hinderer et al 1987) (Fig. 2). These hydroxylases have been characterized as microsomal cytochrome P450 oxygenases (Hinderer et al 1987). The enzymatic data of the hydroxylation reactions and various measurements with molecular inhibitors have shown that there are at least two independent hydroxylases involved in these hydroxylation reactions. Maximum hydroxylase activity is obtained some 16 hours after transfer of cells into medium supplemented with yeast extract followed by a rapid decrease of enzyme activity in the following 24 hours. The transient induction of enzyme activity correlated with the accumulation of phytoalexins where the maximum was observed 24 hours after elicitor treatment.

The next step in pterocarpan biosynthesis is catalyzed by a NADPH:2'-hydroxy-isoflavone oxidoreductase which produces the corresponding isoflavanones (Fig. 2). Essentially as measured for the hydroxylases the oxidoreductase reaches maximum enzyme activity some 16 hours after elicitor treatment. This reduction step provides the correct stereochemistry at carbon atom 3 of the isoflavanone which is required for the configuration of the 6aR:11aR pterocarpan phytoalexins in chickpea (Kessmann and Barz unpublished). In case of medicarpin biosynthesis Fig. 2 shows that there is only one more enzyme necessary to complete the specific biosynthetic pathway. The cell culture system de-

Enzyme	Plant material	Phytoalexin	Induction method	References
Casbene synthetase	Ricinus communis	Casbene[a)]	Rhizopus stolonifer;RsE	Dueber et al., 1978 Moesta and West, 1985
Dimethylallylpyrophosphate: trihydroxypterocarpan dimethylallyltransferase	Glycine max*	Glyceollin[b)]	Pmg, PmgE	Zähringer et al., 1979 Leube and Grisebach, 1983
3,9-dihydroxypterocarpan 6a-hydroxylase	Glycine max*	Glyceollin[b)]	PmgE	Hagmann et al., 1983
S-Adenosyl-L-methionin: bergaptol and S-Adenosyl-L-methionin: xanthotoxol O-methyltransferases	Petroselinum hortense*	Xanthotoxin[c)] Bergapten[c)]	PmgE	Hauffe et al., 1986
S-Adenosylmethionin:(+)6a-hydroxymaackiain O-methyltransferase	Pisum sativum	Pisatin[b)]	Nectria haematococca; $CuCl_2$	Sweigard et al., 1986
Stilbene synthase	Arachis hypogaea*	Resveratrol[d)]	UV-light	Fritzemeier et al., 1983
NADPH: dehydroipomeamarone oxidoreductase	Ipomea batatas	Ipomeamaron[e)]	Ceratocystis fimbriata	Inoue et al., 1984
Dimethylallylpyrophosphate: umbelliferone dimethylallyltransferase	Petroselium hortense*	Psoralen[c)]	AcE	Tietjen and Matern, 1983
Formononetin 2'- and 3'-hydroxylases	Cicer arietinum*	Medicarpin[b)] Maackiain[b)]	Yeast extract	Hinderer et al., in press
NADPH: 2'-hydroxyisoflavone oxidoreductase	Cicer arietinum*	Medicarpin[b)] Maackiain[b)]	Yeast extract; ArE	Tiemann et al., 1987

Table 2: Enzymes specifically involved in phytoalexin biosynthesis. RsE: elicitor from Rhizopus stolonifer (polygalacturonase); Pmg: Phytophtora megasperma f.sp. glycinea; PmgE: elicitor from Pmg; ArE: elicitor from Ascochyta rabiei; AcE: elicitor from Alternaria carthami. a) Diterpene, b) Isoflavonoid (Pterocarpane), c) Furanocoumarine, d) Stilbene, e) Furanosesquiterpene, *: enzyme present in elicitor treated cell cultures of the plant

scribed will hopefully allow the isolation and characterization of this terminal enzyme step.

CONCLUSIONS

The recent investigations on phytoalexin formation and accumulation in plant cell suspension cultures have demonstrated that such provide elegant means to study the enzymology and the metabolic regulation of induction processes in plant cells. Furthermore, such studies will lead to a better understanding of the potential of higher plants to form new compounds which can only be synthesized under specific conditions. The natural product chemist will find the system valuable to detect new biologically active compounds and the biochemist will find this field of high interest because one can study the metabolic regulation of a selected pathway due to the specific induction by a biotic elicitor compound.

ACKNOWLEDGEMENTS

The financial support by Deutsche Forschungsgemeinschaft and Fonds der Chemischen Industrie is gratefully acknowledged.

REFERENCES

Ayabe S, Udagawa K, Iida Y, Yohsikawa T (1987) Regulation of retrochalcone biosynthesis: activity changes of O-methyltransferase in the yeast extract-induced Glycorrhiza echinata cells. Plant Cell Reports 6:16-19

Bailey JA and Mansfield JW (1982) (eds) Phytoalexins, Blackie, Glasgow, London

Barz W and Koster J (1981) Turnover and degradation of secondary (natural) products. In: Conn EE (ed) The biochemistry of plants, Vol. 3, Academic Press, New York, London, Toronto, Sydney, San Francisco, pp 35-84

Barz W, Reinhard E and Zenk W (1987) (eds) Plant tissue culture and its biotechnological application, Springer Verlag, Berlin, Heidelberg and New York

Barz W and Ellis BE (1981) Plant cell cultures and their biotechnological potential. Ber. Deutsch. Bot. Ges. 94:1-26

Barz W, Daniel S, Hinderer W, Jacques U, Kessmann H and Tiemann K (1986) Metabolism and enzymology of isoflavone malonylglucosides and pterocarpan phytoalexins in Cicer arietinum. Planta medica 5:420

Barz W, Köster J, Weltring K.-M. and Strack D (1985) Recent advances in the metabolism and degradation of phenolic compounds in plants and animals. In: Van Sumere CF and Lea PJ (eds) Annual Proceedings of the Phytochemical Society of Europe, Vol. 25, Oxford University Press, Oxford pp 307-349

Berlin J (1984) Plant cell cultures - a future source of natural products? Endeavour 8:5-8

Brooks CJW and Watson DG (1985) Phytoalexins. Natural Product Report 2:427-459

Daniel S, Kessmann H, Jacques U and Barz W (1986) Elicitor-stimulated phytoalexin metabolism and changes of enzyme activities in chickpea cell cultures. Biol. Chem. Hoppe-Seyler 367 (suppl.):201

Darvill AG and Albersheim P (1984) Phytoalexins and their elicitors - a defense against microbial infection in plants. Ann. Rev. Plant Physiol. 35:243-312

Dewick PM (1975) Pterocarpan biosynthesis: chalcone and isoflavone precursors of demethylhomopterocarpin in Trifolium pratense. Phytochemistry 14:979-982

Dewick PM (1977) Biosynthesis of pterocarpan phytoalexins in Trifolium pratense. Phytochemistry 16:93-97

Dewick PM and Ward D (1977) Stereochemistry of isoflavone reduction during pterocarpan biosynthesis: an investigation using deuterium nuclear magnetic resonance spectroscopy. J.C.S. Chemm. Comm. 211:338-339

Dixon RA (1986) The phytoalexin response: elicitation, signalling and host gene expression. Biol Rev 61:239-291

Dueber M, Adolf W and West CA (1978) Biosynthesis of the diterpene casbene. Partial purification and characterization of casbene synthetase from Ricinus communis. Plant. Physiol. 62: 598-603

Ebel J, Ayers AR and Albersheim P (1976) Host-pathogen interactions XII. Response of suspension cultured soybean cells to the elicitor isolated from Phytophtora megasperma var. sojae, a fungal pathogen of soybeans. Plant Physiol. 57:775-779

Ebel J (1986) Phytoalexin synthesis: biochemical analysis of the induction process. Ann Rev Plant Pathol 24:235-264

Fritzmeyer KH, Rolf CH, Pfau J and Kindl H (1983) Coordinate induction by UV light of stilbene synthase, phenylalanine ammonia-lyase and cinnamate 4-hydroxylase in leaves of vitaceae. Planta 151:48-52

Funk C, Gugler K and Brodelius P (1987) Increased secondary product formation in plant cell suspension cultures after treatment with a yeast carbohydrate preparation (elicitor). Phytochemistry 26(2):401-405

Hagmann ML, Heller W and Grisebach H (984) Induction of phytoalexin synthesis in soybean. Stereospecific 3,9-dihydroxypterocarpan-6a-hydroxylase from elicitor induced soybean cell cultures. Eur. J. Biochem. 142:127-131

Hahn MG, Bonhoff A and Grisebach H (1985) Quantitative localization of the phytoalexin glyceollin I in relation to fungal hyphae in soybean roots infected with Phytophthora megasperma f. sp. glycinea. Plant Physiol 77:591-601

Hauffe KD, Hahlbrock K and Scheel D (1986) Elicitor-stimulated furanocoumarin biosynthesis in cultured parsley cells: S-adenosyl-L-methionine bergaptol - and S-adenosyl-L-methionine xanthoxol-O-methyltransferases. Z Naturforsch. 41c:228-229

Hinderer W, Köster J and Barz W (1986) Purification and properties of a specific isoflavone 7-O-glucoside-6"-malonate malonylesterase from roots of chickpea (Cicer arietum L.). Arch Biochem Biophys 248(2):570-578

Hinderer W, Flentje U and Barz W (1987) Microsomal isoflavone 2'- and 3'-hydroxylases from chickpea (Cicer arietinum L.) cell suspensions induced for pterocarpan phytoalexin formation. FEBS - Lett. 214:101-106

Hösel W, Frey G and Barz W (1975) Über den Abbau von Flavonolen durch pflanzische Peroxidasen. Phytochemistry 14:417-422

Ingham JL (1982) Phytoalexins from the Leguminosae. In: Bailey JA and Mansfield JW (eds) Phytoalexins Blackie, Glasgow and London, pp 21-80

Inoue H, Oba K, Ando M and Uritani I (1984) Enzymatic reduction of dehydroipomeamarone to ipomeamarone in sweet potato root tissue infected by Ceratocystis fimbriata. Physiol. Plant Pathol 25:1-8

Jaques U, Köster J and Barz W (1985) Differential turnover of isoflavone 7 - O - glucoside - 6" - O - malonates in Cicer arietinum roots. Phytochemistry 24:949-951

Kessmann H and Barz W (1986) Elicitation and suppression of isoflavone and pterocarpan phytoalexin accumulation in plants and cell cultures of chickpea. Biol Chem Hoppe - Seyler 367 (Suppl.):201

Kessmann and Barz W (1987) Accumulation of isoflavone and pterocarpan phytoalexins in cell suspension cultures of different cultivars of chickpea (Cicer arietinum). Plant Cell Reports 6:55-59

Köster J and Barz W (1981) UDP - glucose: isoflavone 7-O glucosyl - transferase from roots of chickpea (Cicer arietinum L.). Arch. Biochem. Biophys. 212(1):98-104

Köster J, Strack D and Barz W (1983) High performance liquid chromatographic separation of isoflavones and structural elucidation of isoflavone 7-O-glucoside 6"-malonates from Cicer arietinum. Planta Medica 48:131-135

Köster J, Bussmann R and Barz W (1984) Malonyl - coenzyme A: isoflavone 7-O-glucoside-6"-malonyltransferase from roots of chickpea (Cicer arietinum L.). Arch Biochem Biophys 234:513-521

Leube J and Grisebach H (1983) Further studies on induction of enzymes of phytoalexin synthesis in soybean and cultured soybean cells. Z Naturforsch 38c:730-735

Matern U, Heller W and Himmelspach K (1983) Conformational changes of Apigenin-7-O-(6-O-malonylglucoside), a vacuolar pigment from parsley, with solvent composition and proton concentration. Eur J Biochem 133:439-448

Moesta P and West CA (1985) Casbene synthetase: regulation of phytoalexin biosynthesis in Ricinus communis L. seedlings. Arch Biochem Biophys. 238:325-333

Nene YL (1982) A review of Ascochyta blight of chickpea. Tropical Pest Management 28:61-70

Robbins MP, Bolwell GP and Dixon RA (1985) Metabolic changes in elicitor treated bean cells (Selectivity of enzyme induction in phytoalexin accumulation). Eur J Biochem 148:563-569

Saxena MC and Singh KB (1984) (eds) Ascochyta blight and winter sowing of chickpea, Kluver Academic Publishers Group, The Hague, Boston, Lancaster

Smith DA and Banks SW (1986) Biosynthesis, elicitation and biological activity of isoflavonoid phytoalexins. Phytochemistry 25(5):979-995

Sweigart JA, Mattews DE and VanEtten H (1986) Synthesis of the phytoalexin pisatin by a methyltransferase from pea. Plant Physiol 80:277-279

Tiemann K, Hinderer W and Barz W (1987) Isolation of NADPH: isoflavone oxidoreductase, a new enzyme of pterocarpan phytoalexin biosynthesis in cell suspension cultures of Cicer arietinum. FEBS-Lett 213(2):324-328

Tietjen KG, Hunkler D and Matern U (1983) Differential response of cultured parsley cells to elicitors from two non-pathogenic strains of fungi. I. Identification of induced products as coumarin derivatives. Eur J Biochem 131:401-407

Tietjen KG and Matern U (1983) Differential response of cultured parsley cells to elicitors from two non-pathogenic strains of fungi. 2. Effects on enzyme activities. Eur J Biochem. 131: 409-413

VanEtten HD, Matthews PS and Mercer EH (1983) (+) Maackiain and (+) medicarpin as phytoalexins in Sophora japonica and identification of the (-) isomers by biotransformation. Phytochemistry 22(10):2291-2298

Weigand F, Köster J, Weltzien HC and Barz W (1986) Accumulation of phytoalexins and isoflavone glucosides in a resistant and a susceptible cultivar of Cicer arietinum during infection with Asochyta rabiei. J Phytopatol 115:214-221

Whitaker RJ and Hashimoto T (1986) Production of secondary metabolites. In: Handbook of plant cell culture. Techniques and applications. Vol. 4. Evans DA, Sharp WR and Ammirato PV (eds.), Macmillan, New York, pp 264-286

Yamada Y and Fujita Y (1983) Production of useful compounds in culture. In: Handbook of plant cell culture. Techniques for propagation and breeding. Vol. 1. Evans DA, Sharp WR, Ammirato PV and Yamada Y (eds.), Macmillan, New York, pp 717-728

Zähringer U, Ebel J, Mulheirn LJ, Lyne RL and Grisebach H (1979) Induction of phytoalexin synthesis in soybean. Dimethylallyl-pyrophosphate: trihydroxypterocarpan dimethylallyl transferase from elicitor induced cotyledons. FEBS - Lett. 101:90-92

IN VIVO CHARACTERIZATION OF NADPH : 2'-HYDROXYISOFLAVONE OXIDOREDUCTASE IN ELICITOR TREATED CHICKPEA CELL CULTURES AND STEREOCHEMICAL ASPECTS OF THE PHYTOALEXINS MEDICARPIN AND MAACKIAIN

H Kessmann, K Tiemann, JR* Jansen, H* Reuscher, G* Bringmann and W Barz

Lehrstuhl für Biochemie der Pflanzen
Westfälische Wilhelms-Universität
Hindenburgplatz 55, D-4400 Münster, FRG

INTRODUCTION

Pterocarpan phytoalexins such as medicarpin, maackiain, phaseollin, pisatin or the various isomers of glyceollin are antimicrobial compounds synthesized by plant species of the Leguminosae in response to stress, primarily caused by fungal infection (Dixon 1986, Ebel 1986).

After infection with the phytopathogenic deuteromycete *Ascochyta rabiei* (Pass.) Lab. (syn. *Mycosphaerella rabiei* Kovachevski) medicarpin and maackiain accumulated in plants of the resistant cultivar ILC 3279 to a much higher level than in the susceptible cultivar ILC 1929 (Weigand et al 1986). Similarly, after treatment with an elicitor from *A. rabiei* the cell culture derived from the resistant cultivar ILC 3279 accumulated the phytoalexins to a much higher degree than the cell culture of the cultivar ILC 1929 (Kessmann and Barz 1986, 1987).

Investigations on medicarpin biosynthesis in $CuCl_2$ treated seedlings of red clover (*Trifolium pratense*) suggested, that the pterocarpan biosynthetic pathway involves a 2'-hydroxylation of the isoflavone formonetin followed by reduction of the 2'-hydroxyisoflavone to the corresponding isoflavanone vestitone (Dewick 1975, Dewick 1977, Smith and Banks 1986).

Two isoflavone hydroxylases (cytochrome P450 monooxygenases) and a NADPH :2'-hydroxyisoflavone oxidoreductase (IFR) have

*Organisch-Chemisches Institut
Westfälische Wilhelms-Universität
Orleansring 23, D-4400 Münster, FRG

NATO ASI Series, Vol. H18
Plant Cell Biotechnology. Edited by M. S. S. Pais et al.

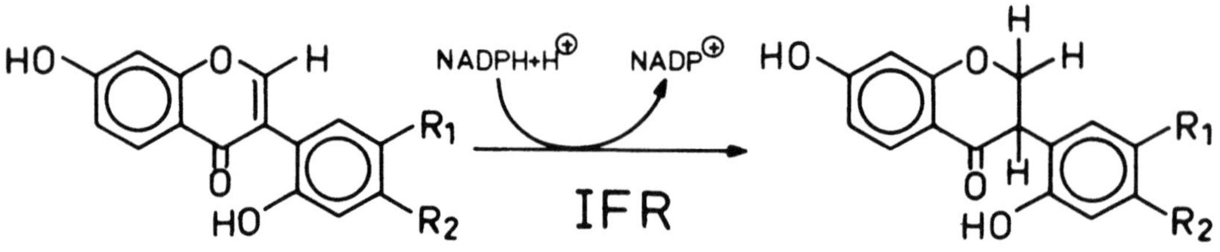

Fig. 1: Reductive step in the biosynthesis of the pterocarpan phytoalexins, catalyzed by the NADPH : 2'-hydroxyisoflavone oxidoreductase (IFR) from chickpea. R_1=H, R_2=OCH_3:2'-hydroxyformononetin; R_1=R_2=O-CH_3-O:2'-hydroxypseudobaptigenin (Tiemann et al 1987)

been characterized as new enzymes involved in pterocarpan phytoalexin biosynthesis in chickpea cell cultures (Hinderer et al 1987, Tiemann et al 1987).

Here, the in vivo characterization of the IFR in elicitor treated cell cultures of both chickpea cultivars and the concomitant accumulation of phytoalexins is presented. In order to study the stereochemistry of the reduction process (Fig. 1) the phytoalexins were purified and their absolute configuration and conformation was determined.

MATERIAL AND METHODS

Cell cultures of chickpea cultivars ILC 3279 and ILC 1929 were grown on modified PRL-4c medium as previously described (Kessmann and Barz 1987). Yeast extract was substituted by glycin (2 mg/l).

Elicitor was prepared from fermenter grown mycelium of A. rabiei essentially as described by Ayers et al (1976) with the exception that proteins were separated by precipitation with trichloracetic acid.

NADPH : 2'-hydroxyisoflavone oxidoreductase was assayed as described recently (Tiemann et al 1987).

In order to isolate the phytoalexins 80 ml cell suspension (ILC 3279, 10-15 g fresh weight) were transferred into 400 ml medium supplemented with 1 g yeast extract (Difco, Detroit, USA; Hinderer et al 1987). 16 hours after transfer cells and medium were separated by filtration. The growth medium was twice extracted with diethylether, the organic phases were pooled and evaporated to dryness. The residue was suspended in water (pH 7) and was immediately transferred to a polyamide column (34x4 cm, MN polyamide SC 6, Machery and Nagel, Düren, FRG). It was eluted with (1 litre each) water, 20% methanol and 100% methanol. The 100% methanol fraction was evaporated to dryness and dissolved in ethylacetate. The phytoalexins were further purified by TLC (Merck silica gel, PLC-plates, 2 mm, or silica gel 60 F_{254}, 0,2 mm). After the first TLC in solvent system 1 (chloroform : methanol = 50:1) the non separated phytoalexins were eluted with methanol. Medicarpin and maackiain were separated in solvent system 2 (n-petan : ether : glacial acetic acid =75:25:3, Strange et al 1985) and eluted with methanol.

Optical rotation was measured on a Perkin Elmer 241 polarimeter.

Proton NMR studies were carried out with a Bruker WM 300 spectrometer.

RESULTS AND DISCUSSION

After treatment with an elicitor from A. rabiei medicarpin and maackiain accumulated in cell cultures of chickpea cultivar ILC 3279 with the maximum 8 hours after elicitor application (Fig. 2 C,D). In the cells a second maximum of phytoalexin accumulation was observed after 72 hours. This may be a result of induction by pectic fragments released after a limited death of cells caused by the phytoalexins (Dixon 1986) though no browning of the cells could be observed.

Concomitant with the elicitor induced pterocarpan formation a 15-fold increase of the specific activity of the IFR was measured (Fig. 2A).

In cell cultures of cultivar ILC 1929, however, phytoalexins accumulated in very low amounts only after elicitor treatment and sometimes they were only barely detectable. It was also shown that these quantitative differences in phytoalexin accumulation are a very stable trait of the two cell culture lines. These differences in phytoalexin accumulation were observed regardless of elicitor preparation (elicitor from A. rabiei; elicitor from Phytophtora megasperma f.sp. glycinea, yeast extract), growth stage of cell culture and medium composition

(Kessmann and Barz 1986, 1987, unpublished data). The IFR was also present in cells of cell culture ILC 1929. In contrast to culture ILC 3279 the elicitor led only to a slight increase in enzyme activity (Fig. 2B).

The correlation of induced enzyme activity and phytoalexin accumulation and further enzymatic data (Tiemann et al 1987) support the assumption that the NADPH: 2'-hydroxyisoflavone oxidoreductase is involved in pterocarpan phytoalexin biosynthesis as postulated by Dewick and associates (Dewick 1975, 1977, Dewick and Ward 1977).

In order to study the stereochemistry of the reduction process the phytoalexins were isolated from the growth medium of culture ILC 3279 and purified by polyamide column and silica gel thin layer chromatography. Phytoalexin accumulation was induced by transfer of 80 ml cell suspension (10-15 g fresh weight) into 400 ml PRL-4c medium supplemented with 1 g yeast extract. Under these conditions the phytoalexins predominantly accumulated in the growth medium (Kessmann and Barz 1987).

The isolated phytoalexins were strongly laevorotatory (medicarpin: -239° +/- 4° (0,3 mg/ml methanol or ethanol), maackiain: -264° +/- 30° (0,27 mg/ml methanol). These data are in agreement with reports from the literature (Ingham 1982) for (-) medicarpin and (-) maackiain. According to Sicherer and Sicherer-Roetman (1980) the absolute configuration of the two phytoalexins can therefore be assigned as 6a R:11a R (Fig. 3).

For this configuration there are two conformations possible as depicted in Fig. 3 (Pachler and Underwood 1967). The proton-proton coupling constants for the 6, 6a and 11a protons in these two conformers should be significantly different (Tab. 1).

Thus, the conformation can be established unambiguously from the experimentally determined coupling constants unless the molecule undergoes rapid conformational changes.

In order to develop a suitable and rapid method to determine the conformation of pterocarpans we have measured the proton NMR spectrum (300 MHz) in solvent mixtures of acetone d_6/benzene d_6 because the decisive coupling constants could not be determined by proton NMR spectroscopy in acetone d_6 only (Pachler and Underwood 1967, Strange et al 1985).

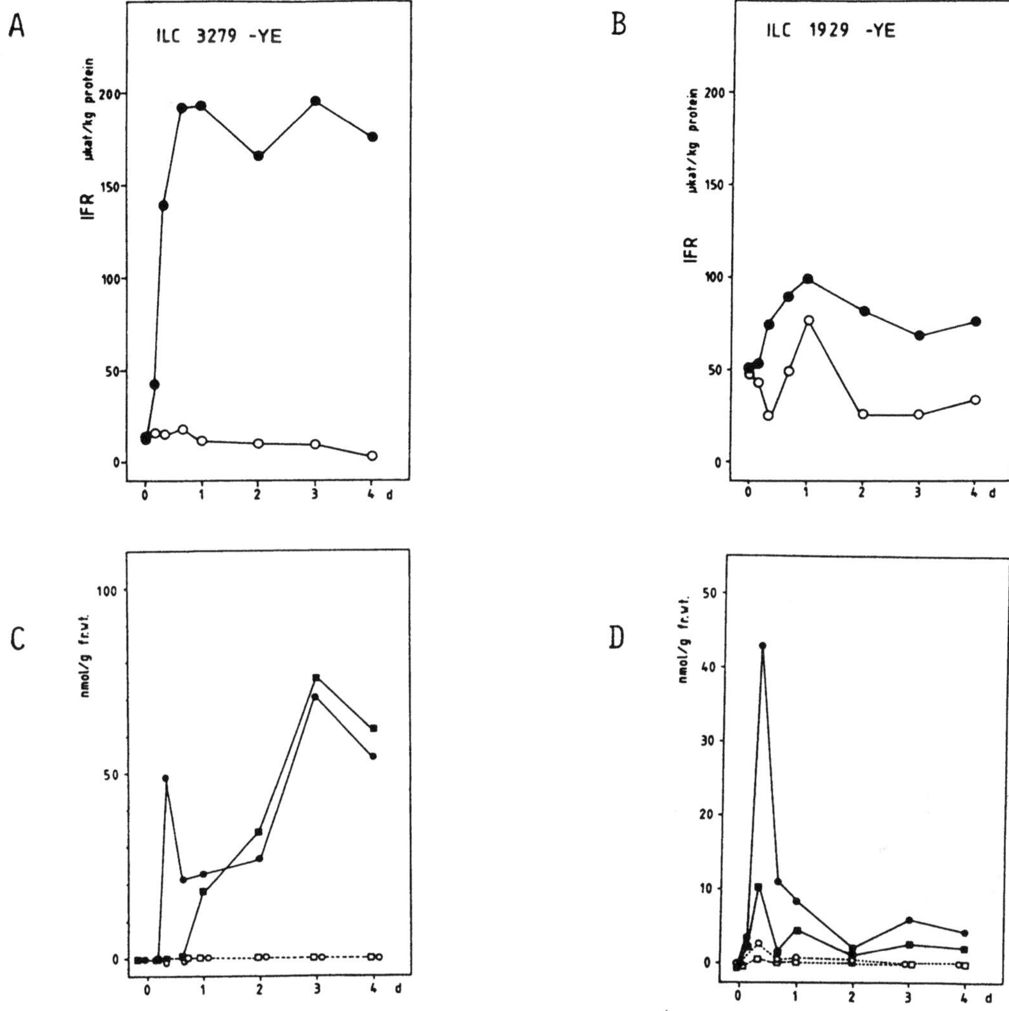

Fig. 2. Kinetics of induction of NADPH : 2'-hydroxyisoflavone oxidoreductase (IFR) activity and of phytoalexin accumulation in chickpea cell cultures ILC 3279 and ILC 1929 after elicitor treatment. -YE, growth without yeast extract.
The elicitor of A. rabiei was applicated on day 3 of growth cycle (5 mg/40 ml culture, n=3)
A: Specific activity of IFR in culture ILC 3279 after elicitor treatment (•) and in non induced cultures (o).
B: Specific activity of IFR in culture ILC 1929 after elicitor treatment (•) and in non induced cultures (o). Under these conditions the cell culture accumulated only 1-10 nmol medicarpin and maackiain per g fresh weight (cells and medium).
C: Accumulation of medicarpin (•) and maackiain (■) in cells of cell culture ILC 3279. Controls (dotted lines) were given water.
D: Accumulation of medicarpin (•) and maackiain (■) in the medium of cell culture ILC 3279 after elicitor treatment. Controls (dotted lines) were given water.

Comparison of the experimentally determined coupling constants with the expected coupling constants (Pachler and Underwood 1967) clearly shows (Tab. 1) that the benzofurobenzopyran ring system of medicarpin exists in conformation A (Fig. 3) with the six membered ring in a staggered half-chair joint to the planar five membered ring.

These data are in line with results from Pachler and Underwood (1967) about the conformation of maackiain.

Proton NMR studies of other naturally occuring pterocarpan derivatives indicate that these compounds also exist in conformation A as shown in Fig. 3 (for references see Pachler and Underwood 1967).

Some stereochemical features of the reduction process during medicarpin biosynthesis were investigated by Dewick and Ward (1977). It was shown that in Trigonella foenum-graecum 2'-hydroxy (2-^{2}H)formononetin (Fig. 1) was an excellent precursor of medicarpin.

Proton NMR studies revealed that the deuterium atom from this percursor was located in the equatorial position at carbon atom 6 of medicarpin (Fig. 3).

The stereochemistry of pterocarpan formation in chickpea can thus be interpreted that the NADPH : 2'-hydroxyisoflavone oxireductase catalyses a trans addition of hydrogen to the double bond of 2'-hydroxyisoflavones during medicarpin and maackiain biosynthesis.

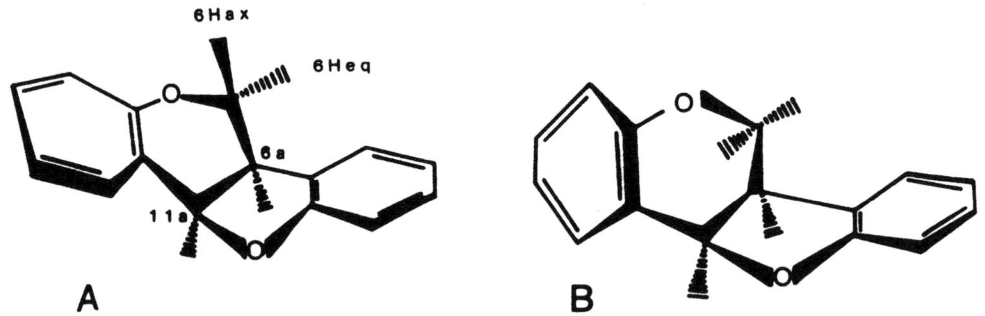

Fig. 3: The two possible conformations of (6a R:11a R)-6a, 11a-dihydro-6H-benzofuro (3,2c) (1) benzopyran (Pachler and Underwood 1967).

COUPLING PROTONS	CONFORMATION A DIHEDRAL ANGLE	COUPLING CONSTANT	CONFORMATION B DIHEDRAL ANGLE	COUPLING CONSTANT	EXPERIMENTALLY DETERMINED COUPLING CONSTANTS
6a - 11a	0°	7 - 9	0°	7 - 9	7
6a - 6H ax	180°	10 - 12	60°	2 - 3	10,5
6a - 6H eq	60°	4 - 5	60°	2 - 5	4,9

Table 1. Expected coupling constants (J, Hz) for the vicinal protons of the heterocyclus of the two possible conformations (Fig. 3, Pachler and Underwood 1967) and the experimentally determined coupling constants for medicarpin. The coupling constants were determined in a solvent mixture of acetone d_6/benzene d_6 (65:35).

ACKNOWLEDGEMENTS

Financial support by Deutsche Forschungsgemeinschaft is gratefully acknowledged.
We thank Professor Ebel, Freiburg, for a generous gift of elicitor from _Phytophtora megasperma_ f.sp. _glycinea_.

REFERENCES

Ayers AR, Ebel J, Valent B, Albersheim P (1976) Hostpathogen interactions X. Fractionation and biological activity of an elicitor isolated from the mycelial walls of _Phytophtora megasperma_ var. _sojae_. Plant Physiol 57:760

Dewick PM (1975) Pterocarpan biosynthesis: chalcone and isoflavone precursors of demethylhonopterocarpen in _Trifolium pratense_. Phytochemistry 14:979-982

Dewick PM (1977) Biosynthesis of pterocarpan phytoalexins in _Trifolium pratense_. Phytochemistry 16:93-97

Dewick PM, Ward D (1977) Stereochemistry of isoflavone reduction during pterocarpan biosynthesis: an investigation using deuterium nuclear magnetic resonance spectroscopy. J C S Chem Comm 211:338-339

Dixon RA (1986) The phytoalexin response: elicitation, signalling and control of host gene expression. Biol Rev. 61:239-291

Ebel J (1986) Phytoalexin synthesis: biochemical analysis of the induction process. Ann Rev Plant Pathol 12:235-264

Hinderer W, Flentje U, Barz W (1987) Microsomal isoflavone 2'- and 3'-hydroxylases from chickpea (_Cicer arietinum_ L.) cell suspensions induced for pterocarpan phytoalexin formation. FEBS - Lett 214(1):101-106

Ingham JL (1982) Phytoalexins from the Leguminosae. In JA Bailey and JW Mansfield (eds) Phytoalexins, Blackie, New York:21-80

Kessmann H and Barz W (1986) Elicitation and supression of isoflavone and pterocarpan phytoalexin accumulation in plants and cell cultures of chickpea (_Cicer arietinum_). Biol Chem Hoppe-Seyler 367(Suppl.):201

Kessmann H and Barz W (1986) Elicitation and suppression of phytoalexin and isoflavone accumulation in cotyledons of Cicer arietinum L. as caused by wounding and polymeric components from the fungus Ascochyta rabiei. J Phytopatol 117:321-335

Kessmann H and Barz W (1987) Accumulation of isoflavones and pterocarpan phytoalexins in cell suspension cultures of different cultivars of chickpea (Cicer arietinum). Plant Cell Reports 6:55-59

Köster J, Zuzok A and Barz W (1983) High performance liquid chromatography of isoflavones and phytoalexins from Cicer arietinum. J Chromatogr. 270:392-395

Pachler KGR and Underwood WGE (1967) A proton magnetic resonance study of some pterocarpan derivatives. Tetrahedron 23:1817-1826

Sicherer CAXGF and Sicherer-Roetman A (1980) The R,S nomenclature of pterocarpan stereochemistry. Phytochemistry 19:485-486

Smith DA and Banks SW (1986) Biosynthesis, elicitation and biological activity of isoflavonoid phytoalexins. Phytochemistry 25(5):979-995

Coocksey CJ and Garatt PJ (1985) Isolation of the phytoalexin medicarpin from leaflets of Arachis hypogaea and related species of the tribe Aeschonomeneae. Z Naturforsh 40c:313-316

Tiemann K, Hinderer W and Barz W (1987) Isolation of NADPH: isoflavone oxidoreductase, a new enzyme of pterocarpan phytoalexin biosynthesis in cell suspension cultures of Cicer arietinum. FEBS-Lett 213:324-328

Weigand F, Köster J, Weltzien HC and Barz W (1986) Accumulation of phytoalexins and isoflavone glucosides in a resistant and a susceptible cultivar of Cicer arietinum during infection with Ascochyta rabiei. J Phytopatol 115:214-221

NICOTINATE CONJUGATE METABOLISM IN PLANT CELL CULTURES

B Upmeier, S Köster, C Otto and W Barz

Biochemistry of Plants
Univeristy of Münster
Hindenburgplatz 55
4400 Münster, Federal Republic of Germany

INTRODUCTION

Nicotinic acid can be regarded as an important connecting link between primary and secondary metabolism in higher plant cells. Being a constituent of the pyridine nucleotide cycle it is a precursor and a degradation product of NAD and NADP which play a decisive role as coenzymes of oxidation-reduction reactions (Waller and Nowacki 1978, Wagner and Wagner 1985).

The pyridine nucleotide cycle also provides components for the pyridine alkaloid metabolism (Waller and Dermer 1981).

The nicotinic acid conjugates trigonelline (N-methyl nicotinic acid) and nicotinate-N-glucoside are known as constituents of higher plants and plant cell cultures (Schütte 1969, Mizusaki et al 1970, Leienbach et al 1975, Leienbach and Barz 1976). For suspension cultured plant cells the rapid conversion of exogenously applied nicotinic acid to either the N-methyl or the N-glucosyl conjugate has been observed (Willeke et al 1979). Due to the fact that the nicotinate moiety can be converted in the pyridine nucleotide cycle, the conjugates are thought to be storage forms of nicotinic acid (Barz 1985). Our recent studies are designed to further elucidate the metabolic role of the nicotinic acid conjugates in plant cell suspension cultures.

MATERIAL AND METHODS

Cell Cultures

Cell cultures of *Glycine max*, *Petroselinum hortense*, *Coffea arabica* and *Coffea robusta* were grown as previously described (Heeger et al 1976, Frischknecht and Baumann 1980).

NATO ASI Series, Vol. H18
Plant Cell Biotechnology. Edited by M. S. S. Pais et al.

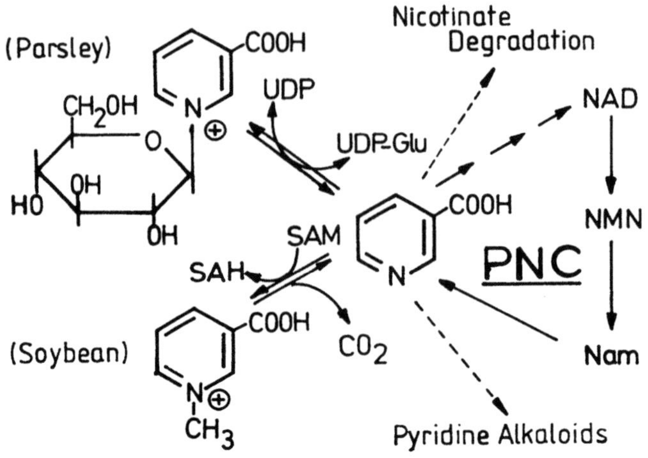

Fig. 1: Enzyme reactions for the formation of nicotinic-acid-N-glucoside and trigonelline in plant cell suspension cultures in relation to the pyridine nucleotide cycle (PNC)

The 15 l airlist bioreactor was inoculated with 150 g fr.wt. of soybean cells from the late logarithmic growth phase.

The application of (U-^{14}C)-aspartate, the collection and measurement of $^{14}CO_2$ followed previous reports (Berlin and Barz 1971).

Quantitative Determination of the Nicotinate Conjugates

Cells were obtained by filtration and homogenized in 80% methanol. The methanolic solution was reduced to a few ml and subjected to ion exchange chromatography on DOWEX 50 WX 8. The conjugates were eluted from the resin with a solution of 2 M NH_4OH. The NH_4OH-fraction was brought to dryness and the residue redissolved in 80% methanol. Identification and isolation of the conjugates were achieved by HPLC with an isocratic elution system of 85% aqueous acetonitrile on a NH_2-column. Quantitation was performed via UV-absorption at 263 nm by external standardization.

Enzyme Assay

The enzyme assay consisted of 50 nmoles nicotinic acid (containing 0.1 μCi (7-^{14}C)-nicotinic acid), 75 nmoles SAM and 100 μl enzyme preparation in a total volume of 150 μl. The assay was incubated for 1 h at 35°C. The enzyme activity was determined by measuring the radioacitivity of the product formed. The product trigonelline was isolated by ion exchange chromatography on DOWEX 1X8.

RESULTS AND DISCUSSION

The plant cell cultures under investigation rapidly convert nicotinic acid to conjugates, alternatively by methylation or glucosylation (see Fig. 1).

Our studies on callus and suspension cultures of two Coffea arabica species show that, in contrast of expectation (Willeke et al 1979), these cultures convert nicotinic acid to trigonelline. The methyl conjugate is formed by Coffea robusta suspension cultures in amounts of about 145-220 nmoles/g fr.wt., whereas in Coffea arabica cultures the trigonelline content increases with the size of cell aggregates in the suspension (70-500 nmoles/g fr.wt.). 100 nmoles/g fr.wt. of trigonelline can be detected in callus culutres of the two plant species.

Cell cultures of Glycine max were previously shown to produce trigonelline (Leienbach et al 1975). The content was now determined to be 5-10 nmoles/g fr.wt. over a growth period of 12 days. The level of trigonelline slightly increased during the stationary phase of growth.

Petroselinum hortense suspension cultures accumulate the nicotinic acid-N-glucoside in amounts of 30-40 nmoles/g fr.wt. at any stage of the growth cycle.

Trigonelline and nicotinic acid-N-glucoside are endogenously produced by soybean and parsley cell suspensions in the same amounts as mentioned above, even when nicotinic acid is omitted from the growth media.

Application of (U-^{14}C)-aspartate to soybean cell cultures led to a small amount of ^{14}C-labelled trigonelline, whereas parsley cells exhibited a considerable incorporation of the label into the N-glucoside indicating a higher metabolic activity of the latter compound. Pulse labelling experiments indicated a turnover of nicotinic acid-N-glucoside with a halflife of 24 h, whereas the demethylation of trigonelline seems to proceed very slowly. Therefore the N-glucoside can rather be considered as a

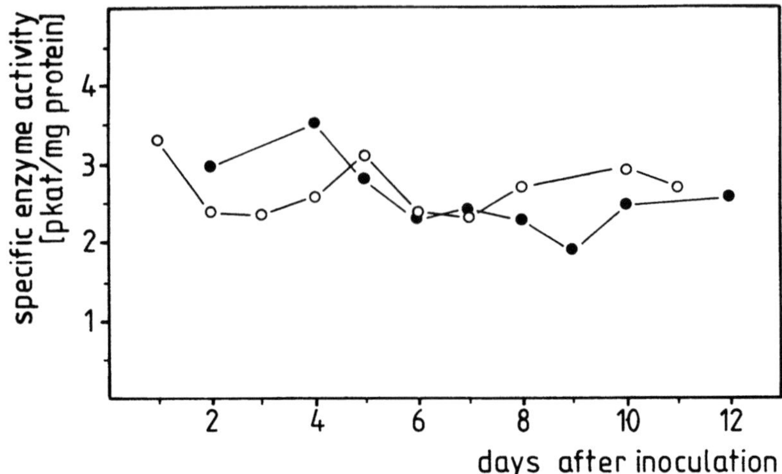

Fig. 2: Development of specific enzyme activity of the nicotinic acid: SAM-N-methyltransferase during growth of soybean cell suspensions in erlenmeyer flasks (o——o) or a 15 l air-lift bioreactor (●——●)

storage form for nicotinic acid. It appears to be much closer linked to the pyridine nucleotide cycle than the methyl conjugate which exhibits the properties of a typical secondary metabolite (Barz 1985).

Recently, studies on trigonelline formation were continued by measuring the enzyme involved. The acitivity of nicotinic acid-N-methyltransferase was determined in soybean cell suspension cultures. Identical values of specific enzyme activity were found when cells were cultivated either in 200 ml erlenmeyer flasks or after scale up in a 15 l airlift bioreactor (2-4 pkat/mg protein). In both cases this value remained constant over the whole growth cycle of the cell cultures (see Fig. 2). In contrast, the specific enzyme activity of the nicotinic acid: UDPG-N-glucosyltransferase in erlenmeyer grown parsley cells significantly increased towards the stationary phase of growth of the cell cultures (Upmeier 1985).

The nicotinic acid:SAM-N-methyltransferase was purified 200 fold from soybean cells by means of ion exchange chromatography and gel filtration. Characterization of the enzyme revealed the following kinetic data Km (nicotinic acid) 78 μM, Km (SAM) 55 μM, Ki (SAH, competitive inhibitor) 95 μM.

The methyltransferase showed a temperature optimum of 40°C and a pH optimum at pH 8.0. Among several pyridine derivatives tested the enzyme exclusively accepted nicotinic acid as a substrate.

REFERENCES

Barz, W (1985) Metabolism and Degradation of Nicotinic Acid in Plant Cell Cultures. In: Neumann, KH, Barz, W and Reinhard, E (eds). Primary and Secondary Metabolism of Plant Cell Cultures. Springer-Verlag (1985) Berlin, Heidelberg pp 186-195

Berlin, J and Barz, W (1971) Stoffwechsel von Isoflavonen und Cumöstanen in Zell- und Callussuspensionskulturen von Phaseolus aureus Roxb.. Planta 98:300-314

Frischknecht, PM and Baumann, TW (1980) The Pattern of Purine Alkaloid Formation in Suspension Cultures of Coffea arabica. Plant Med 40:245-249

Heeger, V, Leienbach, KW and Barz, W (1976) Stoffwechsel von Nicotinsäure in pflanzlichen Zellsuspensionskulturen, III; Bildung und Stoffwechsel von Trigonellin. Hoppe Seyler's Z Physiol Chem 357:1081-1087

Leienbach, KW, Heeger V, Neuhann, H and Barz, W (1975) Stoffwechsel und Abbau von Nicotinsäure und ihren Derivaten in pflanzlichen Zellsuspensionskulturen. Plant Med (Suppl.): 148-152

Leienbach, KW and Barz, W (1976) Stoffwechsel von Nicotinsäure in pflanzlichen Zellsuspensionskulturen, II; Zur Isolierung, Charakterisierung und Enzymologie von Nicotinsäure-N-arabinosid. Hoppe Seyler's Z Physiol Chem 357:1069-1080

Mizusaki, S, Tanabe, Y, Kisaki, T and Tamaki, E (1970) Metabolism of Nicotinic Acid in Tobacco Plants. Phytochem 9:549-554

Schütte, HR (1969) Methylierung und Transmethylierung. In: Mothes, K and Schütte, HR (eds). Biosynthese der Alkaloide. VEB Deutscher Verlag der Wissenschaften (1969) Berlin pp 123-167

Upmeier, B (1985) Untersuchungen zum Stoffwechsel von Nicotinsäure-Konjugaten in pflanzlichen Zellkulturen. Master Thesis. University of Münster

Wagner, R and Wagner, KG (1985) The pyridine nucleotide cycle in tobacco. Enzyme activities for the de-novo synthesis of NAD. Planta 165:532-537

Waller, GR and Nowacki, EK (eds) (1978) Alkaloid Biology and Metabolism in Plants. Plenum Press, New York, London

Waller, GR and Dermer OC (1981) Enzymology of Alkaloid Metabolism in Plants and Microorganisms. In: Conn, EE (ed). Biochemistry of Plants, Vol. 7. Academic Press (1981) New York, London, Toronto, Sydney, San Francisco pp 317-402

Willeke, U, Heeger, V, Meise, M, Neuhann, H Schindelmeister, I, Vordemfelde, K and Barz, W (1979) Mutually Exclusive Occurrence and Metabolism of Trigonelline and Nicotinic Acid Arabinoside in Plant Cell Cultures. Phytochem 18:105-110

Alkaloids from Organized *Atropa belladonna* Cultures

Jeffrey Lang, Robert Hamilton, Henrik Pedersen
Department of Chemical and Biochemical Engineering,
and Chee-Kok Chin[1]
Rutgers, The State University of New Jersey
New Brunswick, New Jersey 08902, USA

INTRODUCTION

Unorganized suspension cultures of some higher plants are known to produce various industrially important products, i.e. nicotine from *Nicotiana tobacum* [Al-Abta, 1978] and purine alkaloids from *Coffea arabica* [Frischknecht, 1977, 1980]; however, structurally organized organs or somatic embryos are needed in some cases [Butcher, 1977; Al-Abta, 1978, 1979; Yamada, 1983] for secondary metabolite production. To achieve redifferentiation, media and culture conditions sometimes need to be varied. These changes complicate the selection of optimum production conditions by introducing more manipulative variables. A three stage scheme for production of tropane alkaloids from *Atropa belladonna* is proposed here.

EXPERIMENTAL

Callus is used as the first stage, aggregate suspensions as the second, and finally root induction in liquid media as the last stage. Callus was maintained in a six week culture cycle at 25 C, 70% humidity, in the dark, on 50 ml agar (0.8%, Agar-Agar, US Biochemicals, Cleveland, OH) solidified medium. Two types of

---

[1]Department of Horticulture and Forestry

NATO ASI Series, Vol. H18
Plant Cell Biotechnology. Edited by M.S.S. Pais et al.

callus were examined and evaluated for further use in the cycle. Callus A, root callus from sterile, excised, seedling roots was initiated and maintained on the medium of White [White, 1954] and Wood and Braun (WWB) [Wood, 1961] containing 2% sucrose, 2 mg/l beta-napthalene acetic acid (NAA), and 0.5 mg/l kinetin, pH 5.2. From this, aggregate suspensions were established after at least two subcultures of the initial callus innoculum.

The second type (Callus B) was root callus initiated from sterile seedling roots on Revised Tobacco (RT) medium [Kaul, 1968; Khanna, 1977] but maintained on WWB medium supplemented with 10% coconut water. Callus initiation, in this case, took approximately 3 months but on transfer to WWB medium the growth rate was comparable to that seen in Callus A.

From the two callus lines, roots were established by four methods: 1) Callus A was transferred directly to root induction media (1/2 MS [Murashige, 1962] total strength salts plus 10 mg/l tropic acid in place of NAA and kinetin), 2) Callus B was transferred to root induction medium, 3) Callus A suspension aggregates were harvested at two week intervals and transferred to root induction medium, and 4) Callus B was transferred to embryogenesis medium (EM) [Thomas, 1985; Konar, 1972] for approximately seven days and then into root induction medium.

RESULTS AND DISCUSSION

In cases 1 and 2 where callus was put directly into root induction medium, not more than 30% of the total biomass formed roots. In case 3, 80% of the aggregates formed roots, however the excessively long time needed for aggregate establishment (3 weeks between subcultures) makes it inefficient for further use. The EM step in case 4 caused the yield from callus to increase to 50% of the biomass with the overall process from suspension to roots taking on about 3 weeks (7 days in EM and 2 weeks in root induction). This appears to be the best scheme due to the ease of callus handling and rate of growth.

It has been shown [Hamilton, in prep.] that callus can be grown on autoclaved, polyethylene-bottom, membrane "boats" supported by polyurethane foam pads (1 cm thick) [Conner, 1984] which deliver nutrients by capillary action. By using this method, an 80% increase in growth rate (gms fw/day) can be achieved. Because the alkaloid level in the redifferentiated roots is constant, the amount of product obtained is growth associated therefore making this advantageous.

Precursor studies were done to possibly eliminate the embryogenesis media step and to increase alkaloid level. The addition of 10 mg/l tropic acid to Callus A when innoculated in both root induction medium and normal suspension medium yielded no alkaloid production, reaffirming the fact that redifferentiation is needed for product formation.

Tropic acid concentration was investigated in established seedling root cultures in the range 5 to 15 mg/l with no hormone media as the control. As shown in Figure 1, atropine content increased 10% and scopalamine content increased 112% at 5 mg/l concentration over 15 days of incubation. In the same range of investigation, final root biomass also increased and was maximized at 15 mg/l, approximately 2X that seen in no hormone media.

CONCLUSIONS

The three stage scheme for obtaining tropane alkaloids from redifferentiated root cultures of <u>Atropa belladonna</u> can be used to yield semicontinuous production in 5-6 week cycles. Callus growm and harvested about every two weeks on membrane-bottom "boats" can be transferred to embryogenesis medium for 1 week and finally to root induction medium supplemented with the optimum tropic acid concentration for alkaloid production.

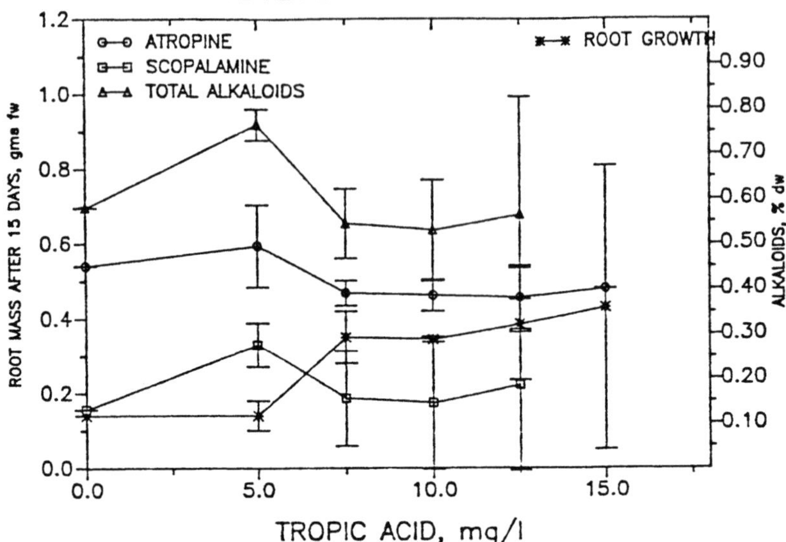

Figure 1. Growth rate and tropane alkaloid content of *Atropa belladonna* root cultures in WWB media supplemented with tropic acid

ACKNOWLEDGEMENTS

This work was supported in part by a research grant from the National Science Foundation.

REFERENCES

S. Al-Abta and H.A. Collin, New Phytol., **80**, 517 (1978).

S. Al-Abta, I.J. Galpin, and H.A. Collin, Plant Sci. Lett., **16**, 129 (1979).

D.N. Butcher, Plant Cell, Organ and Tissue Culture (Springer-Verlag, New York, 1977), p. 688.

A.J. Conner and C.P. Meredith, Plant Cell Tiss. Org. Cult., **3**, 59 (1984).

P.M. Frischknecht, T.W. Baumann, and H. Wanner, Planta Med., **31**, 344 (1977).

P.M. Frischknecht and T.W. Baumann, Planta Med., **40**, 245 (1980).

R. Hamilton, H. Pedersen, and C.-K. Chin, in preparation.

B. Kaul and E.J. Staba, Lloydia, **31**, 171 (1968).

P. Khanna, G.L. Sharma, and A. Uddin, Ind. J. Exp. Biol., **15**, 323 (1977).

R.N. Konar, E. Thomas, and H.E. Street, Ann. Bot., **36**, 249 (1972).

T.Murashige and F. Skoog, Physiol. Plant, **15**, 473 (1962).

M. Tabata , H. Yamamoto, N. Hiraoka, Y. Marumoto, and M. Konoshima, J. Phytochem., **10**, 723 (1971).

E. Thomas and M.R. Dewey, From Single Cells to Plants (Wykehen, London, 1985).

P.R. White, The Cultivation of Animal and Plant Cells (Roland, New York, 1954).

H.N. Wood and A.C. Braun, Proc. Nat. Acad. Sci. U.S.A., **47**, 1907 (1961).

Y. Yamada and Y. Fujita, Handbook of Plant Cell Culture (Macmillan, New York, 1983).

CULTIVATED PLANT CELLS: AN ENZYME SOURCE FOR ALKALOID FORMATION

J. Stöckigt and H. Schübel

Lehrstuhl für Pharmazeutische Biologie, Universität München, Karlstr. 29, D-8000 München 2, FRG

ABSTRACT

Plant cell cultures have been proved in the past to be an efficient source for the isolation of novel enzymes. Therefore, cultivated plant cells are a valuable tool für the investigation of the biosynthesis of secondary metabolites. Three examples in the alkaloid field are discussed: the enzymatic formation of the isoquinoline alkaloid berberine in the family Berberidaceae, the biosynthesis of the indole alkaloid ajmaline in *Rauwolfia serpentina* Benth. and the in vitro preparation of the dimeric alkaloids voafrine A and B.

KEY WORDS

Ajmaline, Berberidaceae, berberine, biosynthesis, enzymes, *Rauwolfia serpentina*, voafrine A and B.

INTRODUCTION

A detailed understanding of the enzymatic formation of secondary plant products is in general limited, on the one hand, by the low levels of the involved enzymes and, on the other hand, by difficulties in a complete recovery of the active enzymes from corresponding parts of the plants. The first limitation can be significantly reduced by using cell culture systems capable of producing the desired secondary metabolites instead of the differentiated plant. In cell suspension cultures, for instance, the level of enzyme activities of the secondary metabolism exceeds very often that of the plant. Consequently, product synthesis in cultivated cells proceeds much faster than in plants. For the second limitation, represented by low plant enzyme re-

NATO ASI Series, Vol. H18
Plant Cell Biotechnology. Edited by M.S.S. Pais et al.

covery, there does not seem to be a solution. Obviously enzyme isolation procedures have to be optimized in each single case. However, the in vitro cell systems are in this respect also superior to the plant as demonstrated by the enormous success in cell-free biosynthetic research over the last decade. Work with cell suspension cultures revealed for the first time substantial insight into the biosynthetic pathways of different groups of natural plant products, e.g. flavonoids (Hahlbrock and Grisebach 1975), heteroyohimbine alkaloids (Stöckigt 1980; Zenk 1980) and quinolizidine alkaloids (Wink and Hartmann 1982).

Furthermore, abundant knowledge has been accumulated over recent years on the biogenesis of pharmaceutical useful alkaloids, when plant cell culture methodology was applied. This knowledge could be one prerequisite if a complete exploitation of the biochemical potential of a cell culture is intended, for instance from the point of view of industrial application. The present communication discusses in this light two biosynthetic routes leading to the isoquinoline alkaloid berberine and to the monoterpenoid indole alkaloid ajmaline. Both compounds have been applied as pharmaceuticals for a long period of time. Up to now they are still isolated from plants for commercial purpose.

1. THE ENZYMATIC SYNTHESIS OF BERBERINE

The isoquinoline alkaloid berberine is a typical constituent of the family Berberidaceae. The alkaloid is widely used in Far East countries because of its antiseptic and anti-inflammatory properties (Kondo 1976). Berberine is in fact a constituent of more than 20 prescriptions in Japan. Therefore as early as 1972 investigations were performed to identify this alkaloid in cultivated plant cells, e.g. of *Coptis japonica* (Furuya et al. 1972). Tabata et al. (1982) and Yamada and co-workers (1982) were able to select high-producing cell lines of *C. japonica* (Sato and Yamada 1984). Moreover, Fujita et al. (1986) at Mitsui Petrochemical Ind. are presently developing a commercial process for berberine production.

In this context it is of special interest to understand the cellular formation of berberine, of its precursors and related alkaloids including the involved intermediates and enzymes of berberine biosynthesis.

Broad investigations of the biogenesis of the isoquinoline alkaloids, which represent the largest group of alkaloids, were carried out many years ago under in vivo conditions (reviewed by Shamma and Moniot 1978). However, the description of individual biosynthetic steps and complete sequences was only attainable after detection of the appropriate enzymes. Recent cell-free work resulted in the complete elucidation of the biosynthetic route to berberine (Zenk et al. 1985). It is indeed the only pathway in alkaloid biosynthesis, for which all the involved enzymes are now known in detail.

The pathway is initiated by the formation of (S)-norlaudanosoline from 3.4-dihydroxyphenylacetaldehyde and dopamine. This reaction is catalyzed by the enzyme norlaudanosoline synthase (Rüffer et al. 1981; Schumacher et al. 1983). In a Pictet-Spengler type reaction the synthase combines dopamine and substituted phenylacetaldehydes with preference for the 3.4-dihydroxyaldehyde. In contrast to previous assumptions (Wilson and Coscia 1975; Scott et al. 1978) the substrate specificity of this enzyme indicates that norlaudanosoline-1-carboxylic acid, which could arise from dopamine and dihydroxyphenylpyruvate condensation, has no biosynthetic relevance. Whereas this synthase exhibits a relatively high substrate and stereospecificity, the next enzyme in the berberine formation is rather unspecific.

The S-adenosylmethionine (SAM)-dependent (R), (S)-norlaudanosoline-6-O-methyltransferase shows a surprisingly low stereo- and regio-selectivity. Both, (S)- and (R)-norlaudanosoline are accepted as substrates by the enzyme to the same extent. Moreover, O-methylation of both isomers takes place at C-6 and C-7 with significant preference for the C-6 position (80%) (Rüffer et al. 1983). (R,S)-norlaudanosoline-1-carboxylic acid is, however, only slightly methylated [5% in comparison to (S)-norlaudanosoline], a fact which also points to the negligible biosynthetic role of this acid.

At the stage of (S)-6-O-methylnorlaudanosoline a further methyl group is introduced by a methyltransferase leading to (S)-norreticuline. The SAM-dependent transferase (also named norreticuline synthase), being highly selective, exclusively trans- for the substrate with (S)-configuration.

The enzyme following this step in the berberine sequence gives rise to the formation of (S)-reticuline, again via methylation (N-methylation). This methyltransferase is, according to the first transferase, not highly specific. Both isomeric series of (R)- and (S)-norbenzylisoquinoline alkaloids act as substrates for this enzyme.

(S)-reticuline, thus formed, occupies the role of a branch-point intermediate because a whole range of groups of isoquinoline alkaloids is derived from reticuline (Cordell 1981), e.g. the protoberberines. The first successful enzymatic transformation of reticuline to the berberine-type structure [(S)-scoulerine] was described by Rink and Böhm (1975).

The enzyme which has been named the berberine bridge enzyme (BBE) has been recently purified to homogeneity and characterized (Steffens et al. 1985). The ring closure catalyzed by BBE is oxygen-dependent and 1 mol each of (S)-scoulerine and hydrogen peroxide is formed during the reaction. As demonstrated by the substrate specificity of this enzyme, BBE clearly belongs to the group of highly specific proteins. After the four-ring skeleton has been closed, generating the tetrahydroberberine system (scoulerine), it is again a stereospecific methyltransferase catalyzing the sixth step in berberine biosynthesis. (S)-scoulerine is transformed by this enzyme into (S)-tetrahydrocolumbamine by regiospecific methylation of the 9-OH group (scoulerine-9-O-methyltransferase).

The final reaction in the cell-free biosynthesis of the protoberberine skeleton is the dehydrogenation of tetrahydroberberines, resulting in the aromatization of the C-ring. The involved enzyme is named (S)-tetrahydroprotoberine oxidase (STOX) (Amann et al. 1984). Although the oxidase accepts a wide variety of tetrahydroprotoberberines, it is absolutely specific for substrates with the (S)-configuration. It is this enzyme which provides the last intermediate of the berberine pathway, the alkaloid columbamine (see Fig. 1).

The final enzyme of this sequence then transforms columbamine into berberine. The necessary conversion of a methoxy- to a me-

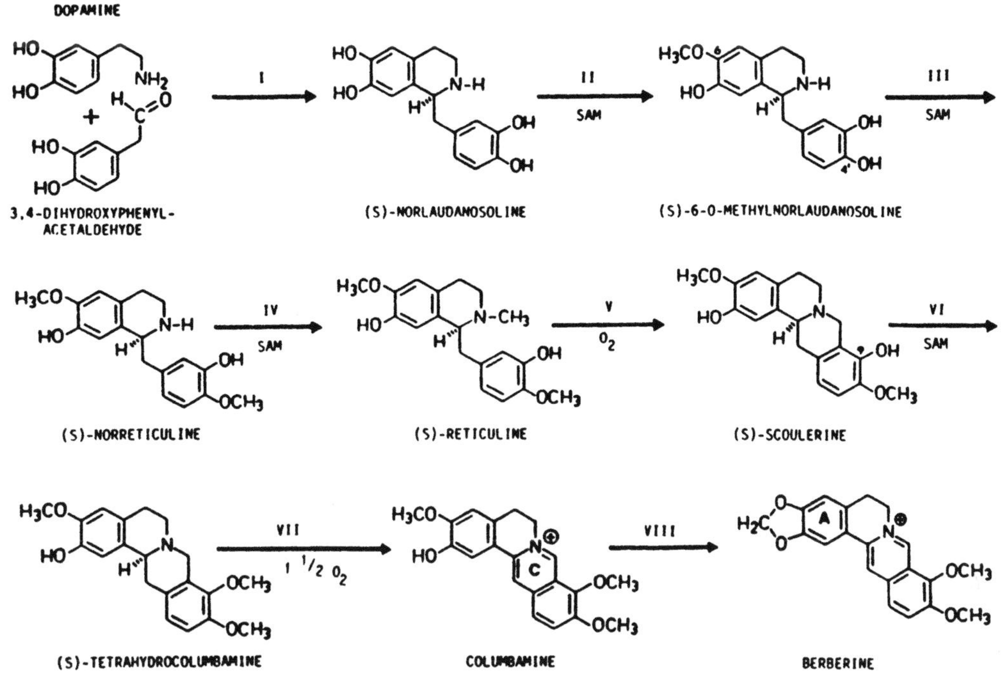

Fig. 1. The enzymatic biosynthesis of berberine. I Norlaudanosoline synthase; II (R), (S)-norlaudanosoline-6-O-methyltransferase; III 6-O-methylnorlaudanosoline-4'-O-methyltransferase; IV (R), (S)-norreticuline-N-methyltransferase; V berberine bridge enzyme; VI (S)-scoulerine-9-O-methyltransferase; VII (S)-tetrahydroberberine oxidase; VIII berberine synthase.
For the revision of the early steps of the reticuline biosynthesis, see Stadler et al. (1987) Tetrahedron Lett. 28:1251-1254

thylenedioxy-group in this reaction has been proved for the first time at the cell-free level. The berberine synthase again was purified to homogeneity and fully characterized (Rüffer and Zenk 1985).

In summary, starting from the precursors, dopamine and dihydroxyphenylacetaldehyde, the complete set of berberine synthesizing enzymes has been isolated and characterized (Zenk 1985; Fig. 1). Simple homogenization of cultivated Berberis cells provides all these biocatalysts and, besides oxygen, the only cofactor needed for the berberine synthesis is the methyl group donor SAM. All these enzymes are soluble, however the berberine bridge enzyme and the oxidase STOX are compartmentalized in specific vesicles. The other enzymes involved occur in the cytoplasm of the cells (Amann et al. 1986). The eight enzymatic steps of berberine formation are performed in a highly organized manner. In addition, the cellular transport and metabolism of the alkaloid precursors, as well as accumulation and storage of these alkaloids, have been recently investigated (Deus-Neumann and Zenk 1984, 1986). The knowledge of these processes, in combination with that of the enzymatic reactions, should help in utilizing the potential of cell cultures to produce the desired berberine.

2. THE ENZYMATIC FORMATION OF THE RAUWOLFIA ALKALOID AJMALINE

Similar to the previously discussed example of berberine, the production of Rauwolfia alkaloids in cultured cells has been a matter of interest for about 10 years. However, the alkaloid formation in such culture systems was only sporadically described for few examples, e.g. for serpentine and ajmaline (Tukhatasinov et al. 1977; Ohta and Yatazawa 1979). The isolated amounts of these indole alkaloids were extremely low. Moreover, nothing was known concerning the involved enzymes of ajmaline biosynthesis as of 5 years ago. Surprisingly, a very limited number of in vivo biosynthetic experiments have been carried out for ajmaline in the past (lit. cited in Rahman and Basha 1983), merely the fact that Rauwolfia plants are difficult to grow under greenhouse conditions. Therefore, knowledge of ajmaline biosynthesis remained almost completely at a hypothetical stage (Koskinen and Lounasmaa 1983).

Ajmaline has been therapeutically in use as an anti-arrhythmic drug since 1959. The alkaloid is commercially obtained by isolation from Rauwolfia roots. Some complicated chemical total syntheses have been developed, which produce ajmaline only with low yields, because the alkaloid structure consists of a six-ring system bearing nine chiral carbon centers. Accordingly, the biosynthesis of ajmaline must also be a complex process catalyzed by a variety of different enzymes. The delineation of this pathway, the description of the single enzyme mechanisms and the characterization of all the enzymes and intermediates involved, would extent our ideas on the biochemical formation of ajmaline. This understanding might be necessary for the regulation and enhancement of alkaloid production in cultivated Rauwolfia cells.

At the beginning of this work, we started with the phytochemical characterization of Rauwolfia serpentina cell suspensions by isolation of the major alkaloids (Stöckigt et al. 1981). Of 12 identified alkaloids 5 belong to the group of the ajmaline type and were structurally and biosynthetically closely related. Therefore, this cell culture fulfilled the prerequisite for a successful investigation of cell-free ajmaline biosynthesis.

The glucoalkaloid strictosidine is an important biogenetic precursor of ajmaline (Nagakura et al. 1979) as for all the other monoterpenoid indole alkaloids (more than 1800 occur naturally). This key intermediate is formed by condensation of tryptamine and the monoterpene secologanin, catalyzed in a stereospecific manner by the enzyme strictosidine synthase (Stöckigt and Zenk 1977). This result, together with the finding that 5-carboxy-glucoalkaloids are not progeners of ajmaline (Stöckigt 1979), ruled out previous assumptions (van Tamelen et al. 1968) of an involvement of a C-5 oxidative decarboxylation in ajmaline biosynthesis.

The second step in the biochemical pathway of ajmaline, the deglucosylation of strictosidine to highly unstable intermediates, is a well-known process. The following reaction, the elaboration of the sarpagan structure from one of these intermediates, has not yet been verified at the enzymatic level. From structural similarities of ajmaline-type alkaloids and from the next enzymes of the sequence it has been proved that the alkaloid polyneuridine aldehyde is one of the stable intermediates at the beginning of the biogenetic route. The enzyme acting on this aldehyde is well characterized from R. serpentina cells. It is an unusually high substrate - specific methylesterase, named PNA-esterase (polyneuridine aldehyde esterase). The immediate enzyme product, polyneuridine aldehyde acid, is extraordinarily unstable (β-ketoacid) and decarboxylates to the previously unknown alkaloid 16-epi-vellosimine. It is this mechanism by which the monoterpenoid carbon skeleton is transformed into a C 9 unit (Pfitzner and Stöckigt 1983). The novel alkaloid 16-epi-vellosimine exhibits the steric requirement for the formation of the characteristic ajmaline-type bond. The enzymatic reaction resulting in the synthesis of the ajmaline skeleton is catalyzed by the enzyme named vinorine synthase (Pfitzner et al. 1986). Vinorine, a constituent of the Rauwolfia cell culture, is an acetylated indolenine alkaloid. The isolation of this compound as well as a series of related acetyl alkaloids led to the concept that an acetylation reaction might be part of the vinorine synthesis. In fact, the vinorine synthase reaction depends fully on the presence of acetyl CoA. The enzyme accepts only a very limited number of substrates and operates exclusively in the presence of acetyl CoA as co-substrate. As discussed later, this acetyl unit has the function of a protecting group, stabilizing the indolenine structure for further enzymatic reactions. The structural comparison of vinorine with ajmaline reveals differences only at the periphery of the molecule. These consist of a hydroxy, N-methyl and acetyl group and two hydrogenated double bonds.

Indeed, feeding of labelled vinorine to the Rauwolfia cell culture produces mainly late intermediates of the pathway and a 16% incorporation into vomilenine (21-OH-vinorine). This experiment verifies that the C-21 hydroxylation follows the vinorine formation in ajmaline biosynthesis. The formed vomilenine is then converted into ajmaline by a crude Rauwolfia enzyme preparation in the presence of NADPH and S-adenosylmethionine. Separation and characterization of the corresponding enzymes point to the following sequence of steps: The indolenine double bond is NADPH-dependent and stereospecifically reduced (vomilenine reductase). The acetyl group, introduced earlier by the vinorine synthase, is finally removed by an acetylesterase. The saturation of the C-19 double bond is accomplished by a second NADPH-dependent reductase. This sequence of reduction steps follows from the substrate specificity of the vomilenine reductase(s). Based on the Km values for different substrates, the most plausible sequence of last steps in ajmaline formation is depicted in Fig. 2a. An N-methyltransferase converting N-norajmaline to ajmaline completes the biogenetic sequence (Stöckigt 1986).

Whereas the discussed enzymes catalyze reactions in the main biosynthetic route, we have isolated several enzymes involved in side reactions of the ajmaline pathway. Two branch-point

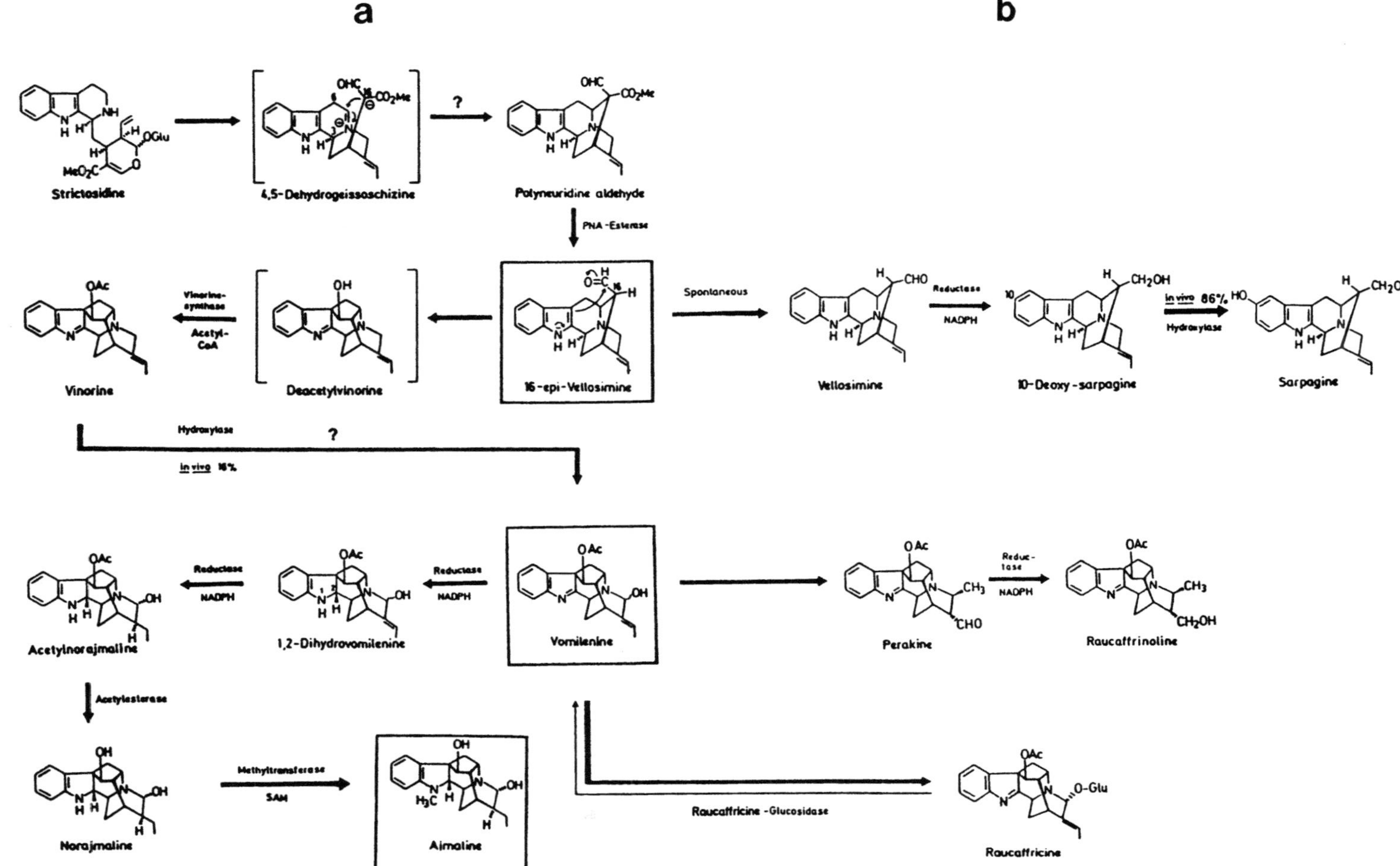

Fig. 2. a The biosynthetic pathway leading to the *Rauwolfia* alkaloid ajmaline. b Side routes of ajmaline biosynthesis

intermediates are of importance: 16-epi-vellosimine and vomilenine. The former enters the route to sarpagine by a spontaneous epimerization leading to vellosimine. This alkaloid is then reduced by vellosimine reductase. The product of this reduction, 10-deoxysarpagine, serves as the direct precursor of sarpagine (Pfitzner and Stöckigt 1984). In vivo feeding of this precursor showed an 86% incorporation rate into sarpagine. This unusually high transformation is easy to explain, because 10-deoxysarpagine cannot be channelled back to the ajmaline route since vellosimine reductase does not catalyze the reverse reaction. This enzyme is inhibited by $NADP^+$. The only biotransformation of deoxysarpagine seems to be its hydroxylation to sarpagine (Fig. 2a).

The second branch-point of the ajmaline route is located at the stage of vomilenine. When this intermediate is incubated with pre-purified enzyme preparations in the presence of NADPH, conversion to the *Rauwolfia* alkaloid raucaffrinoline takes place (Polz et al. 1987). Obviously, an isomerization under formation of perakine must first occur.

Both side routes, formation of sarpagine and raucaffrinoline, do not have a significant influence on the ajmaline production of the cultured *Rauwolfia* cells. Sarpagine can be regarded as a trace compound formed during 3 weeks of cell growth in the low milligram range per liter medium. Raucaffrinoline has not yet been detected in our cultures, whereas ajmaline is produced in yields approx. 0.3 g l^{-1} medium.

In sharp contrast, the third side reaction of the ajmaline pathway probably diminishes the cellular ajmaline yield effectively. The intermediate vomilenine is channelled out of the pathway by glucosylation. The produced glucoalkaloid raucaffricine accumulates in the suspension culture in amounts of up to 1.2 g l^{-1} medium (Schübel et al. 1986). It is interesting to note that raucaffricine is a typical constituent of *R. caffra*. The alkaloid has not been isolated from other *Rauwolfia species*. However, in *Rauwolfia* cultures we found the compound in all species so far tested with a maximum yield in *R. serpentina*. It seems that under the growth conditions of the cell suspensions, the pathway to ajmaline is "de-regulated", ending preferentially at the stage of raucaffricine. If in future experiments an inhibition of raucaffricine biosynthesis can be achieved, the real "bottleneck" of ajmaline formation would be recognizable. In the case that high raucaffricine synthase activity is the only limiting factor of ajmaline production, the in vivo inhibition of the enzyme would theoretically result in an ajmaline production rate of 1 g l^{-1} medium. Such conditions, however, still need to be verified. The occurrence of the enzymes described here is taxonomically restricted to those plant families, genera and species which synthesize the appropriate alkaloids. However, we also found the following contradictory example. Recently, we isolated an enzyme from leaves of *Catharanthus roseus* G. Don, which efficiently dimerizes the *Aspidosperma* alkaloid tabersonine. The pre-purified enzyme obtained from 500 g fresh tissue allows the dimerization of 1 g tabersonine, corresponding to a specific activity of 0.15 nkat mg^{-1} protein (Fahn et al. 1987). From the formed dimers the alkaloids voafrine A and voafrine B can easily be prepared by reduction, thus offering a novel access to these rare compounds. Both voafrines were previously detected in cultivated cells of *Voacanga africana* Stapf, being the first isolation of dimeric indole alkaloids from cell suspension cultures (Stöckigt et al. 1984). Whereas the original source of these alkaloids allowed an isolation only in the low milligram range, the voafrines can now be obtained at the gram scale by employing the *C. roseus* enzyme. Although *C. roseus* is the most investigated plant with respect to the formation of indole alkaloids, dimeric tabersonines, however, have never been found in this plant. Therefore we are presently engaged in clarifying the biosynthetic significance of this dimerization process.

Acknowledgements. The Fonds der Chemischen Industrie and the Deutsche Forschungsgemeinschaft, Bonn- Bad Godesberg (SFB 145) are acknowledged for financial support.

REFERENCES

Amann M, Nagakura N and Zenk MH (1984) (S)-Tetrahydroprotoberberine oxidase the final enzyme in the protoberberine biosynthesis. Tetrahedron Lett 25:953-954

Amann M, Wanner G and Zenk MH (1986) Intracellular compartmentation of the two key enzymes of berberine biosynthesis in plant cell cultures. Planta 167:310-320

Cordell GA (1981) Introduction to alkaloids, a biogenetic approach. John Wiley and Sons, New York Chichester Brisbane Toronto

Deus-Neumann B and Zenk MH (1984) A highly selective alkaloid uptake system in vacuoles of higher plants. Planta 162:250-260

Deus-Neumann B and Zenk MH (1986) Accumulation of alkaloids in plant vacuoles does not involve an ion-trap mechanism. Planta 167:44-53

Fahn W, Kaiser V, Schübel H, Danieli B and Stöckigt J (to be published) Helv Chim Acta

Fujita Y and Tabata M (1986) Secondary metabolites from plant cells. Pharmaceutical application and progress in commercial production. In: Somers DA, Gengenbach BG, Biesboer DD, Hackett WP and Green CE (eds) 6th International congress plant tissue and cell culture. University of Minnesota, p 2

Fukui H, Nakagawa K, Tsuda S and Tabata M (1982) Production of isoquinoline alkaloids by cell suspension cultures of Coptis japonica. In: Fujiwara A (ed) Proc. 5th intl. cong. plant tissue and cell culture. Maruzen, Tokyo, p 313

Furuya T, Synono K and Ikuta A (1972) Isolation of berberine from callus tissue of Coptis japonica. Phytochemistry 11:175

Hahlbrock K and Grisebach H (1975) Biosynthesis of flavonoids. In: Mabry TJ, Mabry H (eds) The flavonoids. Chapman and Hall, London, p 866

Kondo Y (1976) Organic and biological aspects of berberine alkaloids. Heterocycles 4:197-219

Koskinen A and Lounasmaa M (1983) The sarpagine-ajmaline group of indole alkaloids. In: Herz W, Grisebach H and Kirby GW (eds) Progress in the chemistry of organic natural compounds. Springer, New York, p 267

Nagakura N, Rüffer M and Zenk MH (1979) The biosynthesis of monoterpenoid indole alkaloids from strictosidine. J Chem Soc Perkin Trans 1:2308-2312

Ohta S and Yatazawa M (1979) Growth and alkaloid production in callus tissues of Rauwolfia serpentina. Agric Biol Chem 43: 2297-2303

Pfitzner A and Stöckigt J (1983) Characterization of polyneuridine aldehyde esterase, a key enzyme in the biosynthesis of sarpagine/ajmaline type alkaloids. Planta Med 48:221-227

Pfitzner A, Krausch B and Stöckigt J (1984) Characteristics of vellosimine reductase, a specific enzyme involved in the biosynthesis of the Rauwolfia alkaloid sarpagine. Tetrahedron 40:1691-1699

Pfitzner A, Polz L and Stöckigt J (1986) Properties of vinorine synthase - the Rauwolfia enzyme involved in the formation of the ajmaline skeleton. Z Naturforsch 41c:103-114

Polz L, Schübel H and Stöckigt J (1987), to be published

Rahman AU and Basha A (1983) Biosynthesis of indole alkaloids. Clarendon Press. Oxford

Rink E and Böhm H (1975) Conversion of reticuline into scoulerine by a cell free preparation from Macleaya microcarpa cell suspension cultures 49:396-399

Rüffer M and Zenk MH (1985) Berberine synthase, the methylene dioxy group forming enzyme in berberine synthesis. Tetrahedron Lett 26:201-202

Rüffer M, El-Shagi H, Nagakura N and Zenk MH (1981) (S)-Norlaudanosoline synthase: the first enzyme in the benzylisoquinoline biosynthetic pathway. FEBS Lett 129:5-9

Rüffer M, Nagakura N and Zenk MH (1983) Partial purification and properties of S-adenosylmethionine: (R), (S)norlaudanosoline-6-O-methyltransferase from Argemone platyceras cell cultures. Plant Med 49:131-137

Sato F and Yamada Y (1984) High berberine producing cultures of Coptis japonica. Phytochemistry 23:281-285

Sato S, Endo T, Hashimoto T and Yamada Y (1982) Production of berberine in cultured Coptis japonica cells. In: Fujiwara A (ed) Proc. 5th intl. cong. plant tissue and cell culture. Maruzen, Tokyo, p 319

Scott AI, Lee SL, Hirata T and Culver MG (1978) Biosynthesis of plant products using cell free systems from tissue cultures. Rev Latinoam Quim 9:131-138

Schübel H, Stöckigt J, Feicht R and Simon H (1986) Partial purification and characterization of raucaffricine β-D-glucosidase from plant cell suspension cultures of Rauwolfia serpentina Benth. Helv Chim Acta 69:538-547

Schumacher HM, Rüffer M, Nagakura N and Zenk MH (1983) Partial purification and properties of (S)-norlaudanosoline synthase from Eschscholtzia tenuifolia cell cultures. Planta Med 48: 212-220

Shamma M and Moniot JL (1978) Isoquinoline alkaloid research 1972-1977. Plenum, London New York

Steffens P, Nagakura N and Zenk MH (1985) Purification and characterization of the berberine bridge enzyme from Berberis beaniana cell cultures. Phytochemistry 24:2577-2583

Stöckigt J (1979) Non-involvement of 5α-carboxystrictosidine and -vincoside in the biosynthesis of sarpagine- and ajmaline-type alkaloids. Tetrahedron Lett 28:2615-2618

Stöckigt J (1980) The biosynthesis of heteroyohimbine-type alkaloids. In: Phillipson JD and Zenk MH (eds) Indole and biogenetically related alkaloids. Academic Press, London New York Toronto, p 113

Stöckigt J (1986) Enzymatic biosynthesis of indole alkaloids: ajmaline, sarpagine, vindoline. In: Rahman AU and Le Quesne PW (eds) New trends in natural products chemistry 1986, studies in organic chemistry. Elsevier Science Publishers B.V., Amsterdam, p 497

Stöckigt J and Zenk MH (1977) Strictosidine (isovincoside): the key intermediate in the biosynthesis of monoterpenoid indole alkaloids. Chem Commun 646-648

Stöckigt J, Pfitzner A and Firl J (1981) Indole alkaloids from cell suspension cultures of *Rauwolfia serpentina* Benth. Plant Cell Rep 1:36-39

Stöckigt J, Pawelka KH, Tanahashi T, Danieli B and Hull WE (1984) Voafrine A and voafrine B, new dimeric indole alkaloids from cell suspension cultures of *Voacanga africana* Stapf. Helv Chim Acta 66:2525-2533

Tukhatsinov MK, Kiryalov NP, Butenko RG, Grushivitskii IV, Grurevich IY and Nikitina IK (1977) Study of the stem culture of *Rauwolfia serpentina* tissue. Khim Prir Soedin 1:122-123

Van Tamelen EE, Haarstad VB and Orvis RL (1968) Hypohalite induced oxidative decarboxylation of α-amino acids. Tetrahedron 24:687-704

Wilson ML and Coscia CJ (1975) Studies on the early stages of *Papaver* alkaloid biogenesis. J Am Chem Soc 97:431-432

Wink M and Hartmann T (1982) Physiological and biochemical aspects of quinolizidine alkaloid formation in cell suspension cultures. In: Fujiwara A (ed) Proc. 5th intl. cong. plant tissue and cell culture. Maruzen, Tokyo, p 333

Zenk MH (1980) Enzymatic synthesis of ajmalcine and related indole alkaloids. J Nat Prod 43:438-451

Zenk MH (1985) Enzymology of benzylisoquinoline alkaloid formation. In: Phillipson JD, Roberts MF and Zenk MH (eds) The chemistry and biology of isoquinoline alkaloids. Springer, Berlin Heidelberg New York, p 240

Zenk MH, Rüffer M, Amann M and Deus-Neumann B (1985) Benzylisoquinoline biosynthesis by cultivated plant cells and isolated enzymes. J Nat Prod 48:725-738

GLUCOSE REPRESSION AND cAMP EFFECT ON ENZYME ACTIVITIES IN CALLUS TISSUES OF NICOTIANA PLUMBAGINIFOLIA

Alessandra Bottacin[1] and Giovanni Cacco[2]

ABSTRACT

Tissue cultures of *Nicotiana plumbaginifolia* were used to study the effect of glucose on the activity of glutamate dehydrogenase, malate dehydrogenase, lactate dehydrogenase and alcohol dehydrogenase. GDH and MHD were repressed while LDH and ADH were not affected by 2% glucose in the culture medium. The low glucose level (0.5%) or the addition of cAMP to the culture medium containing 2% glucose restored the initial level of GDH while they were ineffective on MDH activity.

KEY WORDS: Callus culture, carbon catabolite repression, cyclic AMP, *Nicotiana plumbaginifolia*.

INTRODUCTION

High glucose concentrations in the culture medium repress the synthesis of some inducible enzymes in bacteria and animal tissues. This phenomenon, known as carbon catabolite repression, may be reversed by cyclic nucleotide 3':5'-monophosphate (Pastan and Perlman 1970; Tomkins 1975). Recently, such a mechanism has been discovered also in photosynthetic microorganisms (Bressan et al. 1980; Handa et al. 1981) and in higher plants (Tassi et al. 1984). The present work provides additional, indirect information about the occurrence of catabolite repression by glucose

[1]Università di Padova, Dipartimento di Biotecnologie Agrarie, Via Gradenigo 6, 35131 Padova, Italy

[2]Università di Reggio Calabria, Istituto di Chimica Agraria e Forestale, Reggio Calabria, Italy

NATO ASI Series, Vol. H18
Plant Cell Biotechnology. Edited by M.S.S. Pais et al.

and the counteracting effect of cyclic AMP (cAMP) in higher plants. The responses of some inducible enzymes, glutamate dehydrogenase (GDH), malate dehydrognease (MDH), lactate dehydrogenase (LDH) and alcohol dehydrogenase (ADH) were assayed in tobacco callus cultured under heterotrophic conditions.

MATERIAL AND METHODS

Callus tissue of <u>Nicotiana plumbaginifolia</u> were cultured in liquid M/S medium (Murashige and Skoog 1962) containing 0.5% (low) and 2% (high) glucose concentrations. The plates were incubated at 25°C on a rotatory shaker (40 rpm) in the dark. After 4 days of culture some calli were submitted to enzyme assays. The remaining calli were transferred to fresh media according to the following protocol:

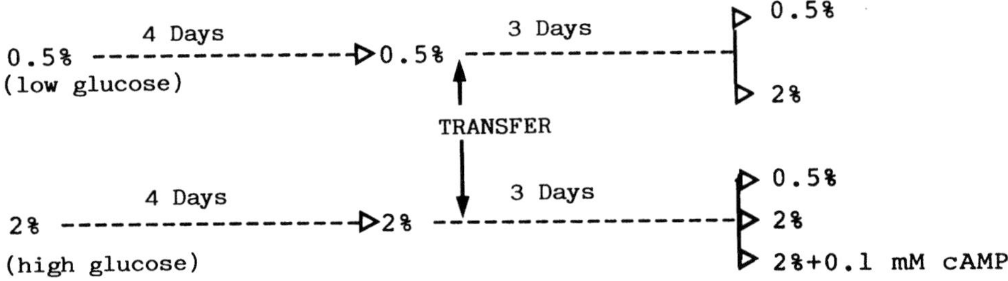

After an additional 3 days of culture, calli were collected and assayed for the recovered enzyme activities. The crude extract was obtained by grinding, with pestle and mortar, 700 mg callus in 5 ml buffer (150 mM Tris-HCl, pH 7.8 and 5 mM dithiothreitol). The enzyme activities were measured spectrophotometrically at 340 nm and 30°C, by the oxidation of NADH in the case of GDH: EC 1.4.1.2 (Groat and Vance 1981), MDH:EC 1.1.1.37 (Quieroz 1969), LDH:EC 1.1.1.27 (Hanson and Jacobsen 1984) and by reduction of NAD, in the case of ADH:EC 1.1.1.1 (Hanson and Jacobsen 1984). Protein was determined by the method of Bradford (1976).

RESULTS AND DISCUSSION

At increasing glucose concentration (0.5-4%) in the culture medium, callus growth was reduced, the rate of protein synthesis was unaffected (data not shown), GDH and MDH decreased, LDH and

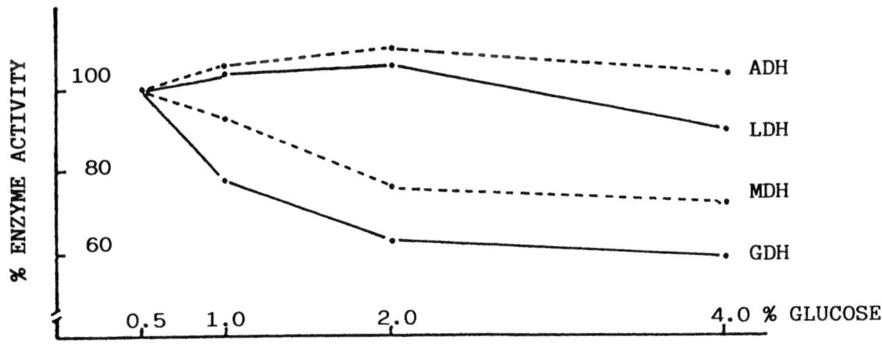

Fig. 1. Relative specific activity of GDH, MDH, LDH and ADH as a function of glucose concentration in the culture medium

ADH did not change (Fig. 1). In order to verify whether the enzymes were subjected to inactivation or repression by glucose, we assayed the enzyme activities in calli grown at low (0.5%) and high (2%) glucose in the liquid medium for 4 days and in calli transferred for an additional 3 days from a high to low glucose level and vice versa, or supplemented with cAMP GDH and MDH showed a 27% repression by high glucose concentration (Table 1).

Table 1. Enzyme activities assayed after 4 and 7 days of culture in 0.5% and 2% glucose (Specific activities are given in nmol min^{-1} mg^{-1} prot. ± SD for GDH, LDH and ADH and in µmol min^{-1} mg^{-1} prot. ± SD for MDH. *P 0.05; **P 0.01; ns, not significant)

	0.5% Glucose	2% Glucose
4 Days		
GDH	30.46±2.06	27.26±2.84*
MDH	1.84±0.11	1.59±0.21*
LDH	41.02±2.99	39.76±1.74ns
ADH	342.13±31.52	349.18±18.23ns
7 Days		
GDH	36.55±2.49	26.85±1.18**
MDH	1.71±0.15	1.24±0.04**
LDH	51.54±3.05	55.67±2.16ns
ADH	231.93±18.62	245.43±19.60ns

GDH activity was restored to the initial level by the addition of glucose only (0.5%) or cAMP to the high glucose medium (Fig. 2). The de-repression of GDH by a low glucose level or by cAMP observed in callus tissues is in accordance with data obtained on single cell cultures of *Asparagus* (Tassi et al. 1984). Since no direct effect of glucose on in vitro GDH activity was observed, it seems likely that carbon catabolite repression may affect some inducible enzyme in higher plants as previosuly demonstrated in heterotrophic or in simple photoautotrophic organisms (Bressan et al. 1980; Handa et al. 1981; Pastan and Perlman 1970). Upon exposure to a low glucose level or to cAMP, no recovery of repressed MDH activity was observed in calli (Fig. 2). A decreased MDH activity was also observed in glucose cultured yeast, and the addition of cAMP restored the enzyme activity only in the case of cytoplasmic MDH but not in the mitochondrial fraction (Takeda 1981). In our experiments a crude extract of callus tissue was used for enzyme assays, so that MDH activity represented a mixture of cytoplasmic and mitochondrial enzymes. Therefore, the glucose-induced inactivation of MDH observed in calli might be due either to carbon catabolite repression or to irreversible inactivation due to a proteolytic regulatory mechanism. The absence of any significant effect of glucose on the inducible cytoplasmic LDH and ADH enzymes is in agreement with the general statement that inducibility is not a sufficient requirement for the expression of the catabolite repression mechanism (Magasanik 1961). In fact, only some inducible enzymes are subjected to carbon repression as previously observed in yeast for glycolysis enzymes (Takeda 1981).

CONCLUSIONS

Several instances show that glucose represses the synthesis of some inducible enzymes in higher plants and this effect is reversed by cAMP. Data reported here indicate that only some inducible enzymes are affected by carbon catabolite repression in tissue culture of tobacco.

Acknowledgements. The authors are very grateful to Prof. G. Ferrari for revising the manuscript.

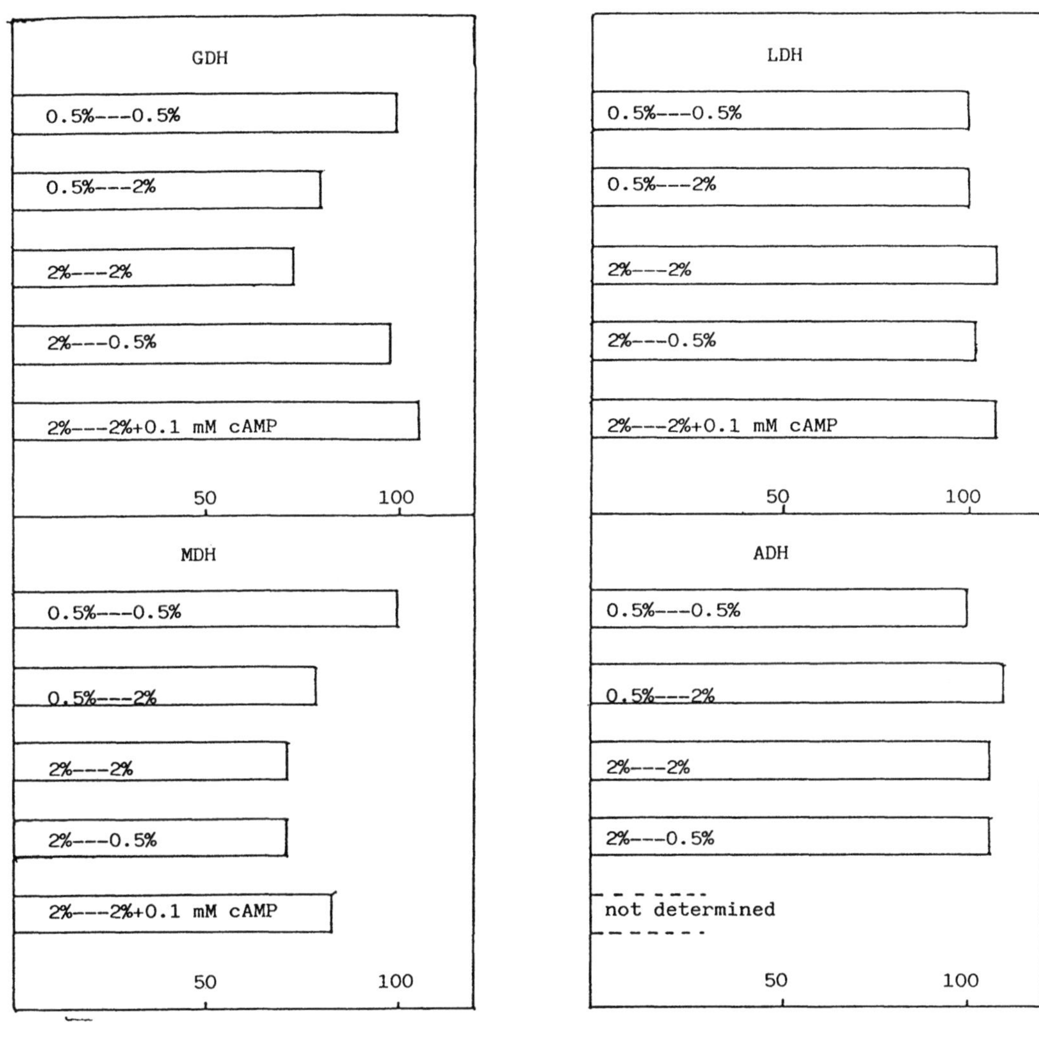

Fig. 2. Recovery of enzyme activities assayed in callus transferred to fresh media containing glucose or cAMP as reported in the figures. Data are expressed as the percentage of the control (0.5% glucose)

REFERENCES

Bradford MM (1976) A rapid and sensitive method for quantitation of microgram quantities of protein using the principle of protein-dye binding. Anal Biochem 72:248-254

Bressan RA, Handa AK, Quader H, Filner P (1980) Synthesis and release of cyclic adenosine 3':5'-monophosphate by _Ochromonas malhamensis_. Plant Physiol 65:165-170

Groat RG, Vance CP (1981) Root nodule enzymes of ammonia assimilation in alfalfa (Medicago sativa L.). Plant Physiol 67: 1198-1203

Handa AK, Bressan RA, Quader H, Filner P (1981) Association of formation and release of cyclic AMP with glucose depletion and onset of chlorophyll synthesis in Poterioochromonas malhamensis. Plant Physiol 68:460-463

Hanson AD, Jacobsen JV (1984) Control of lactate dehydrogenase, lactate glycolysis and a-amylase by O_2 deficit in barley aleurone layers. Plant Physiol 75:566-572

Magasanik B (1961) Catabolite repression. Cold Spring Harbor Symp Quant Biol 26:249-257

Murashige T, Skoog F (1962) A revised medium for rapid growth and bioassays with tobacco tissue culture. Physiol Plant 15:473-497

Pastan I, Perlman R (1970) Cyclic adenosine monophosphate in bacteria. Science 169:339-344

Quieroz O (1969) Photoperiodisme et activite enzymatique (PEP carboxylase et enzyme malique) dans les feuilles de Kalanchoe blassfeldiana. Phytochemistry 8:1655-1663

Takeda M (1981) Glucose-induced inactivation of mitochondrial enzymes in yeast Saccharomyces cerevisiae. Biochem J 198: 281-287

Tassi F, Restivo FM, Puglisi PP, Cacco G (1984) Effect of glucose on glutamate dehydrogenase and acid phosphatase and its reversal by cyclic adenosine 3':5'-monophosphate in single cell cultures of Asparagus officinalis. Physiol Plant 60: 61-64

Tomkins GM (1975) The metabolic code. Science 189:760-763

LONG-TERM STORAGE OF *VINCA MINOR* CELL CULTURES

M. Caruso, L. Crespi-Perellino, L. Garofano and A. Guicciardi
Farmitalia Carlo Erba, Via dei Gracchi 35, 20146 Milano, Italy

INTRODUCTION

Plant cell cultures are potential sources for the production of useful secondary compounds. The selection of high-producing variant cell lines is the first step for scale-up procedures. Unfortunately the maintenance of these variants is not always easy and reliable.

These cultures are commonly maintained through serial subcultures, either in callus or suspension, which has the disadvantages of high cost, contamination and errors. Also, a number of undesirable genetic changes may occur with time (Hauptmann and Widholm 1982; Zenk 1978). Consequently, a culture with particular yield charactristics may not be readily available if serial passage is the only method of preservation. The production of alkaloids by in vitro cultures can vary among cell lines and in an individual line may be fairly stable; in other cell lines, the alkaloid content has been found to be subjected to gradual change over some years of subculturing (Deus and Zenk 1982).

Instead of repeated selection to maintain a high alkaloid level, cryopreservation would appear to be an alternative to overcome the stability problem (Chen et al. 1984; Seitz et al. 1985).

In most of cases of cryogenic storage, viability and regrowth capacity were tested but there are few reports related to the preservation of biosynthetic capacities in cell suspension cultures after freezing and thawing.

In this work we will deal in particular with the preservation of the biosynthetic capacity of *Vinca minor* after storage.

NATO ASI Series, Vol. H18
Plant Cell Biotechnology. Edited by M.S.S. Pais et al.

MATERIAL AND METHODS

Culture conditions

Cell cultures of Vinca minor were established and maintained in Gamborg B5 medium (Gamborg et al. 1968) supplemented with 1 ml l^{-1} NAA, 20 g l^{-1} sucrose and 7 g l^{-1} agar.

Liquid cultures were grown in 300-ml flasks with 50 ml B5 medium at 23° or 28°C in the dark on a rotary shaker at 120 rpm. The final alkaloid concentration was determined by HPLC chromatography.

Preservation protocol

Serial subcultures: slants were inoculated by spreading 0.5 g of a 20-day-old callus on the surface of an agar slant. The remaining callus was inoculated in the flask to control alkaloid production.

Serial liquid cultures: 5-ml portions from a 7-day-old seed stage were inoculated in three production and one seed stage flasks. The production flasks were collated after an incubation of 14 days at 23°C and then the alkaloid yield was determined. The seed stage was incubated at 28°C for 7 days, then used as a new seeding batch.

Storage under mineral oil: sterile liquid paraffin C. Erba RPE was added to young, but almost completely growh, agar slants in order to avoid any air contact. These slants were maintained in the dark at 10°C for 6 months; each month three slants were spread on a petri dish, isolated and transferred to liquid culture for alkaloid production.

Storage under liquid nitrogen (LN): liquid cultures of Vinca minor, 6-days-old, were added with 0.5 M sorbitol (final concentration) and incubated overnight at 28°C on a rotary shaker. The cells were collected by centrifugation, resuspended in a half volume of cryoprotectant solution (30% sucrose, 10% glycerol) and distributed in ampoules (2 ml). The ampoules were slowly frozen to -80°C overnight and stored under liquid nitrogen at -196°C until use.

For regrowth ampoules were thawed rapidly in a water bath at 40°C, and spread on an agar slant. A 5-ml portion of the same

Fig. 1. Reproducibility of alkaloid yield in Vinca minor. △ Serial cultures in liquid medium every 7 day; ▲ serial subcultures on agar medium every 20 days; □ storage under mineral oil; ■ storage under liquid nitrogen

suspension can be inoculated in a seed stage flask for the subsequent, larger inoculum stage.

RESULTS AND DISCUSSION

Our results (Fig. 1) confirm that we can preserve the growth ability of the cultures in each of the examined methods, however, alkaloid yield is not always maintained. In particular, there was a progressive loss of activity when Vinca minor cultures were serially subcultured in liquid or in agar medium.

In contrast, it was possible to maintain a fairly stable yield close to the initial one, when storage was performed under a mineral oil layer in liquid nitrogen.

Storage under mineral oil may be preferred because it is easy to perform and gives good results on regrowth of cultures of different plant species (Augerau et al. 1986).

Cryostorage under LN requires a great deal of effort to obtain reproducible results. Each step has to be checked and optimized for each different plant species. Even with this method we experienced some failure. Nevertheless, cryostorage becomes a necessity when a large stock of readily available inoculum is required for optimization of a scale-up process.

In fact, cultures from LN grew at about the same rate as the standard inoculum stage and can be immediately used for subsequent, larger inoculum stages, without lowering alkaloid yield.

REFERENCES

Augerau JM, Courtois D, Petiard V (1986), Plant Cell Reports 5: 372

Chen THH, Karta KK, Leung NL, Kurz WGW, Chatson GB and Constabel F (1984), Plant Physiol 75:726

Deus B and Zenk MH (1982), Biotech Bioeng 24:1965

Gamborg OL, Miller RA and Ojime K (1968), Exp Cell Res 50:151

Hauptmann M and Widholm JM (1982) Plant Physiol 70:30

Seitz U, Reuff I and Reinhart E (1985) In: Neumann et al (eds) Primary and secondary metabolism of plant cell cultures, Springer Berlin:323 (1985)

BIOTRANSFORMATION OF SYNTHETIC AND NATURAL COMPOUNDS BY PLANT CELL CULTURES

A.W. Alfermann[1] and E. Reinhard[2]

INTRODUCTION

The large biochemical potential of plant cells cultivated in vitro for performing specific biotransformations on particular natural or synthetic substrates resulting in more complex and, from a pharmaceutical point of view, more useful products, has been well recognized in recent years. The reactions observed up until now include reductions, oxidations, hydroxylations, epoxidations, glycosylations as well as esterifications. Various phenols, coumarins, alkaloids, terpenoids, steroids and cardenolides have been used as substrates (for reviews see Furuya 1978, Reinhard and Alfermann 1980, and Kurz and Constabel 1985, among others).

This review will concentrate on some newer data showing that it is possible to produce new compounds not yet found in nature at all or at least not in the particular plant species used. If one produces the natural product of the plant species, then certain prerequisites have to be fulfilled for the reaction to be of biotechnological interest. The product yields must be especially high and, of course, it must be a plant-specific reaction, which means that it is impossible to perform the same reaction by chemical or microbiological methods on a large scale.

[1]Institut für Entwicklungs- und Molekularbiologie der Pflanzen, Universität Düsseldorf, Universitätsstraße 1, D-4000 Düsseldorf FRG

[2]Pharmazeutisches Institut der Universität Tübingen, Auf der Morgenstelle 8, D-7400 Tübingen, FRG

NATO ASI Series, Vol. H18
Plant Cell Biotechnology. Edited by M.S.S. Pais et al.

PRODUCTION OF NEW COMPOUNDS

As early as 1980, Veliky and co-workers reported that cell suspension cultures of the carrot were able to hydroxylate digitoxigenin and later also gitoxigenin and oleandrigenin (Veliky et al. 1980; Jones and Veliky 1981) to the appropriate 5-hydroxy-derivatives. Whereas 5-hydroxydigitoxigenin was already known in nature, both other derivatives were new compounds. Unfortunately, none of these cardiac aglycones is of medicinal importance.

In the following example no unknown compounds are produced; they are only new for the species involved. However, by using this technique enough material for intensive biological testing can be produced.

Relatively few plant constituents are used in cancer treatment. Recently, the pyridocarbazole alkaloid 2-methyl, 9-methoxyellipticinium was introduced to clinical treatment. This type of alkaloid is found in _Ochrosia_ species (_Apocynaceae_) and can be synthesized by tissue cultures of _O. elliptica_ in the same amounts as by the differentiated plant (Kouadio et al. 1984a). One problem in that field is that it is difficult with chemical as well as microbiological methods to produce different derivatives of the ellipticine molecule to modify the biological activity of the alkaloid. For this reason Chenieux and co-workers (Kouadio et al. 1984b) investigated the biotransformation of ellipticine by various cell cultures of different plant species. They were able to show that special strains from one single _Choisya ternata_ (_Rutaceae_) plant were able to transform the added ellipticine. Biotransformation of the cytotoxic ellipticine could only be achieved with agar cultures in which the ellipticine had been mixed with the agar. In suspension, _Choisya_ cell growth was reduced by the added ellipticine. Obviously, the agar-cultivated cells excrete an enzyme which performs this interesting biotransformation reaction. The biotransformation product isolated was identified as 5-formyl-ellipticine (Fig. 1), which could also be isolated in small quantities from _Strychnos dinklagei_ plants (Michel et al. 1980). It is not yet known whether 5-formylellipticine has a better medicinal activity than 2-methoxy-ellipticinium.

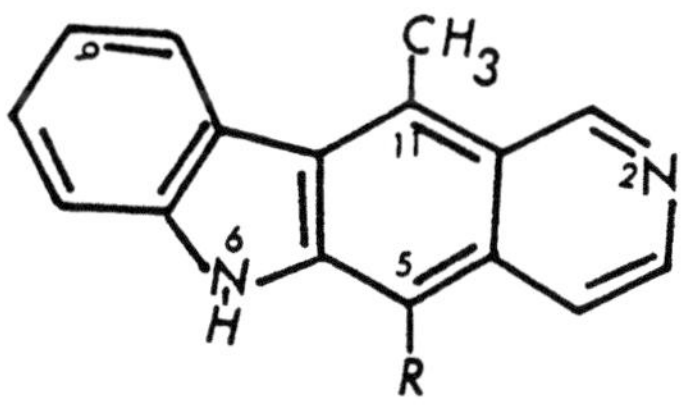

Fig. 1. Chemical structure of ellipticine ($R = CH_3$) and 5-formyl-ellipticine ($R = CHO$)

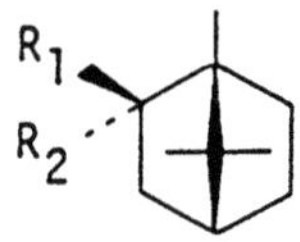

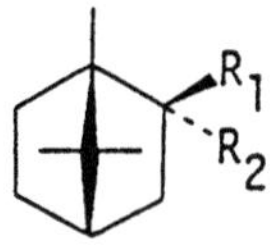

1a: R_1=H, R_2=OH
2a: R_1=OH, R_2=H
4a: R_1,R_2=O

1b: R_1=H, R_2=OH
2b: R_1=OH, R_2=H
4b: R_1,R_2=O

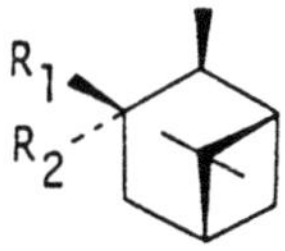

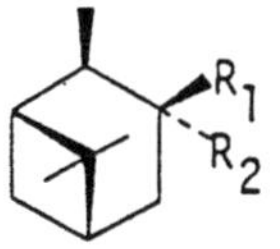

3a: R_1=H, R_2=OH
5a: R_1,R_2=O

3b: R_1=H, R_2=OH
5b: R_1,R_2=O

Fig. 2. Chemical structures of (+)-borneol (1a), (-)-borneol (1b), (-)-isoborneol (2a), (+)-isoborneol (2b), (+)-isopinocampheol (3a), (-)-isopinocamphenol (3b), (+)-camphor (4a), (-)-camphor (4b), (-)-isopinocamphone (5a), and (+)-isopinocamphone (5b). Compounds 1a and 2a are transformed by tobacco cells to 4a, compound 3a to 5a. Compounds 1b, 2b and 3b respectively, remain practically untouched

PRODUCTION OF KNOWN COMPOUNDS

The main advantage of the plant cell or the enzyme as compared with chemical synthesis is stereospecificity. This is demonstrated in Fig. 2 (Suga et al. 1983). Tobacco cell cultures, for example, are able to discriminate between the different enantiomeres and bicyclic monoterpene alcohols. Only (1R, 2S, 4R)-(+)-borneol was oxidized to (1R, 4R)-(+)-camphor, whereas its enantiomere (-)-borneol was oxidized only to a very minor extent. A similar enantiomeric selection was observed for the biotransformation of the enantiomeric pairs of isoborneol and isopinocampheol. These results demonstrate the possible application of plant cell cultures or of enzymes isolated from them for the

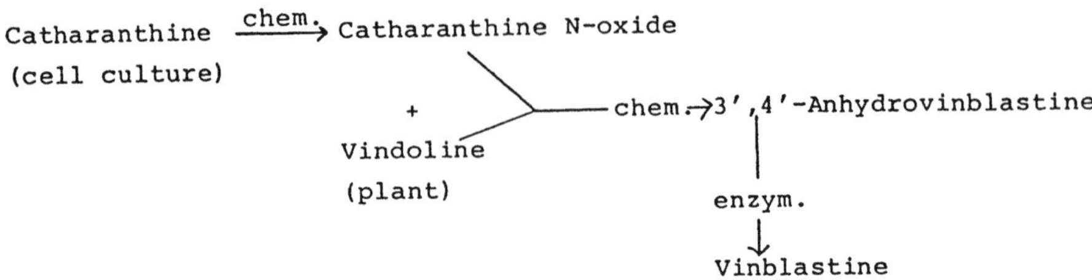

Fig. 3. A possible combined chemical and biological approach to production of vinblastine as investigated in different laboratories (see note added in proof 1)

optical resolution of organic compounds, as in the monoterpene compounds mentioned above. On the other hand, it demonstrates the advantage of using plant cells or enzymes for a combined approach to the synthesis of medicinal compounds by chemical and biological means, as proposed recently (Deus and Zenk 1985).

A very interesting example for the biotransformation of natural compounds is shown in Fig. 3. The dimeric indole alkaloids vinblastine and vincristine are used in the treatment of leukemia. Tons of dried leaves of <u>Catharanthus roseus</u> have to be extracted to isolate only a few grams of both compounds. Therefore, it would be interesting to use a combined chemical and biological approach to synthesize these compounds. Vindoline and catharanthine, available from plants or tissue cultures respectively, are each coupled by chemical means to anhydrovinblastine. A further biotransformation to both target compounds by cell or enzymes of <u>Catharanthus roseus</u> was investigated in several laboratories (Kutney et al. 1983; MacLauchlan et al. 1983; Endo et al. 1986).

Cardiac glycosides can be transformed with quite high yields using cell cultures of <u>Digitalis lanata</u>. The reason for our interest in that field is that <u>Digitalis lanata</u>, the main source of digoxin, always contains quite large amounts of digitoxin. Although both compounds are used in treatment of heart diseases, there is a higher demand for digoxin. Digoxin differs form digitoxin only in an additional hydroxyl function. If it were possible to hydroxylate digitoxin or its derivatives at C-12, then it would be possible to increase the yield of digoxin from the medicinal plant. Digitoxin would seem to be the most appropriate substrate to use because it is produced during the technical isolation of digoxin from the plant. Unfortunately, however, purpureaglycoside A is the main biotransformation product, not digoxin (see note added in proof 2). Therefore, we have tested a series of digitoxin derivatives, of which β-methyldigitoxin proved to be the most promising (Fig. 4). Specialized cell strains could be isolated after screening a number of cell lines which transform β-methyldigitoxin to β-methyldigoxin at a high transformation rate (Heins 1978; Hu 1984).

Those strains of Digitalis with a high hydroxylating capacity were tested in different types of bioreactors and various culture parameters were manipulated to improve product yields. Up to about 0.8 g l^{-1} of β-methyldigoxin during 20 days of culture in a 20-l reactor could be achieved. In contrast to results with Morinda (Wagner and Vogelmann 1977), Digitalis cells grow very well in normally stirred tank reaktors as well (Helmbold 1979; Spieler et al. 1985). Similar results were reported recently for Lithospermum (Fujita et al. 1982) and Coleus cells (Ulbrich et al. 1985). This means that the fermentation industry does not have to install totally new fermenter systems for industrial cultivation of plant cells.

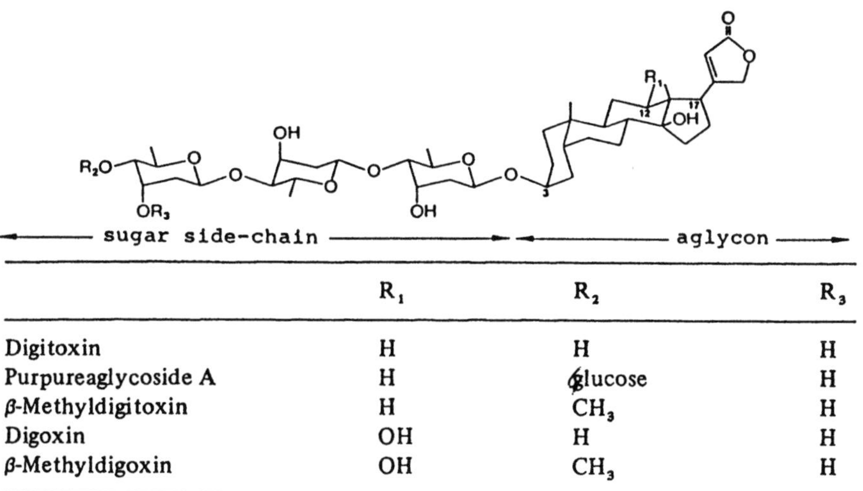

	R_1	R_2	R_3
Digitoxin	H	H	H
Purpureaglycoside A	H	glucose	H
β-Methyldigitoxin	H	CH_3	H
Digoxin	OH	H	H
β-Methyldigoxin	OH	CH_3	H

Fig. 4. The chemical structure of different natural and semisynthetic cardiac glycosides

More recently we were able to install a 200-l pilot plant at Tübingen and test our cell lines in a setup more like that used in the fermentation industry. In principle, the same product yields as in the laboratory fermenter were achieved. In industry, our cell lines were tested in a 1 m^3 stirred tank reactor. The product yields were only slightly lower as in the laboratory fermenter (Wahl 1985).

There are two main reasons why plant cell cultures are not widely used in industry. First, in most cases the interesting compounds are not produced in sufficiently high amounts such that practical application would be considered. Second, the plant cell cultures multiply very slowly. Therefore, to reduce the costs for the long period of scaling-up by using the cells during a prolonged production phase, we have tested a semi-continuous (repeated batch) cultivation of *Digitalis* cells for biotransformation of β-methyldigitoxin. Thus, 500 g of β-methyldigoxin were produced during a production time of more than 3 months in the 200-l pilot plant. Such a technique may help to reduce the production costs of such a process substantially (Alfermann et al. 1985). Nevertheless, since the product yields of this process are still too low, practical application is already economic. The enzyme system of this interesting biotransformation reaction has been characterized recently (Peterson and Seitz 1985), and not only the cells (Alfermann et al. 1980) but also the immobilized enzyme can be used for biotransformation of cardiac glycosides (Petersen et al. this Vol.).

PERSPECTIVES

The product yields of our process for biotransformation of β-methyldigitoxin have been increased by a factor of 50 during the last 10 years when calculated on a per day and fermenter volume basis. Although the yields are still too low for practical application the results presented clearly show that productivity of plant cell culture systems can be increased decisively by various manipulations. This sustains our optimism that besides production of shikonin other plant cell cultures will be used for biotechnological production of natural compounds in the future.

Acknowledgements. The work discussed on the biotransformation of cardiac glycosides has been sustained by a research grant of the Federal Ministry of Research and Technology, Bonn.

Notes added in proof:

1) Effective coupling of catharanthine and vindoline by various enzyme preparations to 3', 4'-anhydrovinblastine as well as an increased product yield of vinblastine has been published recently (Kutney JP 1987) Studies in Plant Tissue Culture. The Synthesis and Biosynthesis of Indole Alkaloids. Heterocycles 25:617-640; Goodbody A, Endo T, Vukovic J, Misawa M (1988) The Coupling of Catharanthine and Vindoline to Form 3', 4'-Anhydrovinblastine by Haemoproteins and Haemin. Planta Medica 54:210-214; Goodbody AE, Endo T, Vukovic J, Kutney JP, Choi LSL, Misawa M (1988) Enzymatic Coupling of Catharanthine and Vindoline to Form 3', 4'-Anhydrovinblastine by Horseradish Peroxidase. Planta Medica 54:136-140; Endo T, Goodbody A, Vukovic J, Misawa M (1988) Enzymes from *Catharanthus roseus* cell suspension cultures that couple vindoline and catharanthine to form 3', 4'-anhydrovinblastine. Phytochem. 27: 2147-2149; Endo T, Goodbody A, Vukovic J, Misawa M (1987) Biotransformation of anhydrovinblastine to vinblastine by a cell-free extract of *Catharanthus roseus* cell suspension cultures. Phytochem. 26:3233-3234)

2) Using a two-stage culture system digitoxine can be hydroxylated effectively at C-12 to deacetyllanatoside C and digoxin by cell cultures of *Digitalis lanata* (Kreis W, Reinhard E 1988) 12β-Hydroxylation of digitoxin by suspension-cultured *Digitalis lanata* cells: Production of Deacetyllanatoside C using a two-stage culture method. Planta Medica 54:143-148.

REFERENCES

Alfermann AW, Schuller I, Reinhard E (1980) Biotransformation of cardiac glycosides by immobilized cells of *Digitalis lanata*. Planta Medica 40:218-223

Alfermann AW, Spieler H, Reinhard E (1985) Biotransformation of cardiac glycosides by *Digitalis* cell cultures in airlift reactors. In: Neumann KH, Barz W, Reinhard E (eds). Primary and secondary metabolism of plant cell cultures. Springer, Berlin Heidelberg New York:316-322

Deus B, Zenk MH (1985) Gewinnung von Pflanzeninhaltsstoffen mit Hilfe pflanzlicher Zellkulturen. In: Pflanzliche Zellkulturen, BMFT-Statusseminar Jülich 1985, Bundesministerium für Forschung und Technologie, Bonn:91-110

Endo T, Goodbody A, Vukovic J, Chapple C, Misawa M, Choi LSL, Kutney JP (1986) Enzymatic synthesis of 3', 4'-anhydrovinblastine by cell free extracts from cultured Catharanthus roseus cells. In: Somars DA, Gengenbach BG, Biesboer DD, Hackett WP, Green CE (eds). VI International Congress of Plant Tissue Culture, Book of Abstracts, University of Minnesota:143

Fujita Y, Tabata M, Nishi A, Yamada Y (1982) New medium and production of secondary compounds with the two-staged culture method. In: Fujiwara A (ed). Plant tissue culture 1982, Japan. Assoc Plant Tissue Culture, Tokyo:399-400

Furuya T (1978) Biotransformation by plant cell cultures. In: Thorpe T (ed). Frontiers of plant tissue culture 1978, Int Assoc Plant Tissue Culture, Calgary (Canada):191-200

Heins M (1978) Selektion 12β-hydroxylierender Zellstämme von Digitalis lanata und Umwandlung von β-Methyldigitoxin durch Fermenterkulturen. Doctoral Thesis, University of Tübingen

Helmbold U (1979) Versuche zur Steigerung der 12β-Hydroxylierung von β-Methyldigitoxin durch Fermenterkulturen von Digitalis lanata Zellstämmen. Doctoral Thesis, University of Tübingen

Hu Z (1984) Untersuchungen zur Biotransformation von speziellen Zellstämmen von Digitalis lanata. Doctoral Thesis, University of Tübingen

Jones A, Veliky IA (1981) Biotransformation of cardenolides by plant cell cultures. II. Metabolism of gitoxigenin and its derivatives by suspension cultures of Daucus carota. Planta Medica 42:160-166

Kouadio K, Chenieux JC, Rideau M, Viel C (1984a) Antitumor alkaloids in callus cultures of Ochrosia elliptica. J Nat Products 47:872-874

Kouadio K, Rideau M, Ganser C, Chenieux JC, Viel C (1984b) Biotransformation of ellipticine into 5-formyl ellipticine by Choisya ternata strains. Plant Cell Reports 3:203-205

Kurz WGW, Constabel F (1985) Aspects affecting biosynthesis and biotransformation of secondary metabolites in plant cell cultures. CRC Critical Reviews in Biotechnology 2:105-118

Kutney JP, Aweryn B, Choi LSL, Honda T, Kolodziejzyk P, Lewis NG, Sato T, Sleigh SK, Stuart KL, Worth B, Kurz WGW, Chatson KB, Constabel F (1983) Studies in plant tissue culture: the synthesis and biosynthesis of indole alkaloids. Tetrahedron 39:3781-3795

McLauchlan WR, Hasan M, Baxter RL, Scott AI (1983) Conversion of anhydrovinblastine to vinblastine by cell-free homogenates of Catharanthus roseus cell suspension cultures. Tetrahedron 39:3777-3780

Michel S, Tillequin F, Koch M (1980) L'Oxo-17 Ellipticine, Nouvel Alcaloide de Strychnos dinklagei. Tetrahedron Lett 21: 4027-4030

Petersen M, Seitz HU (1985) Cytochrome P-450-dependent digitoxin 12β-hydroxylase from cell cultures of Digitalis lanata. FEBS Lett 188:11-14

Reinhard E, Alfermann AW (1980) Biotransformation by plant cell cultures. Adv Biochem Engin 16:49-83

Spieler H, Alfermann AW, Reinhard E (1985) Biotransformation of β-methyldigitoxin by cell cultures of Digitalis lanata in airlift and stirred tank reactors. Appl Microbiol Biotechnol 23:1-4

Suga T, Hirata T, Hamada H, Futatsugi M (1983) Enantioselectivity in the biotransformation of bicyclic monoterpene alcohols with the cultured suspension cells of Nicotiana tabacum. Plant Cell Reports 2:186-188

Veliky IA, Jones A, Ozubko RS, Przybylska M, Ahmed FP (1980) 5β-Hydroxygitoxigenin, a product of gitoxigenin produced by Daucus carota culture. Phytochem 19:2111-2112

Wagner F, Vogelmann H (1977) Cultivation of plant tissue cultures in bioreactors and formation of secondary metabolites. In: Barz W, Reinhard E, Zenk MH (eds). Plant tissue culture and its biotechnological application. Springer, Berlin Heidelberg New York:245-252

Wahl J (1985) Adaption konventioneller Fermenter zur Züchtung von Pflanzenzellen zum Zwecke der Gewinnung von Naturstoffen. In: Pflanzliche Zellkulturen, BMFT-Statusseminar Jülich 1985, Bundesministerium für Forschung und Technologie, Bonn:35-43

GROWTH AND SUBSTRATE UTILIZATION OF CELL SUSPENSION CULTURES OF *CUCUMUS SATIVUS* ON MILK WHEY BASED MEDIA

Alfons Callebaut, Jean-Claude Motte, Michel Hoenig, Willy De Cat and Hervé Baeten

Instituut voor Scheikundig Onderzoek, Museumlaan 5, B-1980 Tervuren, Belgium

INTRODUCTION

Sucrose and glucose are the most effective carbon sources in supporting high growth rates and biomass yields in plant cell cultures. Media based on waste material such as milk whey and molasses have been tested also (Fowler and Stepan-Sarkissian 1985), but no growth data in these media have been reported (Berlin and Sasse 1985; Fowler 1982).

A prerequisite for using milk whey (a by-product of the cheese-making industry) in plant cell culture is the ability to metabolize lactose. Lactose-utilizing plant cells have been cultured (Hess et al. 1979; Chaubet et al. 1981), although the biomass yield and growth rate were considerably lower than in media based on glucose or fructose (Fowler 1982). Recently, we reported that a non-embryogenic suspension culture of cucumber grows equally well on lactose as on sucrose (Callebaut et al. 1987). In this work we report on the growth of cucumber cells in media based on milk whey permeate.

MEDIA COMPOSITION

Suspension cultures of cucumber are routinely subcultured on Murashige-Skoog (MS) medium as described in Gross et al. (1981) and Callebaut et al. (1987). MS-L medium has the same composition, with lactose (30 g l^{-1}) replacing sucrose. The milk whey medium is based on milk whey permeate (MWP), which is obtained from milk whey after recovering the proteins by ultrafiltration. MW-medium is made as follows: 23.5 g MWP (Lacprodan permeat, Danmark); micro-elements of MS-medium (KC Biological, Kansas, USA); 100 ml of a stock solution containing 500 mg l^{-1} $CaCl_2$,

NATO ASI Series, Vol. H18
Plant Cell Biotechnology. Edited by M. S. S. Pais et al.

16.5 g l^{-1} NH_4NO_3 and 19 g l^{-1} KNO_3; vitamins and hormones as in MS medium. The MW medium is autoclaved and the precipitated proteins are removed by centrifugation, whereafter the medium is filter-sterilized. This basal MW medium is modified by adding sugars (MW-SU, MW-Glu, etc.) before autoclaving or by doubling the MWP amounts (MW-D).

GROWTH OF CUCUMBER CELL CULTURES ON MS, MS-L AND MW MEDIA

Sugar utilization of cucumber cells in MS medium is shown in Callebaut et al. (1987). Sugar utilization in MS-L medium is shown in Fig. 1, together with the growth curve. In both media maximum fresh and dry weight, as well as the growth rate are comparable, at least when, upon subculture, the dilution factor in MS-L media is not too low. Upon dilution factors of 1:10 the lag period, during which there is no lactose utilization, increases. Figure 1 also shows that no lactose is metabolized in the MW medium; also, there is no growth and the cells are dying. Cation and anion analyses of the MW medium showed that every necessary macro- and micro-nutrient element is present.

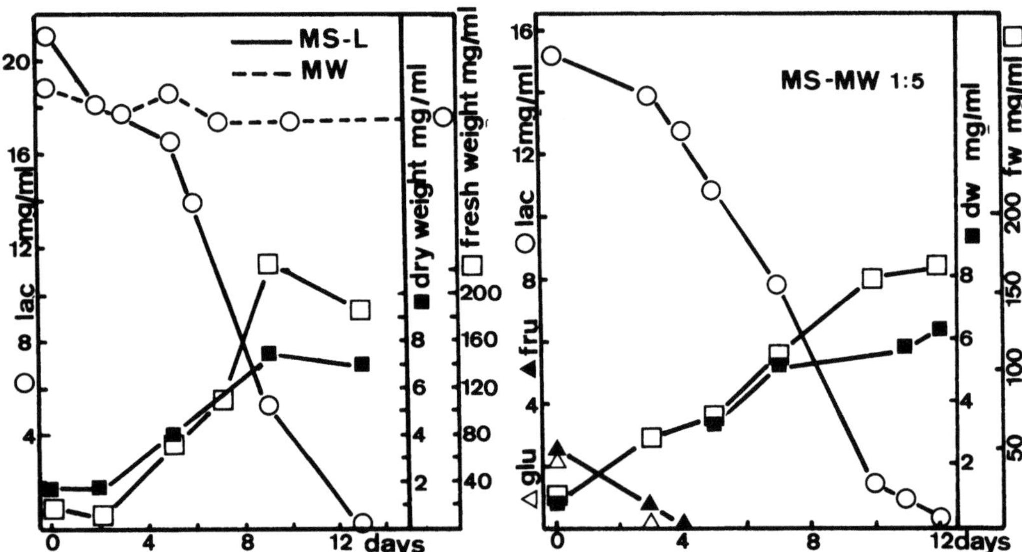

Fig. 1. Sugar utilization and growth in MS-L media (—) and in MW medium (---)

Fig. 2. Sugar utilization and growth in MS-MW (1:5) medium

Mixing MS and MW media results in normal growth and similar sugar utilization patterns as in MS or MS-L media (Fig. 2). The sucrose in the medium is dosed as its hexose components as described in Callebaut et al. (1987).

GROWTH OF CUCUMBER CELLS ON MW MEDIA SUPPLEMENTED WITH SUGARS

MW media supplemented with 24 g l^{-1} glucose, fructose or 12 g l^{-1} sucrose support growth of cucumber cells. Maximum fresh and dry weights are 50% increased compared with MS media. This can be explained by the higher sugar concentration (± 50%) and the high amounts of PO_4^- in milk whey permeate. The cells also grow in an MW-D medium (47 g l^{-1} MWP) supplemented with 12 g l^{-1} sucrose (Fig. 3). Preliminary experiments show that the concentration of sucrose can be lowered to 4 g l^{-1} or even lower without growth reduction.

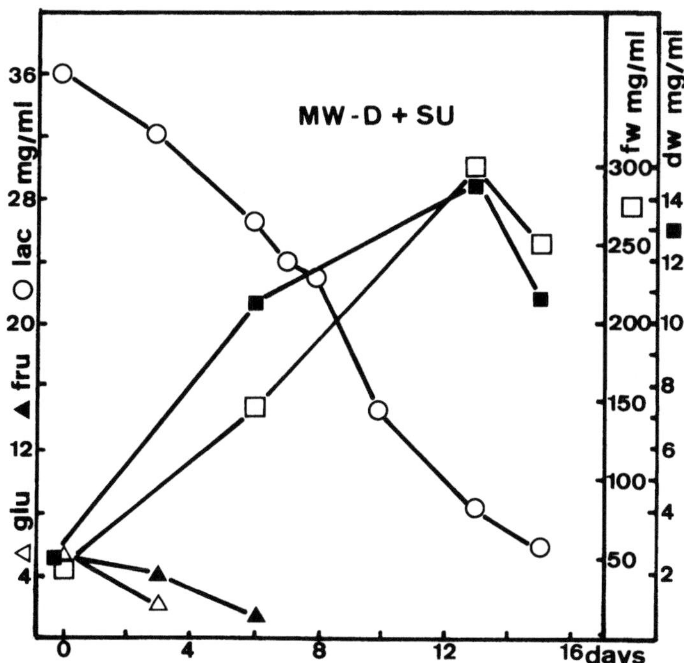

Fig. 3. Sugar utilization and growth in MW-D medium supplemented with 12 g l^{-1} sucrose

Glucose is always preferentially metabolized in media with several sugars, as observed before (Callebaut et al. 1987). In media with fructose and lactose, there is less preference for the hexose. Our results also support an uptake of the intact lactose molecule, in contrast to sucrose which seems to be hydrolyzed extracellularly.

Supplementing the MW medium with 24 g l^{-1} galactose also induces growth, but the growth rate is decreased. The lactose is metabolized after depletion of the medium in galactose; the lactose utilization rate is lower. Adding lactose to an MW medium does not induce growth or sugar utilization.

β-Galactosidase activity, assayed with p-nitrophenyl-β-D-galactopyranoside as substrate, was comparable for the medium, cell homogenates and intact cells of cultures grown in MS, MS-L, MS-MW and MW media. Previous work showed that the enzyme activity could not be correlated with growth on lactose media (Callebaut et al. 1987).

CONCLUSIONS

Our results show that cucumber cells can be grown in suspension culture in a medium based on milk whey permeate, a cheap waste product. Next to milk whey permeate, the medium has to be supplemented with an inorganic N-source and a second carbon source. The most effective carbon sources are sucrose, glucose, fructose and to a lesser extent also galactose. Sucrose is effective in concentrations of 0.4% or even lower. Lactose is completely ineffective. Although we supplemented the MW medium with the microelements and the vitamins of the MS medium, our analyses and literature data (Moulin and Garzy 1984) suggest that these can at least partially be omitted.

Our experiments suggest that the lack of growth in a medium based only on milk whey as carbon source is due to the impossibility of utilizing lactose in this medium. The second carbon source seems to induce the lactose uptake and/or the lactose hydrolyzing capacity of the cells.

REFERENCES

Berlin J and Sasse F (1985) Selection and screening techniques for plant cell cultures. In: Fiechter A (ed). Advances in biochemical engineering/biotechnology 31: plant cell culture. Springer, Berlin Heidelberg New York, p. 99-132

Callebaut A, Motte J-C and De Cat W (1987) Substrate utilization by embryogenic and non-embryogenic cell suspension cultures of Cucumis sativus L. J Plant Physiol 127:271-280

Chaubet N, Pétiard V and Pareilleux A (1981) β-Galactosidases of suspension-cultured Medicago sativa cells growing on lactose. Plant Science Letters 22:369-378

Fowler MW (1982) Substrate utilisation by plant-cell cultures. J Chem Tech Biotechnol 32:338-346

Fowler MW and Stepan-Sarkissian G (1985) Carbohydrate source, biomass productivity and natural product yield in cell suspension cultures. In: Neumann K-H, Barz W and Reinhard E (eds). Primary and secondary metabolism of plant cell cultures. Springer, Berlin Heidelberg New York Tokyo, p.66-73

Gross KC, Pharr DM and Locy RD (1981) Growth of callus initiated from cucumber hypocotyls on galactose and galactose-containing oligosaccharides. Plant Science Letters 20:333-341

Hess D, Leipoldt G and Illg RD (1979) Investigation on the lactose induction of β-galactosidase activity in callus tissue cultures of Nemesia strumosa and Petunia hybrida. Z Pflanzenphysiol 94:45-53

Moulin G and Galzy P (1984) Whey, a potential substrate for biotechnology. In: Russell GE. Biotechnology and genetic engineering reviews vol 1. Intercept, Newcastle upon Tyne, p. 347-374

CHANGES IN ANION AND SUGAR CONTENT IN LIQUID MEDIA DURING *IN VITRO* CULTURE OF CELLS FROM CEREALS, SUGARCANE AND TOBACCO

Ulrich Schmitz and Horst Lörz

Max-Planck-Institut für Züchtungsforschung, Egelspfad,
D-5000 Köln 30, FRG

INTRODUCTION

Our aim is the improvement of cereal cell cultures with regard to plant regeneration from cultured cells and protoplasts. To achieve this objecitve we modify the composition of media, baded on investigations of uptake and release of components in cell suspension cultures. This work was started with assays of sugars (glucose, fructose and sucrose) and anions (chloride, nitrate, phosphate and sulfate) in graminaceous and tobacco cell suspension cultures.

Abbreviations: MS=Murashige/Skoog medium (Murashige and Skoog 1962); CC=CC medium (Potrykus et al. 1979); N6=N6 medium (Chu et al. 1975); CS=CS medium (Lörz et al. 1983).

I) INVESTIGATIONS OF SUGAR CONSUMPTION

Cell suspensions, cultured under standard conditions, were pooled in one vessel. After shaking, the medium was removed and 5 g of the cells were incubated in each of five Erlenmeyer flasks, together with 50 ml of fresh medium (for sugar content of fresh media see Table 1). Every 24 h these cultures were assayed for cell growth and sugar content in the medium (Fig. 1). Under sterile conditions the medium was removed from the cells and the volume of medium as well as the cell mass remaining in the Erlenmayer flask were measured. Additionally, 10 µl of conditioned medium was removed for enzymatic assays of fructose, glucose and sucrose. The remaining medium was returned to the cells into the Erlenmeyer flasks and the suspensions were cultured for the next 24 hours.

NATO ASI Series, Vol. H18
Plant Cell Biotechnology. Edited by M.S.S. Pais et al.

Table 1. Sugar content of fresh media. Sugars were assayed enzymatically in coconut water. These data led to the calculation of sugar content in rice (CC) medium (10% coconut water).

(mol m^{-3})	Tobacco (MS)	Sugarcane (CS)	Rice (CC)	Coconut water
Fructose	0	0	1.4	14.4
Glucose	0	0	0.8	8.2
Sucrose	87.6	87.6	70.1	118.2

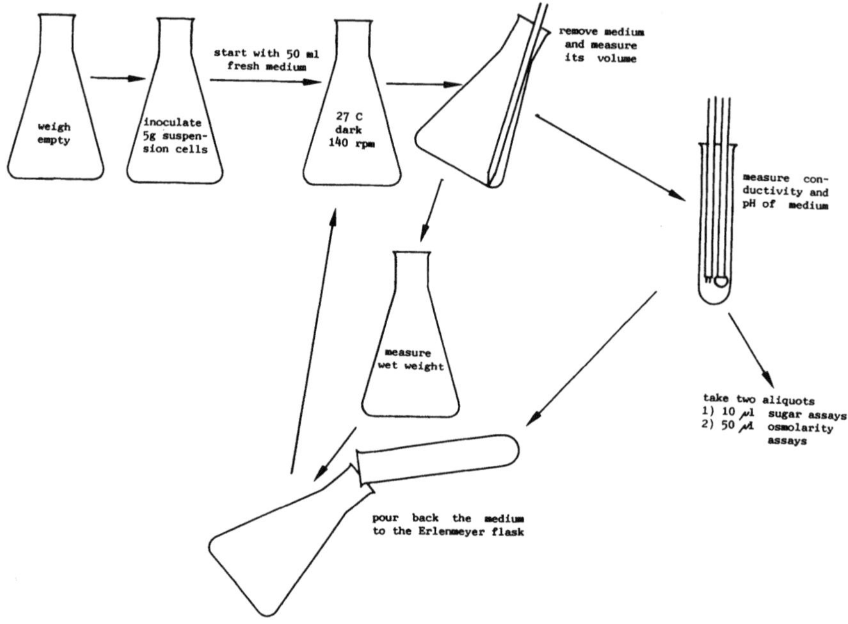

Fig. 1. Scheme for assaying sugars in suspension cultures. This cycle was repeated every 24 h. Sugars were tested with a Boehringer enzyme kit.

In all suspension cultures investigated, we observed a decrease in sucrose concentration in the first days of culture, and glucose and fructose appeared in the medium during this culture period. These results lead to the conclusion that sucrose is hydrolyzed in the medium as has been shown previously for carrot suspensions (Kanabus et al. 1986), pea stem protoplasts (Singh and Maclachlan 1986), corn roots (Giaquinta et al. 1983) and sugarcane suspensions (Komor et al. 1981). In tobacco suspension cultures, at any time the concentration of fructose is higher than the concentration of glucose (Fig. 2). In all graminaceous

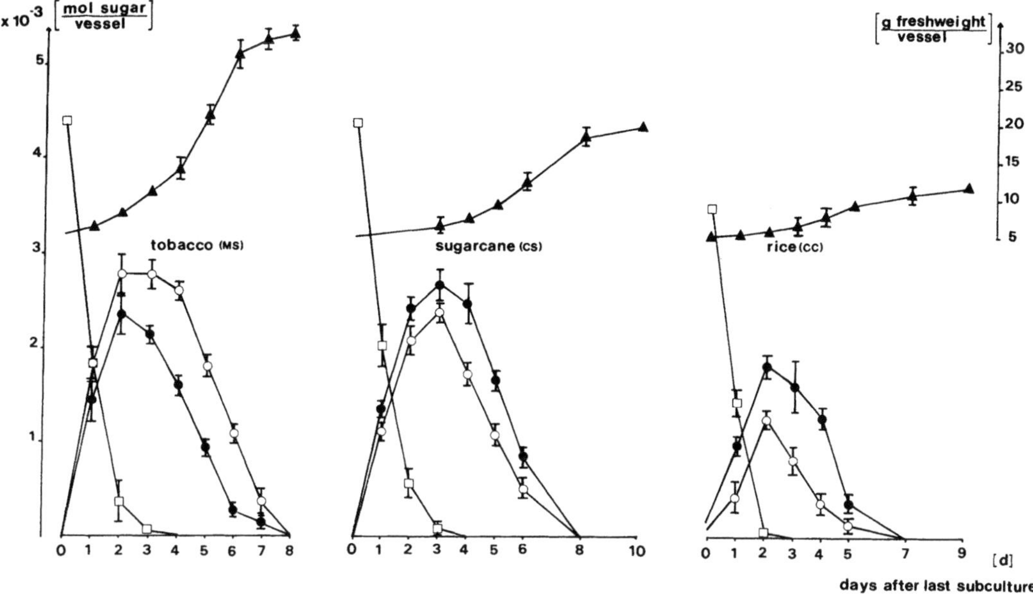

Fig. 2. Sugar content of conditioned media of sugarcane (CS), rice (CC) and tobacco (MS). ▲: wet weight of cells; □: sucrose; o: fructose; ●: glucose

cultures tested, the situation is reversed. Since carrot suspensions (Kanabus et al. 1986) showed the same behavior as tobacco suspensions, the difference between graminaceous and dicotyledoneous suspension cultures may be general.

II) INVESTIGATIONS OF ANION CONSUMPTION

Table 2. Anion content of fresh media. All media contained basal anions, but rice (CC) and maize (MS) media were supplemented with 10% coconut water. Anion content of coconut water was measured by HPLC.

($mol\ m^{-3}$)	Tobacco (MS)	Maize (MS)	Rice (CC)	Rice(N6)	Coconut water
Chloride	6	11.42	13.42	2.2	54.2
Nitrate	39.4	39.4	20	28	0
Phosphate	1.2	1.92	1.72	2.9	7.2
Sulfate	1.6	3.74	3.24	4.4	21.2

Cell suspensions, cultured under standard conditions, were pooled in one vessel. After shaking, the medium was removed and suspension cultures were started from each pooled suspension (for anion

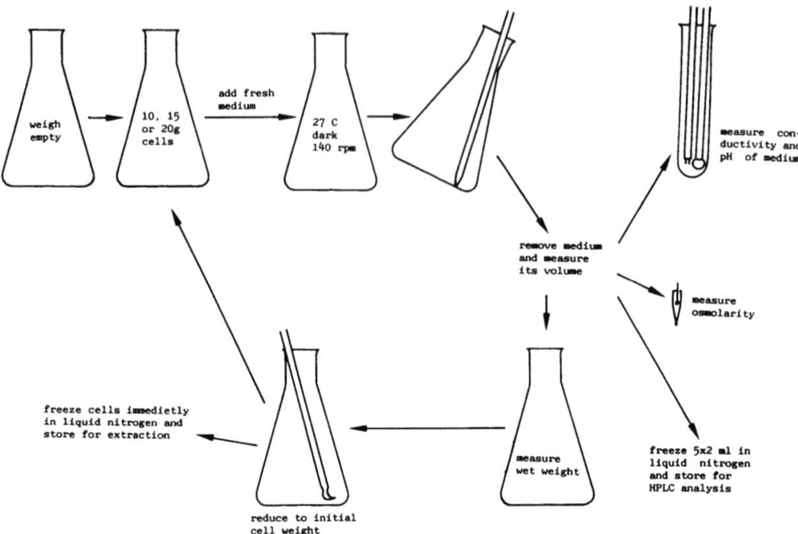

Fig. 3. Subculture cycle for anion analysis. Subcultures were made at a defined time during the day for each suspension culture.

content of fresh media see Table 2). For maize (MS) cultures, suspensions were initiated with 10, 15 and 20 g fresh weight of cells, with two cultures being started at each initiation weight. For tobacco (MS), rice (N6) and rice (CC) cultures, two suspensions were started with 20 g of cells, one suspension with 15 g and one with 10 g of cells. Tobacco (MS) and maize (MS) suspensions were subcultured in a 2 day/2 day/3 day cycle, while both rice lines were subcultured in a 3 day/4 day cycle. During subculture (see Fig. 3) osmolarity, conductivity, pH, fresh weight of cells and volume of the medium were measured. In addition, samples were taken for high performance liquid chromatography (HPLC) analysis. To date the inorganic anion content in conditioned media has been investigated. Rice (N6) and rice (CC) suspension cultures take up all phosphate of the fresh medium within 3 days of culture. More sulfate is taken up in rice (CC) than in rice (N6) suspensions, although the original sulfate concentration in N6 medium is higher. In both suspension cultures nitrate uptake is limited by the initial nitrate concentration of the fresh medium, when subcultured at exactly the same time on each day of subculture (Fig. 4). A similar pattern of limitation was seen in chloride uptake in rice (N6) cultures, while rice

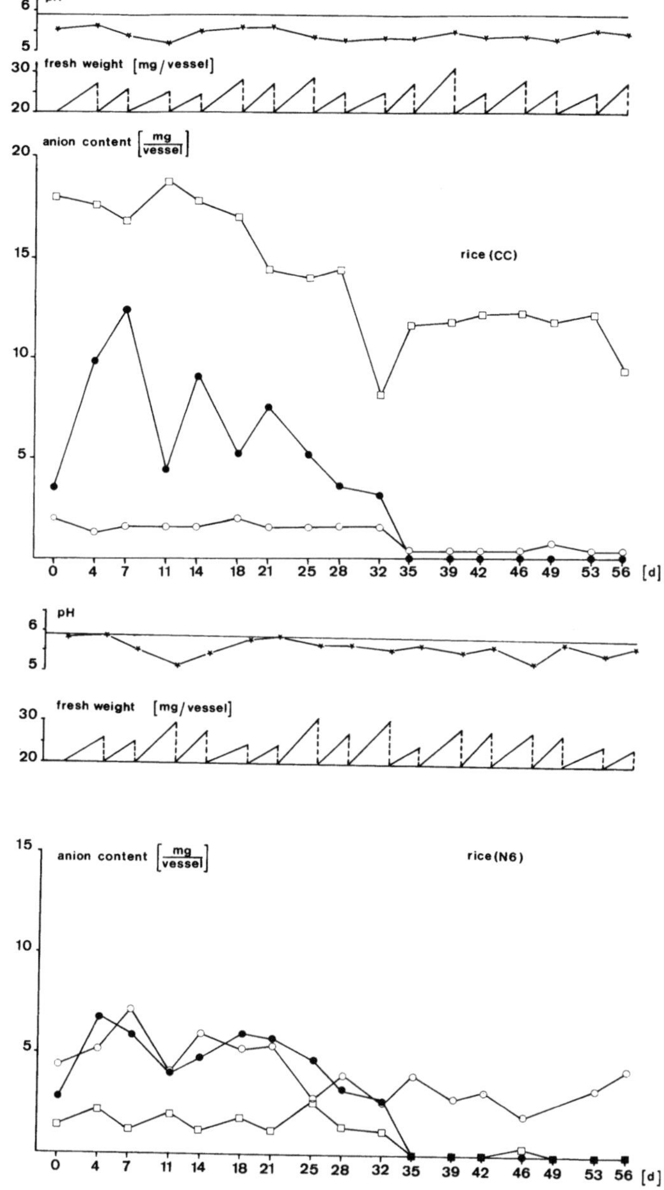

Fig. 4. Anion content, cell growth and pH of rice (CC) and rice (N6) suspension cultures. Wet weight of cells; * pH; □ chloride; ● nitrate; o sulfate

(CC) suspensions took up more chloride, probably because the chloride concentration is higher in the CC medium.

Subculturing according to a precise schedule had two effects. First, more anions were taken up from the medium than in cultures with more random subculture intervals. Second, following a precise subculture regime it was possible to isolate dividing protoplasts from a rice (CC) suspension from which previously only non-dividing protoplasts were obtained.

Acknowledgements. This work was supported by BMFT Project BCT 0365 2-2/3. Suspension cultures were kindly provided by E. Göbel (rice), B. Junker (maize) and D. Barfield (tobacco).

REFERENCES

Chu CC, Wang CC, Sun CS, Hsu C, Yin KC, Chu CY and Bi FY (1975) Establishment of an efficient medium for anther culture of rice through comparative experiments on the nitrogen sources. Sci Sinica 18:659-668

Giaquinta RT, Lin W, Sadler NL and Franceschi VR (1983) Pathway of phloem unloading of sucrose in corn roots. Plant Physiol 72:362-367

Kanabus J, Bressan RA and Carpita NC (1986) Carbon assimilation in carrot cells in liquid culture. Plant Physiol 82:363-368

Komor E, Thom M and Maretzki A (1981) The mechanism of sugar uptake by sugarcane suspension cells. Planta 15:181-192

Lörz H, Larkin PJ, Thomson J and Scowcroft WR (1983) Improved protoplast culture and agarose media. Plant Cell Tissue Organ Culture 2:217-226

Murashige T and Skoog F (1962) A revised medium for rapid growth and bioassays with tobacco tissue cultures. Physiol Plant 15:473-497

Potrykus I, Harms CT, and Lörz H (1979) Callus formation from cell culture protoplasts of corn (Zea mays L.). Theor Appl Genet 54:209-214

Singh R and Maclachlan G (1986) Differential rates of uptake of fructose, glucose and sucrose by pea stem protoplasts. J Exp Bot 37:1164-1169

CONDITIONING OF MEDIA: AN ELABORATE METHOD OF OPTIMIZING INITIAL GROWTH HORMONE CONCENTRATION

R. Wijnsma[1], R. Verpoorte[1], P.A.A. Harkes[2], F. van Iren[2] and H.J.G. ten Hoopen[3]

Biotechnology Delft Leiden (BDL)

[1]Center for Bio-Pharmaceutical Sciences, Dept. Pharmacognosy, University of Leiden, P.O. Box 9502, 2300 RA Leiden, The Netherlands

[2]Department of Plant Molecular Biology, University of Leiden, The Netherlands

[3]Department of Biochemical Engineering, Delft University of Technology, The Netherlands

INTRODUCTION

The conditioning of media is a well-known method for growing plant cells at low density (Street 1977). Stuart and Street (1969) studied the conditioning effect in more detail. They were able to prove a conditioning effect for Acer cell cultures. The critical initial cell density of such cultures could be lowered by a factor of at least 10 by using a conditioned medium. It was also attempted to define the chemical basis of the conditioning effect (Stuart and Street 1971). From experiments measuring the cell numbers after 4 weeks of culture of different initial cell densities on various (conditioned) media, the authors concluded that the carbon course, the vitamins and other growth factors as well as the growth hormones (2,4-D and kinetin) were probably not involved in the conditioning effect. Analysis of a conditioned medium showed the presence of a series of amino acids in small amounts. Addition of these compounds to a non-conditioned medium indeed resulted in a lowering of the critical initial cell density, however, not to the same level as a conditioned medium. Furthermore, a volatile factor was proven ro be involved in the conditioning effect.

Although the authors were able to improve the synthetic medium, it never completely complied with a conditioned medium. In fact, we have to conclude that still little is known about the conditioning effect. To our knowledge no extensive study on

NATO ASI Series, Vol. H18
Plant Cell Biotechnology. Edited by M. S. S. Pais et al.

this phenomenon using various different cell lines has been published.

Further knowledge of the conditioning effect is of interest for improving growth of plant cells at low density (single cells), protoplast regeneration, recovery of cells after cryopreservation and last, but not least, for industrial scale plant cell culture.

A better understanding of the conditioning effect could contribute to a reduction of fermentation time by shortening the lag phase. This prompted us to study the factors involved in the conditioning effect.

To measure a conditioning effect we have chosen to follow the growth directly the first days after subculturing, the duration of the lag phase should be related to the conditioning effect. The shorter the lag phase, the more optimal are the growth conditions for the cells at the relatively low density after subculturing. A further argument for using the lag phase as a parameter is the possibility of translating the results to an industrial process.

MATERIAL AND METHODS

Plant material

A Cinchona ledgeriana Moens cell line was used (Ci R1, 1 l). It was grown on Gamborg B5 medium, containing 1 mg l^{-1} 2.4-D and 0.2 mg l^{-1} kinetin and 4% sugar. Subculturing was every 7th day by adding 350 ml of the culture to 1000 ml of fresh medium. The cultures were grown in the light (Philips TL F33, 1500 lx) on gyrotary shakers (140 rpm) at a temperature of 28°C.

Conditioning of the media

350 ml of a 7-day-old stock-culture was added to 1 liter of B5 medium. After 24 h the cells were filtered off with a G3-glass filter (16-40 µm), all under aseptic conditions using sterile glassware.

Experiments for determining the conditioning effect

The conditioned medium was divided over 250 ml flasks, in each flask 50 ml. Then 10.0 ml of a 7-day-old stock-culture was added to each flask. The growth conditions were as above. Growth was determined by collection of the cell material by filtering and subsequent drying for 24 h at 75°C. All experiments were performed in duplicate, bars in Figs. 2 and 3 represent the two duplicate values.

RESULTS AND DISCUSSION

The first experiments were to determine whether a conditioning of the medium results in a shorter lag phase. As can be seen from Fig. 1 indeed a more rapid increase of the biomass yield (dry weight) is observed in the first 4-5 days in the case of a conditioned medium. Conditioning for 1-2 days proved to be more effective than 3-4 days. In all cases after about 6 days at a biomass concentration of about 2.5 g l^{-1}, the growth in the non-conditioned medium surpasses that in the conditioned medium. The final yields in the controls being about three times as high as that in the conditioned media. Analysis of the sugar contents by means of HPLC showed that during 1 day of conditioning, the total sugar level did not change, i.e. there was no sugar limitation on the growth. Also the uptake of phosphate in this cell line is very slow, excluding this as a possible growth-limiting factor. In a second experiment the heat-stability of the conditioning factor was determined. The conditioned medium was sterilized for 5 min at 120°C. Cells grown on this medium still showed a conditioning effect similar to that obtained with the non-heated conditioned medium.

Changing from sucrose to glucose as carbon source did not influence the conditioning effect.

Considering these results two possibilities can be envisaged for explaining the conditioning effect:

1. Excretion of (a) compound(s) from the cells which promote growth;
2. Uptake (metabolism) of (a) compound(s) from the medium by cells which result in an optimized medium for a subsequent inoculum.

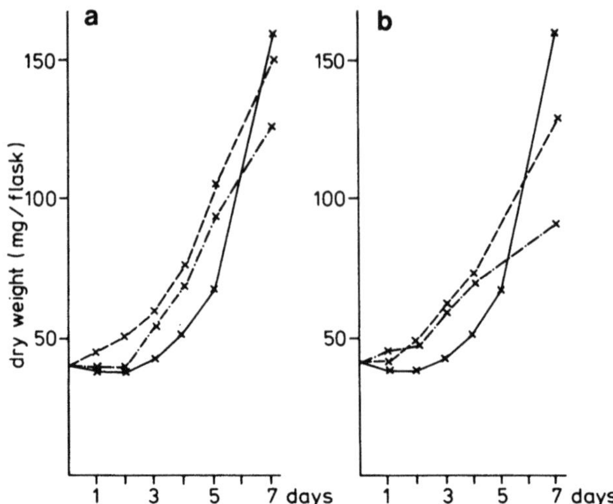

Fig. 1. Dry weight yield of Cinchona ledgeriana cells cultures in media conditioned for 1, 2, 3 or 4 days
a ——: control; — —: 1 day; — . —: 3 days. b ——: control; — —: 2 days; — . —: 4 days

Our further efforts were mainly concerned with the second option. In particular we studied the vitamins and growth hormones as compounds possibly connected with the conditioning effect.

Media without the vitamins and/or growth hormones were conditioned and subsequently these compounds were added in the usual concentrations. In these experiments only the media without the vitamins showed a shortening of the lag phase, i.e. normal vitamin concentrations do not affect the conditioning. A normal growth hormone concentration depressed the conditioning effect. To further prove the involvement of growth hormones in the shortening of the lag phase, the auxin (2,4-D) and/or the cytokinin (kinetin) were left out of the medium. After conditioning the growth was determined. All three types of medium gave similar results, i.e. a shortening of the lag phase (Fig. 2). Apparently the endogenous growth hormones are sufficient for keeping the cells dividing for a few days, however, after about a fourfold increase of the biomass the growth will level off, both in a hormone-free medium, with or without conditioning (e. g. Fig. 3). Therefore, it was concluded that the hormones play an important role in the duration of the lag phase after subculturing. Other experiments with Tabernaemontana cell cultures,

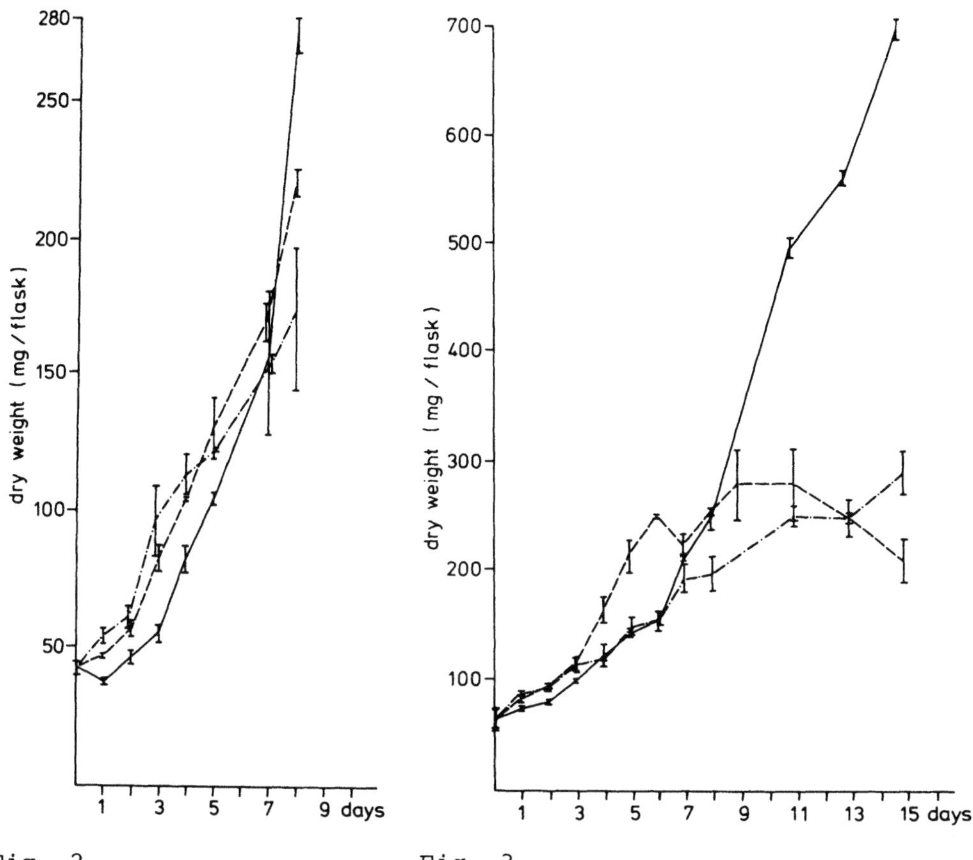

Fig. 2 Fig. 3

Fig. 2. Dry weight yield of Cinchona ledgeriana cell cultures in hormone-free media with or without conditioning: —— control; — — medium without kinetin and 2.4-D; —.— conditioned medium without kinetin and 2.4-D

Fig. 3. Growth of Cinchona ledgeriana cells on medium with 10% of the normal hormone concentration: —— control; — — medium with 10% of the normal hormone concentration; —.— conditioned medium

using the method as reported by Strauss and King (1981, 1984), proved that in fact a concentration of 2.4-D of 1 mg l^{-1} is very close to toxic levels (3 mg l^{-1}) (Van der Heijden et al. unpublished results). To determine the influence of the hormone concentrations on the growth, growth curves were recorded for cell cultures grown on media containing 10, 25 or 50% of the usual growth hormone concentrations (Fig. 3, Table 1).

Apparently the hormones become growth-limiting factors. However, the onset of growth was much faster at the lower growth

Table 1. Growth a *Cinchona ledgeriana* on media with various hormone concentrations

Hormone concentrations	Exponential phase (days after subculturing)	Final biomass yield per flask (dry weight in mg, mean of two)
100 %	6 - 11	750
50 %	5 - 8	710
25 %	3 - 5	400
10 %	2 - 4	275
100 % conditioned	2 - 4	280
0 % conditioned	2 - 4	280

hormone concentrations, e.g. the 10% hormone medium had its exponential growth phase from 2-4 days after inoculation, whereas the control had it from 6-11 days after inoculation. In fact, media with a lower hormone concentration result in a shortening of the lag phase similar to a conditioned medium (Fig. 3).

Recent experiments with *Tabernaemontana* cell cultures showed a clear relationship between the duration of the lag phase and the initial concentration of biomass, also leading to the conclusion that for optimal growth there should be a certain relation between the amounts of certain media constituents and the amount of biomass (Schripsema et al. unpublished results). Preliminary experiments with the recovery of tobacco cells after cryopreservation using hormone- freee media were successful (Van Iren et al. unpublished results).

CONCLUSIONS

For *Cinchona* cell cultures media conditioned for 1-2 day result in a shorter lag phase but also in a reduced biomass yield. Similar results can be obtained by lowering the hormone concentration in the media. Apparently for optimal growth the amounts of hormone in a medium should not be related to the volume but to the amount of biomass present.

REFERENCES

1. Strauss A and King PJ (1981) Physiol Plant 5:123
2. Strauss A and King PJ (1984) Plant cell, Tissue and Organ Culture 3:111
3. Street HE (ed) Plant Tissue and cell culture, Botanical Monographs volume 11. Blackwell Scientific Publications 1977 Oxford
4. Stuart R and Street HE (1969) J. Exp. Bot. 20:556
5. Stuart R and Street HE (1971) J. Exp. Bot. 22:96

PAPAIN INHIBITION BY NEW α-METHYLENE-γ-LACTONE CARBOHYDRATE DERIVATIVES

Amélia P. Rauter[1], Maria S. Pais[1], Christina Duarte[1], Ana Ana Eusébio[1], Lîgia Pinto[1], Christina Simões[1] and Joaquim S. Cabral[2]

INTRODUCTION

Naturally occurring sesquiterpenes with α-methylene-γ-lactone units in its structure are known to exhibit a diversity of biological activities, namely cytostatic and anti-helmintic, and to act as plant growth regulators (Gross 1975; Hoffmann and Rube 1985). The mechanism of its action consists of an irreversible inhibition of sulfhydryl enzymes by a Michael-type reaction, in which the electrophilic methylene group is added to the sulfhydryl group of the enzyme.

α-Methylene-γ-lactone carbohydrate derivatives are a class of compounds which have not been extensively explored in relation to their biological activities. In order to correlate the structure and the stereochemistry with the biological activity, the new carbohydrate derivatives 1 to 5 (Fig. 1) (Rauter et al. 1987) were tested as inhibitors of papain, a plant sulfhydryl enzyme, using azocasein as substrate.

MATERIAL AND METHODS

The sources of the reagents used in these experiments were as follows:

The compounds 1 to 5 were synthesized in the laboratory according to the literature (Rauter et al. 1987); azocasein SIGMA, papain type I SIGMA, tetrahydrofuran p.a. MERCK and sodium phos-

[1]Departamento de Quîmica da Faculdade de Ciências de Lisboa, Universidade de Lisboa, Centro de Engenharia Biológica, INIC, Rua da Escola Politécnica, 1294 Lisboa Codex, Portugal

[2]Laboratório de Engenharia Bioquîmica, Instituto Superior Técnico da Universidade Técnica de Lisboa, Av. Rovisco Pais, 1000 Lisboa Codex, Portugal

NATO ASI Series, Vol. H18
Plant Cell Biotechnology. Edited by M.S.S. Pais et al.

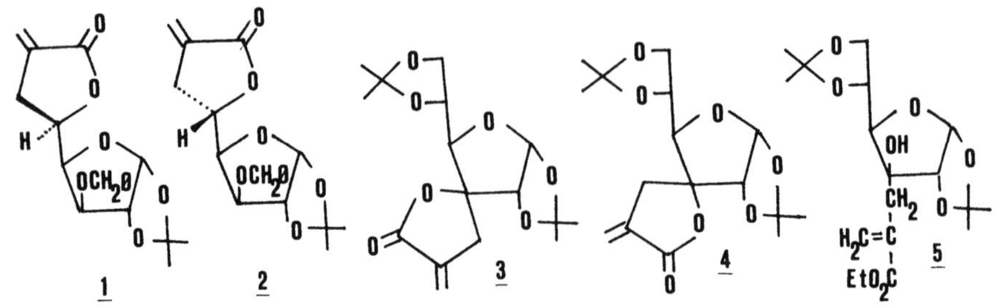

Fig. 1. Carbohydrate derivatives containing α,β-unsaturated carbonyl groups in their structure

phate, trichloroacetic acid and sodium hydroxide of reagent grade.

<u>Assays for determination of papain inhibition by compounds 1 to 5</u>

0.5 ml of a 1% (w/v) papain solution in 0.05 M sodium phosphate solution, pH = 7, containing a determined volume of 1 to 4 mg ml^{-1} of the inhibitor in tetrahydrofuran (final concentrations varying from 1 mg l^{-1} to 40 mg l^{-1}, ca. 10^{-6} M and 10^{-4} M, depending on the molecular weight of the compound used as inhibitor) were added to 0.5 ml of a 2% (w/v) solution of azocasein in the same buffer. Under the standard assay conditions, this solution and the control (prepared as above and containing the same volume of tetrahydrofuran without inhibitor) were shaken at 37°C for different periods of time up to 40 min.

Two other different assays were performed using final inhibitor concentrations of about 10^{-4} and 10^{-3} M and an incubation time of 15 and 30 min at 37°C, before the addition to the azocasein solution.

In all cases the reaction was stopped by adding 5 ml of a 5% trichloroacetic acid solution to each tube, to precipitate the azocasein not hydrolyzed by the enzyme. Each tube was kept at 37°C for 10 min. The unhydrolyzed azocasein was removed by centrifugation at 3000 rpm for 5 min. To 1 ml of the supernatant fraction was added 1 ml of a 2 N sodium hydroxide solution. The adsorbance of these final solutions was determined at 440 nm using a UV spectrophotometer Pye Unicam SP-6.

RESULTS AND DISCUSSION

The effect of the inhibitors on the proteolytic activity of papain was determined by measuring the initial reaction rates. These were given by the slope between the absorbance values and the reaction time.

The inhibition degree of papain was calculated by the following expression:

% inhibition = 100 $(1-a_1/a_0)$;

a_1 = slope of the line corresponding to the assay;

a_0 = slope of the line corresponding to the control.

The inhibition degrees for all the assays performed with compounds 1 to 5 are listed in Table 1. Figure 2 shows the variation of the inhibition degrees with concentration and incubation time for the five compounds tested.

A significant inhibition of papain activity was observed for all the compounds. The most active ones are the least sterically hindered lactones 1 and 2, at 10^{-3} M concentration and with a 30-min incubation time, causing 91.1% inhibition and complete deactivation of papain respectively. Under the same conditions, the spirolactones 3 and 4 originated inhibition degrees of 86.2 and 52.8% respectively. Compounds 1 and 2 are also inhibitors of papain activity at very low concentrations (3.0 x 10^{-6} M) without incubation time, when the other derivatives do not display any inhibition degree.

The difference between the inhibition degrees observed for the pair of diastereomers 3 and 4 is higher than that observed for 1 and 2. This means that the stereochemistry of the more sterically hindered lactones 3 and 4 is more relevant for their activity as inhibitors than the stereochemistry of 1 or 2.

Without incubation time, the most active compounds were 3 and 5. At 10^{-4} M concentration, inhibitions of 27.8 and 30.4% were observed respectively. Again the influence of the stereochemistry in the activity was expressed by the low inhibition values obtained for compound 4, under the same conditions (see Table 1).

Finally, it can be concluded that cyclization is not necessary for inhibition in the carbohydrate derivatives tested,

Table 1. Inhibition of papain by carbohydrates containing α,β-unsaturated carbonyl groups

Compound No.	Concentration (M)	Inhibition degree (%) Incubation time (minutes) 0	15	30
1	$3.0x10^{-6}$	6.0	-	-
	$1.4x10^{-5}$	12.8	-	-
	$1.2x10^{-4}$	17.4	35.6	39.2
	$1.2x10^{-3}$	-	84.0	91.1
2	$3.0x10^{-6}$	6.5	-	-
	$5.8x10^{-5}$	5.0	-	-
	$1.2x10^{-4}$	11.4	35.1	42.5
	$1.2x10^{-3}$	-	84.5	99.2
3	$1.2x10^{-5}$	3.4	-	-
	$6.1x10^{-5}$	7.2	-	-
	$1.1x10^{-4}$	27.8	-	-
	$1.2x10^{-3}$	-	74.7	86.2
4	$3.1x10^{-5}$	2.9	-	-
	$6.1x10^{-5}$	9.9	-	-
	$7.7x10^{-5}$	8.9	-	-
	$1.2x10^{-4}$	-		41.0
	$1.2x10^{-3}$	-	56.6	52.8
5	$1.3x10^{-5}$	0.7	-	-
	$2.7x10^{-5}$	22.0	-	-
	$5.4x10^{-5}$	9.0	-	-
	$6.7x10^{-5}$	25.1	-	-
	$0.9x10^{-4}$	30.4	-	-
	$1.1x10^{-3}$	-	84.1	76.8

because the open-chain α,β-unsaturated ester 5 shows inhibition degrees comparable to those of the α-methylene-γ-lactone 3.

Acknowledgements. This work was partially supported by Junta Nacional de Investigação Científica e Tecnológia under research contract No. 834.86.194.
The authors thank also Prof. Dr. Fernando Fernandes of the De-

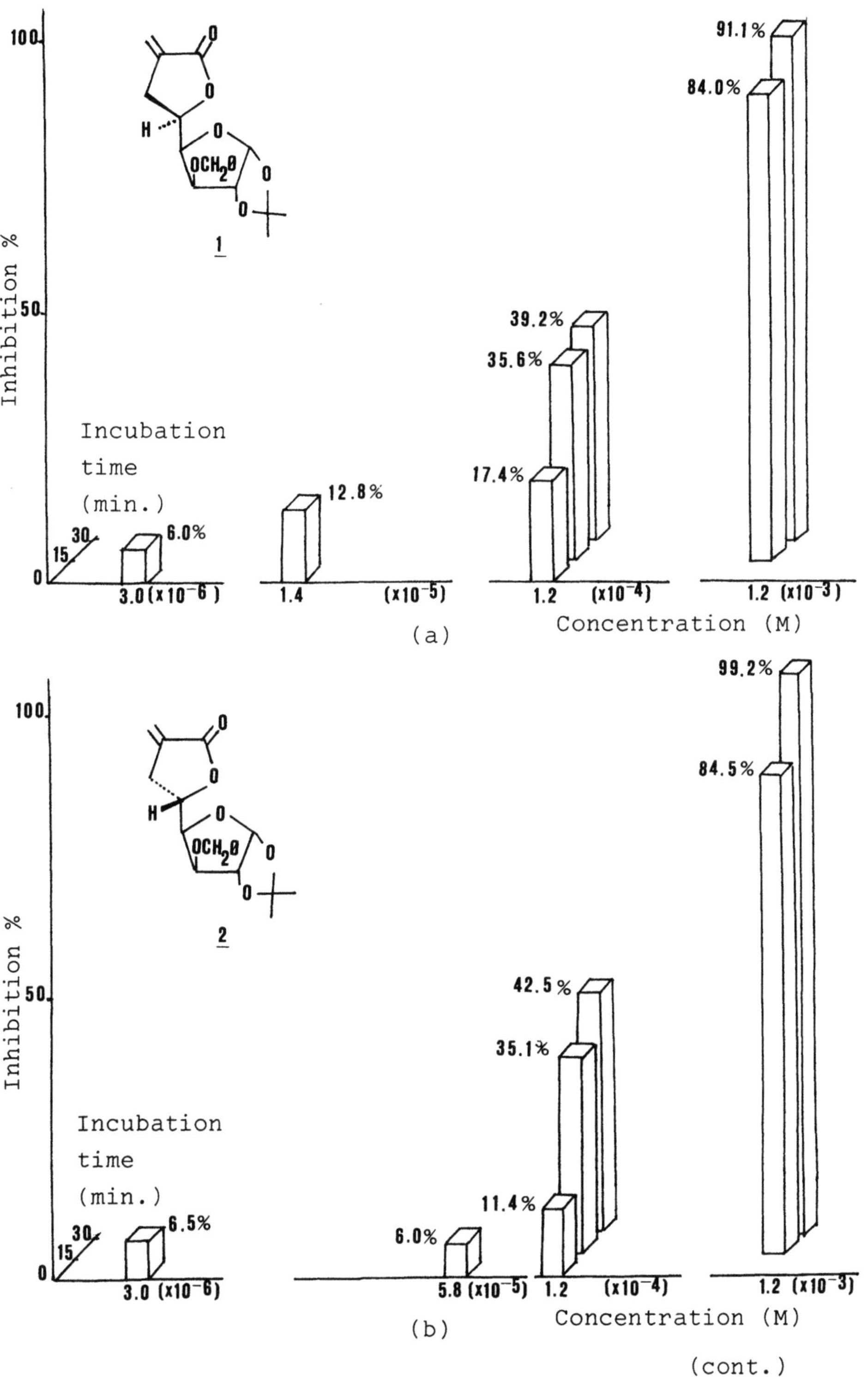

100
50
0
Inhibition %
1
Incubation time (min.)
15
30
6.0%
12.8%
17.4%
35.6%
39.2%
84.0%
91.1%
3.0 (x10⁻⁶)
1.4
(x10⁻⁵)
1.2
(x10⁻⁴)
1.2 (x10⁻³)
Concentration (M)
(a)
100
50
0
Inhibition %
2
Incubation time (min.)
15
30
6.5%
6.0%
11.4%
35.1%
42.5%
84.5%
99.2%
3.0 (x10⁻⁶)
5.8 (x10⁻⁵)
1.2
(x10⁻⁴)
1.2 (x10⁻³)
Concentration (M)
(b)

(cont.)

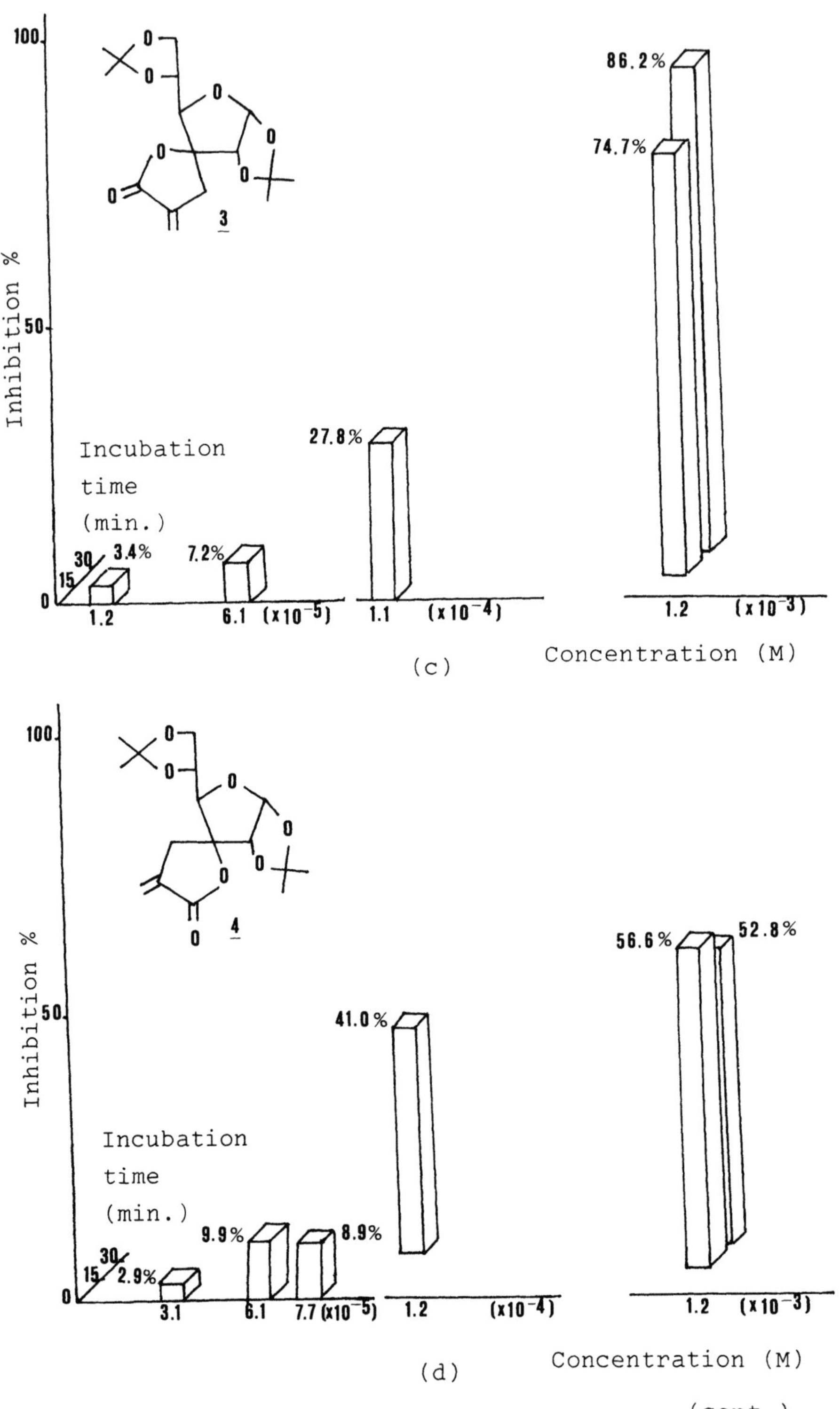
100
50
0
Inhibition %
3
Incubation
time
(min.)
30
15
3.4%
7.2%
27.8%
86.2%
74.7%
1.2
6.1 (x10^-5)
1.1 (x10^-4)
1.2 (x10^-3)
Concentration (M)
(c)
100
50
0
Inhibition %
4
Incubation
time
(min.)
30
15
2.9%
9.9%
8.9%
41.0%
56.6%
52.8%
3.1
6.1
7.7 (x10^-5)
1.2 (x10^-4)
1.2 (x10^-3)
Concentration (M)
(d)

(cont.)

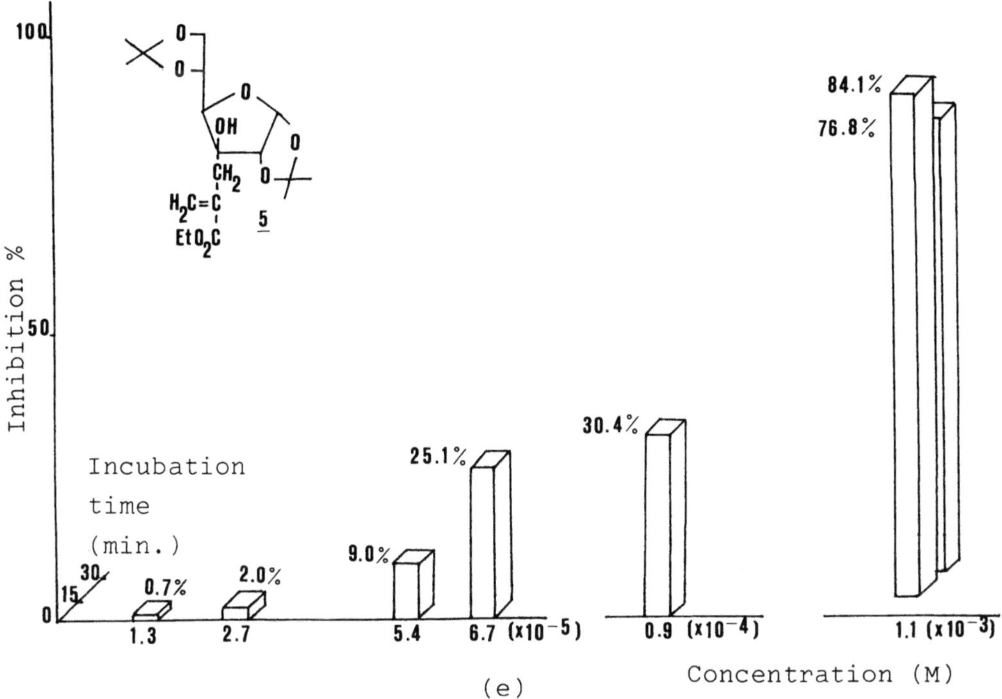

Fig. 2a-e. Effect of compounds 1 to 5 at concentrations in the range of 10^{-6} to 10^{-3}M and incubation time on proteolytic activity of papain

partamento de Quîmica da Faculdade de Ciências da Universidade de Lisboa for providing computer programs used in this work.

REFERENCES

Gross D (1975) Growth regulation substances of plant origin. Phytochem 14:2105-2112

Hoffmann HMR and Rube J (1985) Synthesis and biological activity of α-methylene-γ-butyrolactones. Angew Chem Int Ed Engl 24:99-110

Rauter AP, Figueiredo JA, Ismael I, Pais MS, Gonzalez AG, Diaz J and Barrera JB (1987) Synthesis of α-methylene-γ-lactones in furanosidic systems. J Carbohydr Chem 6:259-272

THE LARGE-SCALE CULTIVATION OF PLANT CELLS

A. Pareilleux

Département de Gênie Biochimique et Alimentaire, UA-CNRS 544,
Institut National des Sciences Appliquées, Avenue de Rangueil,
31077 Toulouse Cédex, France

INTRODUCTION

Higher plants produce a large variety of metabolites, some of which are potentially useful compounds of commercial interest, including pharmaceuticals (steroids, alkaloids, glycosides), flavor and fragrances, sweeteners, etc.; these compounds are mainly secondary metabolites (Berlin 1986; Nickell 1980).

In the last decade, industry has requested factory-type production of plant compounds because of difficulties in marketing field-grown plants linked to uncertainties due to climatic conditions and political disruptions. Therefore the alternative to producing useful compounds by cultured plant cells seems to be attractive and the only culture systems relevant at an industrial scale is the culture in large reactors. While the basic techniques of initiating plant cell cultures is easy, the cultivation of plant cells on a large scale must often be improved.

The main objective of this work is not to speculate on the crucial problem of the expression of secondary metabolism in plant cells or to estimate the actual potential of the cultures (Kurz and Constabel 1985), but rather to assemble some ideas on the present state of the art and to approach on a rational basis the cultivation of plant cells. In consequence, engineering aspects of large-scale cultivation will be introduced, to highlight some of the fundamental problems related to the practical application of plant cell culture technology. As a preamble, two points have yet to be underlined:

Firstly, most of the drawbacks in the field of plant cell culture applications are in fact due to a large gap in the

NATO ASI Series, Vol. H18
Plant Cell Biotechnology. Edited by M. S. S. Pais et al.

knowledge of the physiology, biochemistry (pathways and regulation) and genetics of secondary metabolite formation rather than in strategies and processes available for plant cell cultivation.

Secondly, to date, the only running industrial process of large-scale secondary metabolite production using plant cell cultures is the extraction of shikonin derivatives from *Lithospermum* cells by the firm Mitsui Petrochemicals Industries in Japan (Fujita et al. 1981a,b). However, a few other examples, including the production of berberine by *Thalictrum* cells (Nakagawa et al. 1984), the production of rosmarinic acid by *Coleus* cell suspension cultures (Ulbrich et al. 1985) or the biotransformation of β-methyldigitoxin in β-methyldigoxin by *Digitalis* cells (Alfermann et al. 1983), offer some promising perspectives. This list is not complete and can be extended to other products and cell lines (Anderson et al. 1985, Berlin 1986).

While there is no one-process strategy for all product targets and the ideal bioreactor suitable for large-scale cultivation does not yet exist, various process strategies and cell culture technologies are available (Fowler 1986). They are not basically different from the techniques now used for microbial systems, but they must take into consideration specific features of plant cells.

PROCESS STRATEGIES

Two major considerations are essential for the choice of a process strategy, i.e. a single- or two-stage process, and arise from the knowledge of the physiology (Table 1).

1. The pattern of the synthesis of the product. The relationship between the growth phase and product formation must be considered and the kinetics have to be clearly understood in connection with the influence of the medium conditions on growth and production.

2. The fate of the product following its synthesis, typically compartmentation and accumulation in the vacuole, or spontaneous or induced release into the medium. In the first case biomass has to be harvested from the culture broth, in the second, cell immobilization systems would be of great interest.

Table 1. Process strategies for plant cell cultures with a view to metabolite production (after Fowler 1986)

Kinetic data and physiology	Coupled growth phase and product synthesis. No or little differences in optimal conditions for growth and product formation. Growth medium; minor medium modification or growth limitation.	Separate growth and production phases. Optimal conditions for cell growth and product formation quite different. Production medium; major medium modification and/or severe nutrient limitation.	
Strategy	Single-stage process	Multi-stage (two stage) process	
		1st Stage Biomass production	2nd Stage Product formation
Mode of operation	Batch culture[a]	Batch culture	Batch culture[e]
	Semi-continuous[b]	Continuous/batch	Batch/immobilization[f]
	Continuous culture[c]	Batch/feed-batch	Batch culture[g]
	Feed-batch culture[d]	Continuous	Continuous[h]

	Reactor size	Feed flow	Holding time
One stream[i]	Similar	Similar or different	Similar or different
Two stream[j]	Similar or different	Similar or different	Similar or different

[a]Furuya et al (1984). [b]Dougall et al. (1983). [c]Dougall and Weyrauch (1980). [d]Schiel et al. (1984). [e]Fujita et al. (1982), Mitsui Petrochemical Industries. [f]Reinhard and Alfermann (1980). [g]Alfermann et al. (1983). [h]Alfermann et al. (1985). [i]Ulbrich et al. (1985). [j]Tal et al. (1983).

It can be noted that the productivity of the cell culture, defined as the amount of product per liter and per unit time, is the substantial parameter; then, enhanced productivity can be obtained by increasing the growth rate and the biomass concentration as well as by increasing the rate of product synthesis and its recovery.

When product formation and growth phase are coupled, a single-stage process will be operative. Various modes of operation can be used, batch, semi-continuous, continuous and feed-batch mode; the latter seems to be a useful technique for increasing the productivity of plant cell cultures provided the compound is already produced in the growth medium and the medium constituents affecting product formation have been analyzed. It was successfully applied to cynnamoyl putrescine biosynthesis in *Nicotiana tabacum* cultures by Schiel et al. (1984). Opposing a two-stage process may provide the best approach when the product synthesis is associated with low cell growth rates; severe medium modification and/or nutrient limitation are usually needed. The first stage is directed towards rapid growth and biomass accumulation and the second, which can be optimized for product formation, is the production run. According to the influence of the medium conditions on growth and product formation, a wide range of two-stage process strategies are available with various modes of operation including immobilized cell systems (see Table 1). For continuous mode we have to discern two-stage one stream from two-stage two stream systems; the residence time in each reactor can be similar but, using a different reactor size or feed flow, the dilution rates are generally different in the two reactors, the result being the increase in biomass or product productivity. The potential of these systems as an experimental tool to test the response of plant secondary metabolic pathways to various environmental parameters has to be mentioned. In addition, one interest of continuous cultures concerns the controlled use of secondary metabolite-acting substances, e.g. phytohormones or phosphates, to increase the production with the maintenance of appropriate specific growth rates (King 1976, 1977).

LARGE-SCALE CULTIVATION

Reactor design and scale-up

The specific features of plant cells govern to a large extent the most suitable technology for large-scale cultivation with special attention given to process variables of engineering interest (Table 2).

Table 2. Specific properties of plant cells and process variables of engineering interest

Specific features	Engineering	Interest	Information
Cell size: large Clump size and number Cell wall	Low shear resistance	Mixing	Shear stress Rheology (viscosity) Sedimentation
Oxygen demand: low (Order 1 µmol O_2 $h^{-1}.10^6$ $cell^{-1}$)	Growth rate dependence	Aeration (Interlinked with mixing)	Control of gas exchanges (oxygen transfer, venting rate)
Doubling time: long (order 20-100 h) Complex nutrient requirements	Sensitivity to infection	Sterility	Foaming Surface adhesion

The main requirement of any mass culture system is the need for adequate mixing and efficient mass transfer of nutrients from the liquid phase to the cells with the lowest power input and reduced shear stress, in fact:

1. To achieve a homogeneous culture and cell environment;
2. To promote dispersion of air bubbles for effective oxgenation.

Generally, three different reactor types are considered, with various mixing systems:

1. Those whose mixing is achieved by admission of air, bubble column, airlift reactor (internal recycling), loop reactor (external recycling).

2. Those whose mixing is governed by a rotative impeller and turbine system of various devices (blade, disc, paddle, propeller), which are stirred tank reactors; conventional ratios of geometrical dimensions are normally observed.

3. Mixt reactors which are a combination of the two previous types, draft tube reactor with or without Kaplan turbine for example, etc.

Air is admitted through a number of sparger devices (nozzles, sinters and ceramics with different porosities, annuli with perforations, etc.). Since 1950 a wide variety of shapes and sizes of reactors have been used for plant cell cultivation; for further details see the reviews of Martin (1980) and Fowler (1982). They are not fundamentally different from microbial fermenters and only few special apparatus have yet been developed. For example, Tanaka et al. (1983) tested a rotating drum fermenter and concluded that it was superior to a mechanically agitated one in supplying oxygen under high viscosity and low hydrodynamic stress conditions. It can also be noted that many of the designs, while adequate for laboratory scale, are not suitable for scale-up.

As bioreactors should be evaluated by considering their main technological factors, capacity of oxygen supply, intensity of hydrodynamic stress effects and intensity of mixing and air bubble dispersion, a number of studies have been concerned with the effect of shear on plant cells, the rheological properties of culture broths (viscisity, specific gravity, concentration and size of aggregates) and the fluid regime in fermenters (Kato et al. 1978, Tanaka 1982, Vogelmann 1981, Wagner and Vogelmann 1977). The authors observed that broths of plant cells could be relatively viscous, up to 100 cp at high cell density of about 20 g dry wt. l^{-1}; the culture broth exhibited a pseudoplastic behaviour. The influence of rheological properties on mass transfer coefficients have been studied by Tanaka (1981) in various culture devices. The volumetric oxygen transfer rate k_la was the highest in the airlift reactor at a given hydrodynamic stress intensity, whereas by increasing the cell mass concentration the best k_la values were found in stirred reactors compared to a bubble column or an air-driven reactor. A marked decrease in k_la values with cell growth was mentioned as the result of the increase in viscosity (Kato et al. 1975, Kato et al. 1978, Tanaka 1982). Respective attributes and drawbacks of air-driven, impeller-driven or hybrid reactors have been discussed by several authors (Fowler 1982, Martin 1980, Spier and Fowler 1984, Wagner and Vogelmann 1977); a digest is presented in Table 3. In summary, air-driven systems are suitable particularly when used at smaller scale and low cell densities; they have the advantage of low shear and generally substantial transfer rates and good mixing. However, the mixing could become inadequate for maintaining a homogeneous culture at high cell mass concentrations, thus resulting in inefficient gas transfer and formation of "dead" areas (Kato et al. 1975). Furthermore, due to the problem of overventilation, often mentioned by authors, lowered growth rates can be observed. Stirred reactors are more versatile systems, although there are many indications that the effect of shear on cell viability may vary with the cell line (Fowler 1982).

For the scale-up and for the rational design of new reactor configurations the capacity of the cell line for hydrodynamic stress should be known quantitatively. As there is a lack of information of this kind in the literature, it is not easy to decide whether dissipated energy or linear velocity should be preferred as the scale parameter. Bioreactors are characterized by a number of parameters and various scale-up criteria should be used (for further details see Kossen and Oosterhuis 1985).

Table 3. Respective advantages and drawbacks of airlift or stirred reactors for plant cell cultivation

	Airlift reactor	Stirred reactor
Device	Simple	More complex
Mixing	Correct Problem at high cell density	Efficient
Hydrodynamic shear forces	Low	High
Oxygen transfer	Correct (except at high cell density; often insufficient in large volumes)	Efficient even at high cell density
Overventilation	Possible and frequent	Not frequent
Scaling	Up to medium volumes	Up to large volumes

The ratio stirrer shear/stirrer flow is dependent on both the shape and size of the impeller and is sometimes chosen as the scale-up criterion ($\sim N/D$). Small diameter impellers rotating rapidly generate high shear forces but are not effective systems, whereas large diameter impellers moving at low speeds have a low shear action with a small shear-flow ratio. Another characteristic which relates the shear-flow ratio is the power input and then the power number can be held constant in scaling-up operations ($\sim N^3D^2$). A different parameter is also held equal, i.e. the dissipated volumetric energy; scale-up should be made on the basis of $\sim N^3D^5$, by increasing the diameter of the impeller with a constant impeller speed; however, due to the high shear forces generated, one should prefer a constant agitator tip velocity. Usually P/V is kept more or less constant during scale-up, resulting in similar values of the oxygen transfer rate but considerable increase of turbulent shear. Another rule is the constant value for the ratio air-flow rate/pumping capacity. In summary, due to overlapping of mass transfer and damage effects, the scale-up is not easy but the control of hydrodynamic forces and the determination of critical values of fermentation parameters is a necessity. It can further be noted that air-driven reactors have been favoured by many groups at a medium scale, up to 1500 l for the bubble column used by Kato et al. (1976), while the largest volume cultivation has been performed by Noguchi et al. (1977) in a 20,000 l conventional stirred fermenter. Examples of large-scale culture systems for the growth of plant cells are given in Table 4.

Some physical parameters of plant cell cultures and process variables related to mass cultivation are reported in Table 5.

CULTURE REGIME

Many studies regarding submerged cultures have involved the use of a "batch mode". While a great deal of biochemical information was collected using such a culture regime, restrictions in batch cultures appear clearly due to a constantly changing environment during the growth of cells. Since 1970 numerous semi-continuous and continuous culture methods have been developed. By producing a steady state growth and enabling the distinction between

Table 4. Development of large-scale cultures of plant cells

Year	Vessel configuration	Volume	Plant species	Reference
1959	Conventional tank	9-30	Ginko, Ilex	Tulecke and Nickell 1959
1971	Round-bottom vessel, stirring bar	5	Acer pseudo-platanus	Wilson et al. 1971
1975-1976	Stirred reactor Bubble column	30; 65 360; 1.500	Nicotiana tabacum	Kato et al. 1975, 1976a,b
1977	Conventional airlift with draught	10	Morinda ci-trifolia	Wagner and Vogelmann 1977
1977-1982	Conventional stirred tank	20,000	Nicotiana ta-bacum	Noguchi et al. 1982
1982	Airlift loop reactor	100	Catharanthus roseus	Fowler 1982

Table 5. Physical parameters and process variables in plant cell cultures

Rheology	Viscosity η app.	Broth up to 35-100 cp (increase from 1.2 to 35 cp; Kato et al. 1978) Filtrate increase from 0.9 to 2.2 cp (Kato et al. 1978)	
	Fluid characteristics	Broth: pseudoplastic Filtrate: Newtonian	
Shear	Aggregates tension rupture	$\tau \simeq 2.10^2$ N m^{-2}	
	Cells	$\tau \simeq 10\text{-}50.10^2$ N m^{-2}	
	Admissible agitation speed	50-200 rpm	
	Tip velocity of agitator	0.1-0.5 m s^{-1}	
Aeration and oxygen transfer			
	Oxygen demand up to	10-15 mg g^{-1} h^{-1}	
	Volumetric oxygen transfer rate	10 to 50 h^{-1}	
	Air flow	0.1 - 1 v.v.m	
	Saturation constant	$K_{s_{O_2}}$ 0.2 mg l^{-1}	Pareilleux and Chaubet (1981) Pareilleux and Vinas (1983)
	(Minimal D.O.concentration 0.8-1 mg l^{-1}, i.e. 10-15% saturation)		

effects of growth rate and effects of growth-limiting substances, the chemostat culture method offers great possibilities in identifying the factors influencing cell metabolism; this may provide basic information on the regulation of secondary metabolite formation.

Extensive studies of chemostat cultures of plant cells have been reported (Bertola and Klis 1979, Kato et al. 1976, King 1976, 1977, Kurz 1971, Wilson 1976, Young 1973) and reviewed periodically (Fowler 1977, Martin 1980, Wilson 1980). Various growth-limiting substrates were used, i.e. phosphate, nitrogen, carbon source, etc. In some cases the data conformed to Monod's model which assumes that the relationship between μ (growth rate) and S (concentration of the growth-limiting factor) is in the form of a saturation curve with an affinity constant Ks, and that Y (growth yield) is independent of μ or D (dilution rate). More recent studies have shown that Monod's model is not always adequate, e.g. steady state culture cell densities are not strictly proportional to the phosphate concentration in the medium (Dougall and Weyrauth 1980). The arrangement of the kinetic model developed by Nyholm (1976, 1978) for growth under limitation by "conservative" substrates and that of Monod may be more applicable. From a Lineweaver-Burk plot, $1/\mu$ against 1/S, affinity constants have been measured for various transport systems in batch and continuous cultures; some values are given in Table 6. As previously discussed such considerations are of great interest for the examination of the potential benefit of continuous culture techniques for improving biomass yield and product formation.

KINETIC STUDIES AND PROCESS VARIABLES

Because of the control of some process variables, mass cultivation even at a small laboratory scale, may provide useful information on the kinetics and physiology of plant cells. Much attention has been given to mixing and gaseous transfer.

With regards to oxygen transfer, data are available concerning the influence of various oxygen supply conditions and agitation speeds on the growth kinetics and cell yields. The most significant has been that of Kato et al. (1975) who worked on

Table 6. Affinity constants measured in batch and continuous cultures (after Monod's model)

Cell line	Substrate	Value	References
Acer pseudoplatanus	Glucose	0.1 g l^{-1}	King et al. (1973) King (1976)
	NO_3^-	0.13 mM	Wilson (1976)
	PO_4^{3-}	0.032 mM	Leguay and Guern (1975)
	2-4 D	$3x10^{-3}\mu M$ (estimated)	
Nicotiana tabacum	Glucose	0.25 g l^{-1}	
	Fructose	0.94 g l^{-1}	Kato and Tsuji (1981)
	(Inhibition for fructose)	10 g l^{-1}	
	PO_4^{3-}	0.04 mM	*
	SO_4^{2-}	15 µM	
	A.I.A	1-5 µM	
Sugarcane cells	Glucose	15 mM Sucrose-grown cells	*
	Glucose	1.4 mM sucrose-grown cells)	
	L-arginine	100 µM	
Phaseolus vulgaris	L-arginine	3.5 µM	*

[a] *For other values and reference details as well as information on growth yields see Dougall (1980), Wilson (1980)

tobacco cells grown in a 15-l stirred reactor and a 65-l bubble column. A hyperbolic relationship was observed between the final cell mass concentration and the initial oxygen transfer rate k_la, with a critical value of 10 h^{-1}. In our laboratory we have also been interested in this aspect of plant cell cultivation. Previously we reported that under limiting oxygen conditions growth of Medicago sativa cells proceeded linearly at a rate proportional to the oxygen supply. Relating oxygen supply and cell growth, the oxygen requirement could be determined by means of a mass balance equation with respect to sugar and biomass;

from the derived stoichiometric oxygen coefficient, i.e. 1.3 g O_2 g^{-1} biomass, experimental and calculated productivities in cell mass were found to be in the same range during the oxygen-limited culture. Moreover, the minimum volumetric oxygen transfer coefficient necessary to achieve an entirely exponential growth could be calculated, assuming that the oxygen uptake rate of the cells is covered by the oxygen transfer (Pareilleux and Chaubet 1981). Indeed, increasing the air flow during the time of the culture was shown to be beneficial, thus lowering deleterious effects of CO_2 "stripping off" and maintaining a sufficient dissolved level; this value was fixed by means of an affinity constant determination K_{SO2} according to Monod. Further results on Catharanthus roseus cell mass cultivation confirmed the above described approach. Linear growth phases with reduced conversion yields occurred at low air-flow rates, whereas exponential growth phase and high yields were obtained at sufficient aeration rates. Furthermore, growth yields and oxygen consumption rates were found to be linked by a linear relationship (Pareilleux and Vinas 1983). After Kato and Nagai (1979), it can be observed that the examination of such quantitative relations on the basis of bioenergetics is a useful tool for the determination of optimal growth conditions for plant cell suspensions. The assumption that the oxygen supply may ensure optimal product formation has also been mentioned recently (Spieler et al. 1985, Ulbrich et al. 1985, Yamakawa et al. 1983). Another aspect has recently become apparent in large-scale cultures of plant cells which is of particular importance in air-driven vessels, i.e. in the culture (Martin 1980, Pareilleux and Chaubet 1981, Smart and Fowler 1981, Tanaka 1982).

This facet was recently studied in detail in our laboratory. Kinetic data for the growth of Catharanthus roseus cells in a 10-l stirred reactor under various aeration rates showed a detrimental effect of the gassing rate on the growth characteristics due to CO_2 stripping. At the same aeration flows better growth and enhanced conversion yields were obtained when the CO_2 partial pressure was maintained at a constant level of 20 mbar in the culture (Ducos and Pareilleux 1986).

Some particular problems of plant cell cultures have to be taken into consideration, including foaming and bulking, surface adhesion and sterility. Bulking of the cell mass and foaming are frequently observed in bioreactors at high biomass densities, especially in airlift systems at high aeration rates. To suppress foaming a number of antifoams have been used by workers with frequent deleterious effects on cell growth and physical characteristics of the broth such as extended lag phase, lowered growth rates, biomass yields and carbon conversion and also less efficient gas transfer (Fowler 1982, Smart and Fowler 1981). Surface adhesion on the wall, shaft and probes may occur in later stages of growth, resulting in dry and anoxic zones and death of cells. Some success in reducing this problem was obtained by coating the vessel surface. Close attention to sterile operations is required because of low growth rates. Although many examples of large-scale culutres over extended periods of time (2 months) are described in the literature, some sterility problems may occur. To prevent contamination during cultivation, the addition of antibacterial and antifungal substances is sometimes recommended with some success; since there are some which do not affect cell growth (Matsumoto et al. 1972).

In plant cell cultures maximum growth generally occurs between 25-28°C; in the 20-30°C temperature range the dependence of growth rate is often described by an Arrhenius expression as shown by Kato et al. (1976) for tobacco cell suspensions. The most favourable pH for the growth of plant cells is in the pH range 5 to 6. Due to nitrogen source consumption pH shifts are currently observed during growth which are dependent on the medium consumption. Cultivation under pH stat conditions is possible without decrease in growth rate or carbon conversion (Martin 1980).

CONCLUSION

At the moment, while their potential as a source of natural products has clearly been demonstrated, the successful exploitation of plant cell cultures at an industrial scale is still limited. With a view to the enhancement of product yields these limitations include the lack of knowledge of fundamental cell physiology, biochemistry and genetics. From a biotechnological point of view, due to the extensive information and experience now available, more attention has to be given to process variables of engineering interest for the development and industrial success of plant cell processes.

Acknowledgement. I would like to thank Dr. N. Lindley for his linguistic assistence.

REFERENCES

Alfermann AW, Bergmann W, Figur C, Helmobold U, Schwantage D, Schuller I and Reinhard E (1983) In Mantell SH and Smith H (eds) Plant biotechnology. Cambridge University Press, p.67

Alfermann AW, Spieler H and Reinhard E (1985) In Neumann KH, Barz W and Reinhard E (eds) Primary and secondary metabolism of plant cell cultures, Springer, p. 316

Anderson LA, Philippson JD and Roberst MF (1985) In Fiechter A (ed) Adv Biochem Eng Biotech. Springer, 31:1

Berlin J (1986) In Pape H and Rehm HJ (eds) Biotechnology, microbial products, VCH 4:629

Bertola MA and Klis FM (1979) J Exp Bot 30:1223

Dougall DK (1980) In Staba EJ (ed) Plant tissue culture as a source of biochemicals, CRC Press, p.21

Dougall DK and Weyranth KW (1980) Biotechnol Bioeng 22:337

Dougall DK, Labrake S and Whitten GH (1983) Biotechnol Bioeng 25:569

Ducos JP and Pareilleux A (1986 Appl. Microbiol Biotechnol 25: 101

Fowler MW (1977) In Barz W, Reinhard E and Zenk M (eds) Plant tissue culture and its biotechnological application, Springer, p. 253

Fowler MW (1982) Prog Ind Microbiol 16:207

Fowler MW (1986) Trends in Biotechnol 8:214

Fujita Y, Hara Y, Suga C and Moromoto T (1981a) Plant Cell Reports 1:61

Fujita Y, Hara Y, Ogino T and Suga C (1981b) Plant Cell Reports 1:59

Fujita Y, Tabata M, Nishi A and Yamada Y (1982) In Fujiwara A (ed) Assoc Plant Tissue Culture, Tokyo, p.309

Furuya T, Yoshikawa T, Orihara Y and Oda H (1984) J Nat Prod 47:70

Kato A, Shimuzi Y and Nagai S (1975) J Ferment Technol 53:744

Kato A, Kawazoe M, Iizima M and Shimizu Y (1976) J Ferment Technol 54:82

Kato A, Hashimoto Y and Shoh (1976) J Ferment Technol 54:754

Kato A, Kawazoe S and Shoh Y (1978) J Ferment Technol 56:224

Kato A and Nagai S (1979) Eur J Appl Microbiol Biotechnol 7:219

Kato A, Asakura A, Tsuji K, Ikeda F and Iijima (1980) J Ferment Technol 58:373

Kato A and Tsuji (1981) J Ferment Technol 59:33

King PJ, Mansfield KJ and Street HE (1973) Can J Bot 51:1807

King PJ (1976) Pl Sci Lett 6:409

King PJ (1976) J Exp Bot 27:1053

King PJ (1977) J Exp Bot 28:142

Kossen NWF and Oosterhuis NMG (1985) In Brauer H (ed) Fundamentals of biochemical engineering, VCH 2:571

Kurz WGW (1971) Exp Cell Res 64:476

Kurz WGW and Constabel F (1985) CRC Crit Rev Biotechnol 2:105

Leguay JJ and Guern J (1975) Plant Physiol 56:356

Martin SM (1980) In Staba EJ (ed) Plant tissue culture as a source of biochemicals, CRC Press, p.143

Martin SM (1980) In Staba EJ (ed) Plant tissue culture as a source of biochemicals, CRC Press, p.149

Matsumoto T, Okunishi K, Nishida K and Noguchi M (1972) Agr Biol Chem 36:2177

Nakagawa K, Konagai A, Fukui H and Tabata M (1984) Plant Cell Rep 3:254

Nickell LG (1980) In Staba EJ (ed) Plant tissue culture as a source of biochemicals, CRC Press, p. 235

Noguchi M, Matsumoto T, Hirata Y, Yamomoto K, Katsuyama A, Kato A, Azechi S and Kato K (1977) In Barz W, Reinhard E and Zenk MH (eds) Plant tissue culture and its biotechnological application, Springer, p.85

Nyholm N (1976) Biotechnol Bioeng 18:1043

Nyholm (1978) J Theor Biol 70:415

Pareilleux A and Chaubet N (1981) Eur J Appl Microbiol Biotechnol 11:222

Pareilleux A and Vinas R (1983) J Ferment Technol 61:429

Reinhard E and Alfermann AW (1980) In Fiechter A (ed) Adv Biochem Bioeng, Springer 16:49

Sahai OP and Shuler ML (1984) Biotechnol Bioeng 26:27

Schiel O, Jarhow-Redecker K, Piehl GW, Lehman J and Berlin J (1984) Plant Cell Rep 3:18

Smart NJ and Fowler WF (1981) Biotechnol Lett 3:171

Spieler H, Alfermann AW and Reinhard (1985) Appl Microbiol Biotechnol 23:1

Spier RE and Fowler MW (1984) In Moo-Young M (ed) Comprehensive biotechnology, Pergamon Press 1:301

Tal B, Rokem JS and Goldberg I (1983) Plant Cell Rep 2:219

Tanaka H (1981) Biotechnol Bioeng 23:1203

Tanaka H (1982) Biotechnol Bioeng 24:345

Tanaka H, Nishijima F, Suwa M and Iwamoto T (1983) Biotechnol Bioeng 25:2359

Tulecke W and Nickell G (1959) Science 130:863

Ulbrich B, Wiesner W nd Arens H (1985) In Neumann KH, Barz W and Reinhard E (eds) Primary and secondary metabolism of plant cell cultures, Springer, p. 293

Vogelmann H (1981) In Moo-Young M, Robinson CW and Vezina C (eds) Advances in biotechnology, Pergamon Press 1:117

Wagner F and Vogelmann H (1977) In Barz W, Reinhard E and Zenk MH (eds) Plant tissue culture and its biotechnological application, Springer, p.245

Wilson G, King PJ and Street HE (1971) J Exp Bot 22:177

Wilson G (1976) Ann Bot 40:919

Wilson G (1980) In Fiechter A. (ed) Adv Bioch Eng 13:1

Yamakawa T, Kato S, Ishida K, Katama T and Minoda Y (1983) Agr Biol Chem 47:2185

Young M (1973) J Exp Bot 27:1053

BIOREACTORS FOR PLANT CELL CULTURE

M. L. Shuler

School of Chemical Engineering, Cornell University, Ithaca, New York 14853 U.S.A.

INTRODUCTION

The premise of this article is that the rational design of a bioreactor can be used to direct plant cell culture towards desired physiological states and toward differentiation and organization. Bioreactors can be used as tools to probe stimulus-response in plant cultures and at the commercial level as devices to increase the productivity of a culture for secondary metabolite formation or production of organized tissues. However, before these aspects of bioreactor design can be discussed the reader must understand the essential principles of bioreactor design.

IDEAL REACTOR TYPES

Most bioreactors can be classified as: batch suspension reactors, continuous flow stirred tank reactors (CFSTR), multistage CFSTR's, and immobilized or retained cell systems. Many immobilized systems approximate a plug flow reactor (PFR).

The shake flask is an example of a batch reactor. A nutrient charge and cellular inoculum are mixed, shaken, and allowed to grow. As the cells grow the medium is continually depleted of nutrients and metabolic by-products may accumulate continuously , altering the environment about the cells. This continuously changing environment can cause continuous changes

NATO ASI Series, Vol. H18
Plant Cell Biotechnology. Edited by M. S. S. Pais et al.

in the culture. In a shake flask pH and dissolved oxygen are also changing as a function of the extent and rate of growth. In a more well-controlled bioreactor pH and dissolved oxygen may be held constant, but other nutrients will be changing. The results of batch culture experiments are often sensitive to the initial conditions (e.g. "history" and size of the inoculum).

A CFSTR is a device where a nutrient feed stream or streams is fed continuously and a least one effluent stream is also removed continuously. Microbiologists describe such a device as a "chemostat" and the term is applied to CFSTRs with plant cells (Dougall 1986). Chemostat theory (Herbert et al 1956) demonstrates that at steady-state operation the population -averaged growth rate of the culture (μ) must equal the dilution rate, D, (where D equals the volumetric flow rate into the reactor divided by the reactor volume). Thus the chemostat allows the investigator to independently set the growth rate. At steady-state the culture response has, in principle, no dependence on the inoculum history. The chemostat is an ideal device to probe a cultures response to a variety of stimuli under well-controlled conditions where the environment remains constant.

The application of chemostat theory to plant cell culture requires the investigator to be aware of the basic assumptions in developing the theory. The key assumption is that the contents of the vessel are perfectly mixed. Thus the concentration of cells (and cell types), nutrients, etc. must be the same at any position in the reactor and the same as the concentration in the exit line. Reactors with low agitation are unlikely to keep the plant cells evenly suspended (see Sahai and Shuller 1982) and often wall growth is significant in plant cell reactors. Either condition invalidates the perfect-mixing assumption. The reader should also note that the concentration of all feed components is diluted immediatly to the effluent concentration. If the reaction order is greater than zero (e.g. Monod-type growth kinetics) this dilution will decrease growth rate with respect to what would have been initially possible in the inlet stream. If the stream contains any metabolizable inhibitors (e.g. substrate-inhibition growth kinetics), this dilution effect may increase growth rate.

In a chemostat the cells must be dividing at a sufficiently rapid rate to replace the cells washed out in the effluent stream. Thus any product formation that takes place must be at least partially growth associated.

A multistage CFSTR is better suited to systems where product formation is non-growth associated. In the first stage conditions must be maintained to support cellular replication at rate equal to washout. In the second and subsequent stages replication is no longer vital since the previous reactor acts as a continuous source of new cellular material. Thus conditions in the second and subsequent stages can be used to direct the cells toward forming non-growth associated products. The system can be used to direct the culture through a sequence of physiological states. Riccia(1970) has described a graphical method to predict (at least roughly) the growth and production characteristics of a culture in each reactor of multi-reactor system. Riccia, in fact, successfully applied his technique to spore production from *Baccilus* using a six-stage system with varying reactor sizes.

A multi-stage system approach can also be applied to batch reactors. Usually such a system is used only if the nutritional conditions for growth and product formation are significantly different. If the growth medium contains a compound inhibitory to product formation, the cells and the medium from the first vessel must be physically separated prior to the second stage reactor. However, with the multiple batch reactor system the results will depend on inoculum history, and the cells will be continually adapting to a changing environment.

Two closely related reactor types are either one in which the cells are immobilized or one where cells are retained within the reactor. In an immobilized system the cells are separated from the liquid medium by attachment or adsorption to a solid support, or more commonly entrapment within a gel matrix or between membranes. In a retained cell system the cells remain suspended in the liquid but are not allowed to leave the reactor in the effluent (typically through gravity settling or membrane filtration). Almost invariably these systems are operated with continuous nutrient flow. Often cells are entrapped in beads, and the beads are used to pack a column. Feed enters the column at one end and is withdrawn at the other. Ideally such reactors approach plug flow where there is no mixing. Thus an element of fluid, high in substrate, enters and nutrients are gradually consumed as the fluid element passes through the column. Unlike a CFSTR the concentration varies with position in the reactor and the gradual depletion of nutrient leads to a higher average substrate concentration than in a CFSTR. A multistage CFSTR approximates a PFR since the concentration decreases gradually from one stage to another.

The choice of reactor type depends on the nature of the bio-catalyst. To illustrate the use of these reactor types we must consider the salient features of plant cell cultures.

PLANT CELL CHARACTERISTICS INFLUENCING BIOREACTOR CHOICES

Plant cell cultures are derived from multicellular organisms. When compared to the bacteria, fungi, and yeasts traditionally exploited in commercial scale bioreactors plant cells have differences that make traditional bioreactor designs inappropriate. Plant cells grow slowly, often as aggregates, have complex genetics that are not easily understood or controlled, and are rather sensitive to shear.

The key to understanding these cultures is likely understanding the role cell-to-cell communication plays. In the whole plant cells are in intimate contact with each other and cytoplasms are connected by plasmodesmata. Thus, small molecular weight compounds (<900MW) can move readily from cell to cell (Gunning and Overall 1983). Nutrients diffuse into tissues while metabolic by-products must diffuse out. Compounds transferred via the plasmodesmata or metabolic by-products can act to alter cellular metabolism and development.

Consider a homogenous suspension of single cells. No plasmodesmata exist and all cells are in contact with a liquid environment supplying the same level of nutrient. Any metabolic by-products or cellular degradative products are released directly into the medium and immediatly diluted to a low concentration. Such homogenous single-cell suspensions are difficult to obtain; such cultures have lower growth rates than cultures with cell aggregates although it is difficult to discern whether this reduced growth rate is due to "single-cellness" *per se* or due to the harsh treatments necessary to obtain such cultures (Kubek and Shuler 1978).

Most suspension cultures have a mixture of aggregates over a wide size range. Typically it is observed that cells on the periphery of such aggregates are morphologically distinct from those in the center. The cells on the periphery have higher mitotic indexes than cells in the center of such aggregates or single cells in the same medium and are also more responsive to external stimuli in terms of organogenesis. Indeed it has been shown that a minimum aggregate size is required for organogenesis (Walker et al 1979). In such aggregates diffusional effects will be important for some nutrients (particularly dissolved gases) and for the release of metabolic by-products. In aggregates locally high levels of such by-products could be obtained due to diffusional limitations. Also, in such aggregates the plasmodesmata could form since many aggregates are due to failure of dividing cells to separate rather than clumping of previously formed cells.

Sussex and Clutter (1967) have performed experiments comparing pseudotissue with real tissue. Pseudotissues were formed by concentrating cells from a suspension and entrapping them in a section of dialysis tubing. Diffusional effects would be similar in pseudotissue and real tissue, but the plasmodesmata would not be present to any significant extent in the pseudo tissue. When exposed to the same stimuli the real tissue could differentiate into tracheids while the pseudo-tissue could not. Thus how cells are associated with each other can have a profound effect on biological response.

The reason that such cell to cell interactions may be important in bioreactor design is that high formation rates for secondary metabolites are often associated with cellular differentiation and organization (Tabata et al 1972; Freeman et al 1974; Hashimoto and Yamada 1985; Lindsey and Yeoman 1983). The control of organization, differentiation, and product formation are often interrelated, but in a complex manner not yet well understood. Further high growth rates and high product formation rates are mutually exclusive. Optimal conditions for product formation often suppresses growth and *vice versa*. The incompatibility of growth and product formation greatly complicate reactor strategy. It should be noted that there is no evidence for a causal relationship between slow growth and high levels of secondary metabolism but the weight of the empirical evidence is clearly for high growth rates suppressing product formation.

Another factor influencing cell to cell contact and productivity is liquid shear. Very high shear rates can destroy cells. Traditional reactors often impart sufficient shear to lyse cells. Lysis can be avoided using reactors with specially designed mechanical stirrers or air lift fermentors (Tanaka 1981). However, more subtle shear effects are likely important and these occur at much lower values than necessary for cell lysis. Moderate levels of shear will alter average aggregate size and, if cell to cell communication is important, the resulting mixture of cell types and biosynthetic capabilities. Also, it has been observed that low shear levels can alter animal cell morphology and possibly the number and distribution of chemical receptors on the surface of endothelial cells (Nerem and Levesque 1985). Shear may also have similar effects on plant cells. Shear is of particular importance in bioreactor design, particularly for reactor scale-up. Oldshue (1966) has shown that as traditional bioreactors are scaled up it is impossible to maintain oxygen transfer, extent of mixing, and shear the same. Usually shear must be increased at large scales to maintain sufficient oxygen transfer. Wagner and Vogleman (1977) have examined growth and secondary metabolite formation in *Morinda citrifolia* cells in a variety of reactor sizes and configurations. Dramatic changes in growth and product formation are observed from vessel to vessel.

Further complications in large scale culture are due to genetic or epigenetic (metastable, developmental shifts in cellular competence) instabilities (Dougall 1985). Different bioreactors provide different microenvironments about the cell which can alter epigenetic responses. Since the most rapidly growing variants are often the least productive, CFSTR reactors will magnify genetic instabilities and cause the culture to become less productive.

Another practical problem in reactor design is the slow growth rates of plant cells in culture (ca. 20 to 100 hr doubling times). Such slow growth requires the maintenance of reactor sterility for periods of at least seven to ten days and often much longer. Further, the low growth rates reflect relatively slow internal reaction rates which means the rate per unit reactor volume is low. Low volumetric productivities imply large and costly reactors.

With this overview of the special constraints plant cell cultures may place on bioreactor choices we are ready to discuss reactor choices in the context of plant cell culture and give some examples.

BIOREACTOR CHOICES

Clearly the genetic instability, inhomogeneties in the culture, and the mutual exclusion of growth and product formation mitigate against the use of a single stage CFSTR for large-scale production of metabolites. However, a CFSTR (chemostat) of laboratory scale remains a very useful tool to probe cell physiology. Such systems can probably be used only when the total cell population is kept low since incomplete mixing and wall growth become more severe at high cell concentrations.

Batch reactors are not as sensitive to instabilities and inhomogenities and can be easily adapted to situations where growth and product formation require different environmental conditions. They are also well-suited for fermentations where the product of interest is intracellular. However, batch reactors are inherently less productive than a CFSTR since time is required to clean, sterilize, and refill the reactor between harvest and a long time is required for the culture to grow to a concentration suitable to alter conditions to those favorable for product formation.

If the change to optimal conditions for product formation requires merely the change of pH, temperature, or nutrient addition, then the production phase can easily be accomplished in the same batch reactor as used for growth. However, in many plant systems it often has been found to be advantageous to remove the growth medium as it contains hormones and nutrients suppressing product formation. After removing the cells from the spent growth medium, they are resuspended in a production medium. In a situation where cells must be removed from a spent growth medium a two-stage batch reactor system has been employed. Recently this two stage batch reaction system has been employed for the commercial production of shikonin from *Lithospermum erythrorhizon* (Tabata and Fujita 1985).

Multi-stage continuous culture systems have also been used with plant cell cultures (Kato et al 1980; Sahai and Shuler 1984; Tal et al 1983). In each of these examples depletion of a key nutrient in the second-stage was responsible for an altered physiological state in the culture. Tal et al (1983) have reported significant diosgenin production in the second stage of a two-staged system while production in a single stage or the first stage of the two-stage system was negligible. Sahai and Shuler (1984) have observed increased productivities for phenolics synthesis in a two-staged system, as compared to either batch operation or a single-staged chemostat with the same total holding time. A two-staged system could also be used with removal of a spent medium and replacement with a production medium but a filtration, sedimentation, or centrifugation step would have to be inserted between the two reactors.

Such separation processes may also be useful in reactors which retain cells. However, shear sensitivity and the need for absolute sterility make continuous cell separation outside of the reactor problematic. Internal cell recycle has been achieved through filtration (Styer 1985) and sedimentation

(Pareilleux and Vinas 1984). Such approaches are workable and potentially allow continuous operation at rather high cell densities. However, the product of interest shoud be one excreted into the medium.

Another method of cell retention is cell immobilization. Typically cells are entrapped by enmeshment in a gel or by enclosure with membranes. The only difference between "immobilization" and "cell recycle" systems is that a form of cell to cell association is forced upon the culture with immobilization.

A number of potential advantages exist for immobilized systems (Shuler and Hallsby 1985) although the simultaneous realization of all of these potentials may prove difficult. These advantages have been detailed as:

-. Continuous operation at high flow rates without concern about cell washout.

-. Problems due to genetic instability can be reduced greatly since cell replication of the immobilized cells is greatly repressed.

-. Contamination is no greater a problem for immobilized plant cell systems than for other immobilized systems such has bacteria. Thus, the impact of the slow growth rates of the plant cells can be reduced.

-. The cells are protected from the liquid shear allowing the engineer to choose fluid mixing characteristics based solely on mass transfer characteristics.

-. The system allows the separation of growth (prior to immobilization) and production. With the immobilized cell system the medium can be optimized solely for product formation. The suppression of all replication is desirable to prevent cell overgrowth and the disruption of the immobilizing matrix.

-. Metabolic inhibitors which normally would accumulate in a batch suspension culture are "dialyzed" away and may allow the expression of metabolic pathways which normally would be supressed.

-. Because of the high cell densities achievable in immobilized systems and the potential to increase specific productivity (amount of product per unit weight of cells per unit time) and because the mixture of cell types is controllable, it should be possible to dramatically improve volumetric productivities (amount of product per unit volume of reactor per unit time) over batch reactors.

-. Cell to cell contact can be controlled (e.g. pseudo-tissue or systems where plasmodesmata are well developed) and the degree of mass transfer resistance can be manipulated as the designer wishes.

In general, immobilized cell reactors give the bioreactor engineer an extra degree of freedom. Spatial heterogeneity can be controllably imposed on the culture.

Immobilized systems have disadvantages; these are maintenance of metabolic vigor while restricting the net growth rate and obtaining product excretion. Although plant cells normally store products intracellularly, excretion can sometimes be obtained by manipulating pH or ionic strength or the use of chemicals such as DMSO (dimethysulfoxide). Further excretion can occur "naturally" to some extent with cells under the stress of immobilization.

Probably the most intriguing prospect for the use of immobilized cell cultures is to use control of mass transfer and cell to cell contact to direct dedifferentiated cultures toward more differentiated and organized forms. Hallsby and Shuller (1986) have shown in tobacco cell cultures that imposition of mass transfer restrictions (nutrient flow parallel to a layer of cells) leads to a primitive form of cell organization while no indications of cellular organization are observed in the same reactor system under identical environmental conditions if the medium is pumped through the cell layer (perfusion) so as to remove any mass transfer limitations. Thus it may be possible to use immobilization to force cells toward organization.

FUTURE THEMES

Several other aspects of bioreactor design could become of importance in the next decade. For example, organ culture may be of significance. For secondary metabolite production the central question may be whether organs (e.g. roots) can be maintained at a point of arrested development and still actively synthesize the compound of interest.

Another emerging opportunity in bioreactor engineering will be the production of "artificial seeds" or plantlets. Current micropropagation techniques are labor intensive and expensive. Labor costs could be cut significantly if cellular organization and development could be controllably induced in liquid suspensions.

On-line extraction of metabolites may simplify recovery of plant products. Further, their removal from the medium could, in principle, increase excretion there by lowering intracellular concentrations and relieving feedback effects on synthesis. In such a situation the on-line extraction could "pull" metabolism in a particular direction.

Other emerging themes are the role of elicitors (Dicosmo and Tallevi 1984) and dissolved gas concentrations on plant cell physiology and development. These factors, when we learn to manipulate them correctly, offer significant opportunities for large changes in yields of desirable products.

ACKNOWLEDGEMENTS

I would first like to acknowledge my former and current students and colleagues who have provided the bases for the ideas described in this paper. They are Dennis Kubek, Om Sahai, Greg Payne, John Pyne, G. Anders Hallsby, Bobby Bringi, Chris Prince, Tom Hirasuna and Masanori Asada. I would also gratefully acknowledge the support of the National Science Foundation (Grant Number ECE-8503183).

REFERENCES

Dicosmo F and Tallevi SG (1984) Trends in Biotechnology **3**:110

Dougall DK (1985) In Zatlin M, Day P and Hollaender A Eds. Biotechnology in Plant Science, Academic Press Inc, New York, p.179

Dougall DK (1986) Biotechnol Bioeng Symp **17**: 737

Freeman GG, Whenham RJ, Mackenzie IA and Davey MR (1974) Plant Sci Lett **3**: 121

Gunning BES and Overall RL (1983) Bio Sci **33**: 260

Hallsby GA and Shuler ML (1986) Biotechnol Bioeng Symp **17**: 739

Hashimoto T and Yamada Y (1985) Planta Medica **47**: 195

Herbert D, Elsworth R and Telling RC (1956) J Gen Microbiol **14**: 601

Kato A, Asakura A, Tsuji K, Ikeda F and Iijima M (1980) Ferment Technol **58**: 373

Kubek DJ and Shuler ML (1978) Can J Bot **56**: 2521

Lindsey K and Yeoman MM (1983) J Exp Bot **34**: 1055

Nerem RM and Levesque MJ (1985) 7th Intl Symp on Artherosclerosis, Melbourne, Australia (1985)

Oldshue SY (1966) Biotechnol Bioeng **8**: 3

EFFECT OF AERATION ON *CYNARA CARDUNCULUS* PLANT CELL CULTURES

E Lima-Costa [1], J M Novais,[2], M S Pais[3], and J M S Cabral[2]

1- Unidade Estrutural de Ciências Exactas

Universidade do Algarve

8000 Faro

Portugal

INTRODUCTION

The species *Cynara cardunculus* is traditionally used in Portugal to produce tasteful and valuable cheese .

Freely suspended or immobilized cells of *C. cardunculus* with high proteolytic activity, might be used as an alternative proteinases source to overcome the seasonal dependency of cheese making.

In large scale biomass production, the plant cell growth depends on the growth medium and operational parameters such as aeration rates and inoculum size (Smart and Fowler 1981; Pareilleux and Vinas 1984).

In our paper we report the influence of inoculum density and aeration rate on biomass production and proteolytic activity.

MATERIALS AND METHODS

Biological material - Cell suspension cultures were obtained by transfer of friable calli to TNO_3^- liquid medium (Tulecke

2- Laboratório de Engenharia Bioquímica, Instituto Superior Téc nico Av. Rovisco Pais, 1000 Lisboa

3- Departamento de Biologia Vegetal,Faculdade de Ciências de Lisboa, R. Escola Politécnica ,1294 Lisboa Codex Portugal

NATO ASI Series, Vol. H18
Plant Cell Biotechnology. Edited by M. S. S. Pais et al.

1966) supplemented with 0.1 mg/l kinetin and 1.0 mg/l 2,4-dichlorophenoxyacetic acid, contained in 250 ml erlenmeyer flasks. Fresh medium was added to the cell suspension every ten days. Homogenous suspensions were obtained after 3-4 subcultures.

Growth measurements - Cell growth was measured by optical density , dry weight, cell number and intracellular protein determinations. Optical density was measured at 578 nm after sample homogenization. Culture medium was used as blank. Dry weight was determined after cell drying for 48 hours at 75 ^{o} C. Cell number was measured by counting the cells of several diluted samples in an hemacitometer Moller D.M. Wedd. These values were correlated with the absorbance at 578 nm.

Proteolytic activity - 100 mg of wet biomass were incubated with 5ml of 3.2% azocasein solution (pH 5.75) for 24 hours at 37°C . Several samples were taken and the unhydrolyzed protein was precipitated with 5% trichloroacetic acid. Optical density of the supernatant was measured at 440 nm.

Intracellular protein determination - The intracellular protein content was determined by the Lowry method (Lowry et al. 1951). Bovine serum albumin was used as standard.

Extracellular phenolics - Phenolics content was assayed either by the Lowry method (Lowry et al 1951) or by ultraviolet analysis at 270 nm (Anselmo et al. 1985).

RESULTS AND DISCUSSION

Effect of inoculum density

As it was verified for shake flasks cultures (Lima Costa et al 1986) the growth rate of cell suspension cultures in fermenter is influenced by the inoculum density. For similar values of aeration rates, a 3.75 fold increase in the specific growth rate was obtained when the initial inoculum concentration was 1.7 fold increased (Table I).

Inoc.Conc. cells/ml	aerat.rate (v.v.m.)	μg biomass (day^{-1})	t_d (h)	μg prot. (day^{-1})
$0.70x10^4$	0.01	0.00	0.0	0.00
$0.87x10^4$	0.16	0.08	207.9	0.21
$1.50x10^4$	0.16	0.30	55.5	0.05

Table I - Effect of inoculum concentration and aeration rate on the specific growth rates.

For high inoculum concentration (1.5 x 10^4 cell/ml) and 0.16 v.v.m. aeration rate, a specific growth rate of 0.30 day^{-1} was obtained. This result corresponds to the lowest doubling time (t^d= 55.5h) obtained for C.cardunculus cell suspensions cultures in TNO^3 $^-$ (Table I). Specific growth rate values in the same range of magnitude were reported by several authors (Pareilleux and Vinas 1984; Kurz et al 1986; Drapeau et al 1986; Lipsky and Chernyak 1983) for different plant cell suspension cultures namely Catharanthus roseus (0.30 - 0.42 day^{-1}) and Dioscorea deltoidea (0.25 day^{-1}). When a low inoculum concentration (0.87x10^4 cell/ml) was used , the specific growth rate determined in terms of protein accumulation was higher than that based on cell number (Table I), in opposition to data obtained for shake flasks cultures (Lima Costa et al 1986). Although the lag phase, in terms of cellular division, is not initiated before day 7, protein synthesis increases since the 2nd day (fig.1). The rate of protein synthesis begins to decrease, while cellular division is still at the exponential phase.

When fermentations were started with a high inoculum concentration (1.5x10^{4}cell/ml),the specific growth rate (in terms of protein accumulation) is lower than that expressed in cell number. Similar results were reported for <u>Acer pseudoplatanus</u> (King et al 1973). These results suggest that, in aeration conditions,protein synthesis is favoured by low density inoculum, this effect being verified only for values below a critical inoculum concentration (1.5x10^4 cell/ml) (fig.1,Table I).

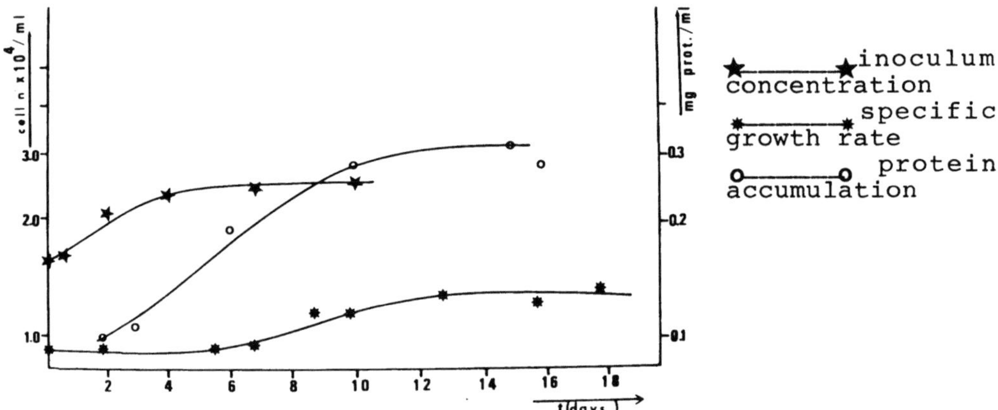

Fig.1. - Influence of inoculum concentration on the specific growth rate at constant aeration rate and protein accumulation obtained for low inoculum concentration

Effect of aeration

Fermentations performed (0.16 v.v.m. aeration rate) with an initial cellular density of 0.87x10^4 cell/ml revealed a specific growth rate 7.3 fold higher than that obtained for shake flasks cultures started with the same initial inoculum concentration (Table I and II).However,for an inoculum density of about 1.5x10^4 cell/ml,the specific growth rate only increased 2 folds. For aeration values of 0.01 v.v.m. and inoculum density of 0.70 x 10^4, no cell growth could be measured (Table I).

Correlations between oxygen consumption, protein synthesis and specific growth rate reported in our results are concordant with those refered for A. pseudoplatanus (Givan and Collin 1968) and Catharanthus roseus cells (Drapeau et al 1986). For Medicago sativa, biomass accumulation is favoured

Inoc. density (cell n° x10^4/ml)	μg (day^{-1})
0.94	0.011
0.98	0.015
1.22	0.158
2.80	0.193

Table II - Effect of inoculum density on specific growth rate obtained for shake flasks cultures

by aeration rates between 0.05 and 0.15 v.v.m. (Pareilleux and Chaubet 1980). Within this range, oxygen seems to be limitant for cell growth which is also in agreement with our results on C. carduncululus. Other authors ,however,have reported that for high aeration rates (0.8-1.6 v.v.m.) cell growth is inhibited probably due to the removal of CO^2 from the fermentation medium by the inlet gas phase (Smart and Fowler 1981; Maurel and Pareilleux 1985). The increase in CO^2 partial pressure seems, in other cases, to induce growth of A. pseudoplatanus cell cultures started with an initial low inoculum density (Gathercole et al 1976).

C.carduncululus shake flasks cultures present high phenolic

accumulation (lmg/ml) in the culture medium. This result is identical to that obtained when fermenter cultures were performed at low initial inoculum concentration and 0.01 v.v.m. aeration rate. However when aeration rate was increased to 0.16 v.v.m., phenolic accumulation did not occur (fig.2). These data may suggest an influence of aeration in phenols accumulation. Some authors have reported that secondary metabolites production increases when cellular growth decreases (Pareilleux and Vinas 1984; Sahai and Schuler 1984; Forrest 1969; Brodelius et al. 1979). Production of phenols has been verified when cells enter the stationary phase (Forrest 1969) or when toxic percursors are added to the culture medium (Sahai

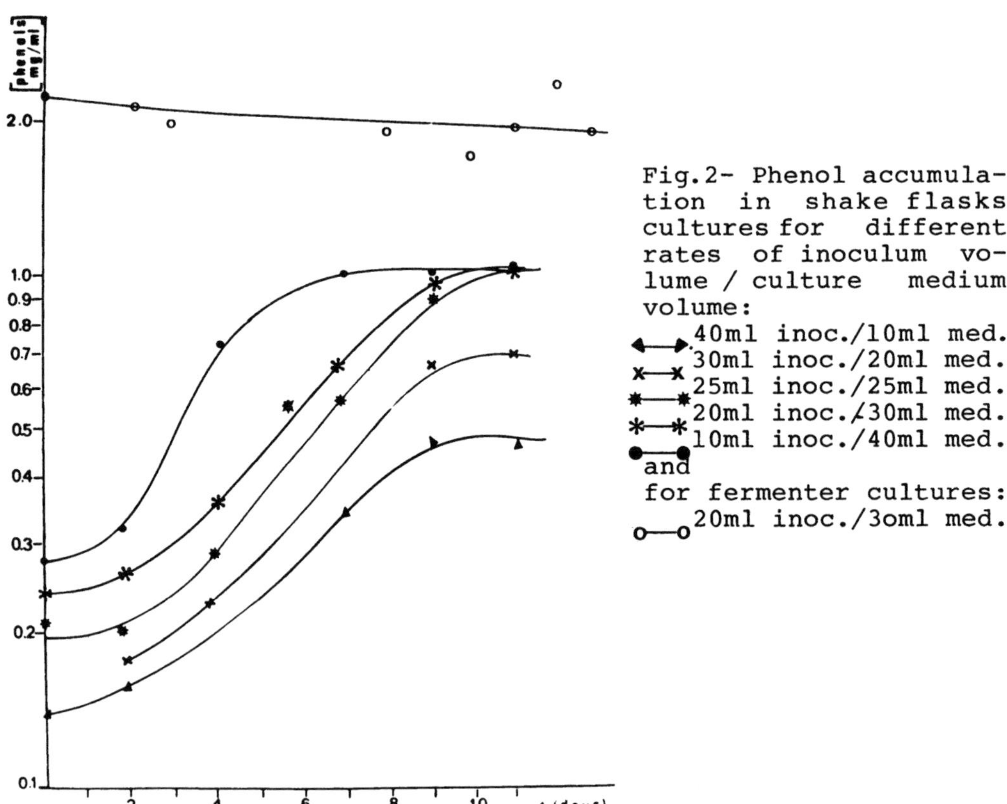

Fig.2- Phenol accumulation in shake flasks cultures for different rates of inoculum volume / culture medium volume:
40ml inoc./10ml med.
30ml inoc./20ml med.
25ml inoc./25ml med.
20ml inoc./30ml med.
10ml inoc./40ml med.
and
for fermenter cultures:
20ml inoc./3oml med.

and Schuler 1984). These results agree with ours concerning the inhibition of phenols accumulation in fermenter conditions, probably due to the great increase in growth rate promoted by aeration. The slow growth accompanied by high accumulation of phenols obtained for shake flask cultures would support this hypothesis.

Proteolytic activity - Fermenter conditions promoted a 4-5 fold increase of proteolytic activity when compared with shake flasks cultures (Table III).

The simoultaneous decrease in phenols production and

Time (days)	Proteolytic Activity (D.O.g^{-1} h^{-1})	
	Shake flasks	Fermenter 0.16v.v.m.
0	0.216	1.250
2	0.040	0.484
6	0.090	1.100
8	0.238	1.060

Table III - Proteolytic activities obtained for shake flasks and fermenter cultures.

the increase in growth rate and protein production may indicate a competition between primary and secondary metabolism. Otherwise, the TNO^{3-} culture medium containing only 1mg/ml nitrogen source when compared with the Gamborg B^5 medium (100 mg/ml nitrogen source) may constitute a limiting factor for cell growth in shakeflasks, this limitation being overlapped by aeration that probably promotes continuous phenols oxidation.

AKNOWLEDGEMENTS:- This work was partially supported by the research contract nº 520.83.47 from JNICT

REFERENCES

Anselmo AM, Mateus AM, Cabral JMS and Novais JM (1985) Degradation of phenol by immobilized cells of Fusarium flociferum ,Biotechnol.Let.**7(12)**:869-894

Brodelius P, Deus B, Mosbach K and Zenk MH (1979) Immobolized plant cells for the production and transformation of natural products , Febs Lett **103(1)**:93-97

Drapeau D, Blanch HW and Wilke C R (1986)Growth kinetics of Dioscorea deltoidea and Catharanthus roseus in batch culture, Biotechnol Bioeng **XXVIII**:1555-1563

Forrest, GI (1969) Studies on the polyphenol metabolism of tissue cultures derived from the tea plant (Camelia sinensis),Biochem.J. **113**:765-772

Gathercole RWE, Mansfield HE and Street R (1976) Carbon dioxide as an essential requirement for culture sycamore cells, Physiol Plant **37**: 213-217

Givan CV and Collin HA (1968) Studies on the growth in culture of plant cells. II Changes in respiration rate and nitrogen content associated with the growth of Acer pseudoplatanus L. cells in suspension culture,J Exp Bot. **18(55)**:321-331

Hegarty PK, Smart NJ, Scragg AG and Fowler MW (1986) The aeration of C. roseus L. G. Down suspension cultures in airlift bioreactors: The rates on culture growth, J Exp Bot **37(185)** :1911-1920

King PJ, Mansfield KJ and Street HE (1973), Control of growth and cell division in plant cell suspension cultures, Can J Bot **51**:1807- 1823

Kurz WGW , Chatson KB, Constabel F, Kudney JP, Choi LFL, Koldziejcvyk PK, Sleigh SK, Stuart KL and Worts BB (1980) Alkaloid production in Catharanthus roseus cell cultures: initial studies on cell lynes and their alkaloid content, Phytochemistry **19**:2583-2587

Lima-Costa E, Novais J, Cabral JMS and Pais MS (1986) Cultura em fermentador de células de Cynara cardunculus .III Enc. Nac.Biot.:63

Lipsky AK and Chernyak ND (1983) Influence of temperature on Dioscorea deltoidea wall cells in submerged culture. Fiziol Rast **30 (3)**:437-447

Lowry OH, Rosebrough NJ,Fau ALA and Randall RJ (1951) Protein measurement with the Folin phenol reagent, J Biol Chem **193** :265-275

Maurel B and Pareilleux A (1985) Effect of carbon dioxide on the growth of cell suspension of Catharanthus roseus, Biotechnol Let **7(5)** :313-318

Pareilleux A and Chaubet N (1980) Growth kinetics of apple plant cell cultures Biotechnol Let**2(6)**:291-296

Pareilleux A and Vinas R (1984) A study on the alkaloid production by resting cell suspension of C. roseus in a continuous flow reactor, Appl Mic Biot **19**:316-320

Sahai OP and Schuler ML (1984) Multistage continuous culture to examine secondary metabolite formation in plant cell phenolics from Nicotiana tabacum , Biotechnol Bioeng **XXVI** :27-36

Smart NJM and Fowler W (1981) Effects of aeration on large scale cultures of plant cells, Biotechnol Let **3**:171-176

Tulecke W (1966) Continuous cultures of higher plants in liquid medium, Ann N Y Acad Sci **139**:142-173 (1966).

METHODS OF IMMOBILIZATION OF PLANT CELLS

Júlio M. Novais

Laboratório de Engenharia Bioquímica, Instituto Superior Técnico,
1000 Lisboa Portugal

INTRODUCTION

Immobilization of biocatalysts has been defined as the operation leading to their confinement in a well-defined region of space, allowing continuous or successive reuse. It is the latter part of this definition that underlies the most commonly referred advantage of immobilization. Besides this possibility of use in continuous reactors, other advantages which may be obtained are the yielding of a purer product, as the biocatalyst is easily separated from the liquid, and better reaction control.

Immobilization methods have been primarily developed for enzymes and were then applied to microorganisms and only more recently to plant cells. All these biological compounds and living organisms can be considered as biocatalysts and many of the techniques and problems involved are common.

Immobilization of enzymes, developed over the past 25 years, is today a routine technique with many applications mainly in the analytical and industrial fields. The immobilization of microorganisms came later and large-scale applications are still lacking.

It was only natural that sooner or later these methods would be applied to plant cells: this happened in the past 8 to 10 years and the work being done is naturally at the laboratory scale, being the object of several reviews (Morris et al. 1985; Rosevear and Lambe; Brodelius 1985). The main difference in relation to both previous cases is that while enzymes and microorganisms, when free, are in a homogeneous form and the immobilized counterparts act in a heterogeneous system, in the case of plant cells they are already heterogeneous in their natural habitat.

NATO ASI Series, Vol. H18
Plant Cell Biotechnology. Edited by M. S. S. Pais et al.

Applications are certainly possible, but it should be noted that even the applications of plant cell cultures are still scarce, in spite of all the work and all the interest which has been developing.

In terms of plant cells, the possibility of retaining and reusing a large concentration of immobilized cells inside a reactor is of the utmost importance, provided that the preparation keeps its viability and capacity for yielding a secondary product since it is well known that it takes a long time to obtain large quantities of suspended plant cells. Using these methods, better yields per unit volume of the fermenter will be obtained.

Immobilized plant cells can be seen as an intermediate state between a homogeneous suspension culture and the highly structured tissue of the whole plant which is their natural status. Immobilization therefore provides an environment which is probably more favourable to cell maintenance of viability than the free suspended cell culture.

For an industrial or large-scale application, it is reasonable to consider that an immobilization system applied to the plant cells and including not only the method of immobilization but also the reactor where it will be used, has tc have the following characteristics (Morris et al. 1985): simple and capable of operating for long periods under sterile conditions, gentle in order to preserve cell viability, stable and durable in use, and it must provide a highly active and specific system which is low in cost, in terms of immobilization, use and recovery of the product.

GENERAL METHODS OF IMMOBILIZATION OF BIOCATALYSTS

The methods which have been most often used for immobilization of biocatalysts, fall in a small number of categories, e.g. a general classification:

1. Immobilization through linkage

a) To a solid support

i) Physical adsorption

ii) Ionic bond

iii) Metal link

iv) Covalent bond

b) Intermolecular or intercellular bond
2. By entrapment
a) In gels
b) In fibers
c) Microencapsulation
d) Enclosure in membranes
3. Pelletization, flocculation, aggregation

All these methods have been formerly used in the immobilization of enzymes and microorganisms, the first group being used particularly for enzymes and the latter being more appropriate for microorganisms.

An immobilization concept which is not included in the previous classification is one which is particularly applicable to cells and in which they are in free suspensions in the reactor but are separated downstream and recycled to the reactor.

APPLICATION OF IMMOBILIZATION METHODS TO PLANT CELLS

The immobilization methods which have been used for microorganisms can certainly be attempted for plant cells. However, there are some points that clearly should not be forgotten and that deal with the particular characteristics of these cells.

Plant cells are much larger than microorganisms, with diameters reaching 100 μm and volumes up to 10^4 times those of the prokaryotes. However, they are not very strong in terms of resistance to environmental conditions, not only to mechanical stresses but also to shocks of nutritional and osmotic nature and to temperature, pH and oxygen concentrations. These problems may arise not only during the utilization of the cells but also during their immobilization.

One problem in the procedure of immobilizing plant cells is to be able to obtain a large amount of suspended cells. Cell growth after immobilization is discouraged and therefore the cells used must already be in the needed quantities and must be predisposed to the production of the required metabolites.

Secondary products are usually best synthesized in slow-growing, differentiated cells, however, immobilization can induce the expression of the desired metabolism through physical changes in the environment of cells caused, for instance, by proximity. A further problem which has not been fully investigated, concerns the possible need of light for the attainment of certain products. Continuous or periodic exposure to light would then be necessary, this involving further problems in terms of immobilization, as the support would then have to be transparent to light.

Considering the immobilization methods that have been used for other biocatalysts, it is legitimate to think that entrapment methods are naturally those which are most suitable for these biocatalysts.

At first sight, the establishment of a linkage between the cells and the support raises problems which are due particularly to the size and weight of the plant cell. However, in some cases, the linkage is possible provided that the surface of the support is conveniently prepared. After the contact between cells and support, adhesion properties of the former are bound to have a responsibility in maintaining the linkage.

Entrapment methods, being used more frequently, will be reviewed in more detail. Four variations of these methods can be distinguished, namely gel immobilization by polymerization, ionic network formation, precipitation and the use of preformed structures.

a) Gel entrapment by polymerization

This method has been widely used for enzymes and has been applied to microorganisms.

A monomer or a mixture of monomers is polymerized in the presence of a cell suspension which will be entrapped inside the lattice of the polymer (Table 1).

The most common example is polyacrylamide. The method is based on the free radical polymerization of acrylamide in an aqueous solution. As the linear polymers are soluble in water, they have to be insolubilized with bifunctional compounds such as N, N' methylenebisacrylamide.

The free radical polymerization of acrylamide is conducted in an aqueous solution containing the cells and the cross-linking agent. Polymerization is commonly carried out in the absence of oxygen and at lower temperatures (10°C) than those usually used in this polymerization to avoid damage to the cells during the operation. An initiator N, N, N', N' tetramethylethylenediamine (TEMED) is used.

Both the initiation and the cross-linking agents are toxic to the cells and therefore their viability can be lost. This has been the case with plant cells, for instance with *Catharanthus roseus* (Brodelius 1984) and *Silybum marianum* (Cabral et al. 1984).

Table 1. Gel entrapment by polymerization

Species	Immobilization method	Reference
Catharanthus roseus	Polyacrylamide	Brodelius 1984
Silybum marianum	Polyacrylamide	Cabral et al. 1984
Mentha sp.	Prepolymerized polyacrylamide	Galun et al. 1983
Catharanthus roseus	Polysaccharide /polyacrylamide	Lambe and Rosevear 1983
Nicotiana sp.	Polysaccharide /polyacrylamide	Rosevear 1981
Lavandula vera	Polyvinylalcohol	Nakajima et al. 1986

However, it was possible to use prepolymerized, linear, water-soluble polyacrylamide, partially substituted with acrylhydrazide groups. Gelation was effected in the presence of the plant cells under mild conditions by the addition of controlled amounts of a dialdehyde (Galun et al. 1983). Plant cells have also been entrapped in polyacrylamide after being mixed with a viscous polysaccharide solution of alginate or xantham gum. In these cases, they would retain up to 90% of their activity (Rosevear et al. 1982).

The use of a mixture of cells with a prepolymer of polyvinylalcohol with styrylpyridinium groups subsequently irradiated with fluorescent lamps provided an immobilization complex which was judged to be superior to alginate in terms of cell growth and product production (Nakajima et al. 1986).

b) Gel entrapment by ionic network formation

In this case polymerization of polyelectrolytes is obtained by the addition of multivalent ions. The most common method is the entrapment in calcium alginate. This is a non-toxic process in which a sodium alginate solution containing the cell suspension is dropped into a mixture of a counterion solution such as calcium chloride. A uniform, spherical, highly microporous structure results, which retains the cells.

Table 2. Gel entrapment by ionic network formation

Species	Gel	Reference
Digitalis lanata	Alginate	Brodelius et al. 1979 Brodelius et al. 1981 Alfermann et al. 1983
Daucus carota	Alginate	Jones and Veliky 1981 Veliky and Jones 1981
Catharanthus roseus	Alginate Carrageenan, alginate Alginate	Brodelius et al. 1979 Brodelius and Nilsson 1980 Brodelius et al. 1981
Morinda citrifolia	Alginate	Brodelius et al. 1980
Lavandula vera	Alginate	Nakajima et al. 1985
Papaver somniferum	Alginate	Furuya et al. 1984

It has been applied to several types of plant cells with success in terms of retention of viability (see Table 2). The method may be inconvenient over long runs in media containing calcium chelating agents, such as phosphates and certain cations such as Mg^{2+} that may accelerate the disruption of the gel by solubilizing the bound Ca^{2+}.

Kappa-carrageenan is also a suitable matrix for the immobilization of plant cells. Calcium and potassium carrageenates gel at room temperature and are very stable at a pH above 4.5.

c) Gel entrapment by precipitation

Gels may be formed by precipitation of some natural and synthetic polymers by changing one or more parameters in the solution, such as temperature, salinity, pH or solvent.

Several materials can be used to apply this method and examples which have been applied to plant cells have involved thermal variations. This was the case by gelatin, agar, agarose and carrageenan (Table 3). Gelatin, however, is only effective if reticulation with glutaraldehyde is carried out and this is lethal to the viability of the cells.

In the methods involving thermal treatment some disruption of viability can naturally occur.

Table 3. Gel entrapment by precipitation

Species	Gel	Reference
Catharanthus roseus	Agarose Agarose, agar	Felix et al. 1981 Brodelius and Nilsson 1980
Silybum marianum	Agar	Cabral et al. 1984

d) Entrapment in preformed structures

Hollow fiber reactors can be used to immobilize plant cells by entrapment (Table 4). The cells are placed on the shell side of the reactor and nutrient medium is rapidly recirculated through the fibers. This may have important applications in large-scale operations. Low and medium molecular weight nutrients and cellular products pass freely through the membrane, while the cells are held between the fibers (Shuler 1981).

Table 4. Entrapment in preformed structures

Species	Procedure	Reference
Glycine max	Hollow-fiber reactors	Shuler 1981
Petunia hybrida	Hollow-fiber reactors	Jose et al. 1983
Lavandula vera	Urethane prepolymers	Tanaka et al. 1984
Capsicum frutescens	Preformed polyurethane foam	Lindsey et al. 1983, Mavituna and Park 1985
Hop and beetroot	Nylon pan scrubbers	Rhodes et al. 1982
Dioscorea deltoidea	Preformed polyurethane foam	Robertson and Ishida 1986

In other examples, the cells are added to preformed polymerized structures such as polyurethane foam. When cells in suspension are mixed with these materials (usually in the form of spheres or cubes up to 1 cm diameter), they are rapidly incorporated into the network and subsequently grow into the cavities of the mesh and become entrapped either by physical restriction or by attachment to the matrix material.

The mechanism of this involves at first a mechanical entrapment and later, the fixation of the cells due to mechanisms of adsorption and adhesion or even to their natural tendency for aggregation.

These methods of immobilization have several advantages over other methods in that they are simple, cheap, gentle and rapid, and mantain cellular functions.

Other methods of immobilization

a) Chelation

Gelatinous hydrous metal oxides (of Ti IV, Zr IV, Fe III, V III and Sn II) are capable of forming insoluble complexes which retain suspended cells.

This method of immobilization involves the replacement of hydroxyl groups on the surface of the hydrous metal oxide by suitable ligands from the cell, resulting in the formation of partial covalent bonds. Ti IV hydrous oxide is more effective under acidic conditions and Zr IV hydrous oxide more effective at neutral or higher pH. *Silybum marianum* was immobilized by this technique with a large retention of its enzyme activity (Cabral et al. 1984).

b) Covalent linkage

At least in one case a covalent linkage was established for the immobilization of *Solanum aviculare* in polyphenylenoxide beads preactivated with glutaraldehyde.

This method has seldom been used but offers large possibilities as some of the available methods are mild and should not involve a decrease in viability (Jirku et al. 1981).

EFFECTS OF IMMOBILIZATION ON THE KINETICS AND PROPERTIES OF LIVING PLANT CELLS

Immobilized cells exist in a heterogeneous two-phase system and therefore a decrease of catalytic activity over free cells can be verified due to a different partition of substrate and product concentrations between the immobilized and the aqueous phase.

Mass-transfer diffusional effects arising from diffusional resistances to the flow of substrate from the bulk solution to the catalytic sites, and the product diffusion of the reaction back to the bulk of the solution, may operate. They can be internal, i.e. diffusion is inside the immobilized particle, and external, which refers to the mass-transfer effects between the bulk solution and the outer surface of biocatalyst particles.

The presence of these diffusional limitations reduces the efficiency of the system and may alter the cellular properties by creating micro-environments within the immobilization matrix which differ from the bulk solution. However, with plant cells, diffusion limitations may be reduced because of the relatively slow metabolism.

Besides the influence of these factors, other properties of the whole cell, namely the operational stability of the immobilized cell, can change.

This stability can be affected by several factors, such as microbial contamination, cell leakage, cell lysis and the effect of endogenous proteases. However, the stability of immobilized plant cells appears to be greater than that of the corresponding cells in suspension.

Again, very often the products of cell metabolism are intracellular and their attainment from suspended cell cultures is destructive. In the case of immobilized cells, product release has to be induced without destroying the cells. This has been possible in at least one case by permeabilization with dimethylsulphoxide (DMSO) (Brodelius and Nilsson 1983). Suspended cells treated with DMSO tend to lose their physical stability but this is not so problematic with immobilized cells as the matrix will contribute to the retention of the physical shape. However, this problem of permeabilization remains to be solved in most cases of plant cells immobilization.

All these factors must naturally be taken into consideration in the design of a suitable reactor to carry the immobilized cells.

IMMOBILIZATION OF PROTOPLASTS

To conclude, a short reference is due on the immobilization of protoplasts.

Particularly, all that which has been stated with regards to plant cells can be applied to protoplasts, except that they are still weaker from the viewpoint of a mechanical stress, since they lack cell walls.

However, this problem can, in certain cases, be solved by immobilization; immobilized protoplasts are bound to have an increased mechanical resistance which is conferred to them by the strength of the entrapping matrix (Linse and Brodelius 1984; Cabral et al. 1984).

CONCLUSION

In conclusion, plant cell and protoplast immobilization is becoming an established technique and although the development of bioreactors and the choice of systems has been slow, there are no real problems which may hinder utilization, apart from the need to increase the effort in research both from scientists and from the technologists who need support from the investors and the industrialists.

The applications may not yet be ready but it is certain that they will come and the utilization of plant cells will play an important part in the production of fine chemicals.

REFERENCES

Alfermann AW, Bergmann W, Figur C, Helmhold U, Schwantag D, Shuler I and Reinhard E (1983) In Mantell SH and Smith H Eds. Plant biotechnology, Cambridge Univ Press pp 67-74

Brodelius P (1984) Annals N Y Acad Sci 434:382-393

Brodelius P (1985) Trends in biotechnology 3:280-285

Brodelius P, Deus B, Mosbach K, Zenk MH (1979) Febs Lett 103: 93-97

Brodelius P, Deus B, Mosbach K, Zenk MH (1981) European Patent Application 80850105.0

Brodelius P, Deus B, Mosbach K, Zenk MH (1980) In Weetall HH and Royer GP Eds. Enzyme engineering, Plenum Press 5:373-381

Brodelius P and Nilsson (1980) Febs Lett 122:312

Cabral JMS, Fevereiro P, Novais JM and Pais MS (1984) Annals N Y Acad Sci 434:501-503

Felix HR, Brodelius P and Mosbach K (1981) Analyt Biochem 116: 462-470

Furuya T, Yoshikawa T and Taira M (1984) Phytochemistry 25:999-1001

Galun E, Aviv D, Dantes A and Freeman A (1983) Planta Med 49: 9-13

Jirku VT, Macek T, Vanerr T, Krumphanzl V and Kubanek V (1981) Biotechnol Lett 3:447-450

Jones A and Velicky IA (1981) Eur J Appl Microbiol biotechnol 13:447-450

Jose W, Pederson H and Chin CK (1983) Annals N Y Acad Sci 413: 409-412

Lambe and Rosevear (1983) In Proceedings of Biotech' 83, London: 565-567

Lindsey K, Yeoman MM, Black GM and Mavituna F (1983) Febs Lett 155:143-149

Linse L and Brodelius P (1984) Annals N Y Acad Sci 434:487-490

Mavituna F and Park JM (1985) Biotechnol Letters 7:637-640

Morris P, Scragg AH, Smart NJ and Strafford A (1985) In Dixon Ed. Plant cell culture, a practical approach, IRL Press pp 127-167

Nakajima H, Sonomoto K, Usui N, Sato F, Yamada Y, Tanaka A and Fukui S (1985) J Biotechnol 2:107-117

Nakajima H, Sonomoto K, Morikawa H, Sato F, Ichimura K, Yamada Y and Tanaka A (1986) Appl Microbiol Biotechnol 24:266-270

Rhodes MJC (1982)

Robertson GH and Ishida BK (1986) In The World Biotechnol Report 1986 2(6):83-90, On line

Rosevear A (1981) Europ Pat Applic 8130400

Rosevear A (1983) In: Topics in enzyme and fermentation biotechnology 7:13

Rosevear A and Lambe CA (1985) In Flechter A Ed. Advances in biochemical engineering, biotechnology, Springer 31:37-58

Shuler ML (1981) Annals N Y Acad Sci 369:65-80

Tanaka A, Sonomoto K and Fukui S (1984) Annals N Y Acad Sci 434:479-482

Veliky IA and Jones A (1981) Biotechnol Lett 3:551-554

Tanaka A, Sonomoto K and Fukui S (1984) Annals N Y Acad Sci 434:479-482

Veliky IA and Jones A (1981) Biotechnol Lett 3:551-554

CHARACTERIZATION AND IMMOBILIZATION OF DIGITOXIN 12β-HYDROXALASE FROM CELL CULTURES OF DIGITALIS LANATA EHRH.

M. Petersen[1], A.W. Alfermann[1] and H.U. Seitz[2]

INTRODUCTION

For many years the biotransformation of digitoxin or β-methyldigitoxin to digoxin or β-methyldigoxin (Fig. 1) by cell cultures of Digitalis lanata has been studied (Reinhard 1974). This reaction is of pharmaceutical interest, since cardiac glycosides, mainly of the C-series, are among the most frequently used drugs against congestive heart failure. The enzyme catalyzing the 12β-hydroxylation was described for the first time by Petersen and Seitz (1985). It could be shown that this enzyme is another cytochrome P-450-dependent monooxygenase from higher plants. Among the best-studied cytochrome P-450-containing enzymes from plants are the cinnamic acid 4-hydroxylase (Russell 1971; Benveniste et al. 1977, and many others), the flavonoid 3'-hydroxylase (Formann et al. 1980; Larson and Bussard 1986) or the monoterpene hydroxylase (Meehan and Coscia 1973; Madyastha et al. 1976).

Immobilization of plant cells or enzymes for production of secondary metabolites is a challenging task. Ca-alginate for immobilization of plant cells producing different secondary products was first used by Brodelius et al. (1979). It was also used to immobilize a mammalian cytochrome P-450-dependent enzyme from liver microsomes (Ibrahim et al. 1986).

[1]Institut für Entwicklungs- und Molekularbiologie der Pflanzen, Universität Düsseldorf, Universitätsstraße 1, D-4000 Düsseldorf, FRG

[2]Institut für Biologie I, Universität Tübingen, Auf der Morgenstelle 1, D-7400 Tübingen, FRG

NATO ASI Series, Vol. H18
Plant Cell Biotechnology. Edited by M.S.S. Pais et al.

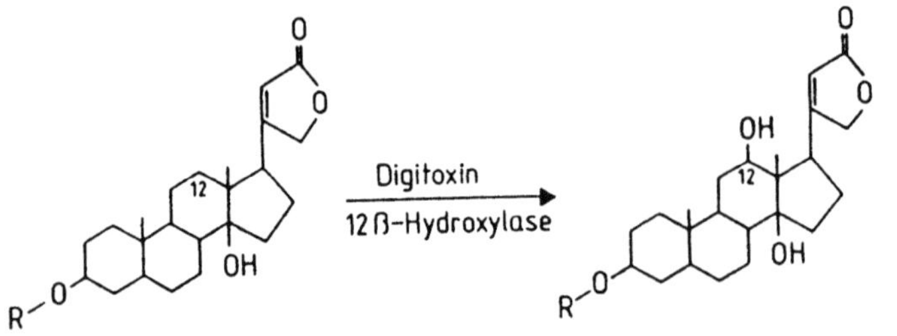

Fig. 1. 12β-hydroxylation by digitoxin 12β-hydroxylase:

R=H digitoxigenin → digoxigenin
R=Dx-Dx-Dx digitoxin → digoxin
R=Dx-Dx-Dx-methyl β-methyldigitoxin → β-methyl-digoxin

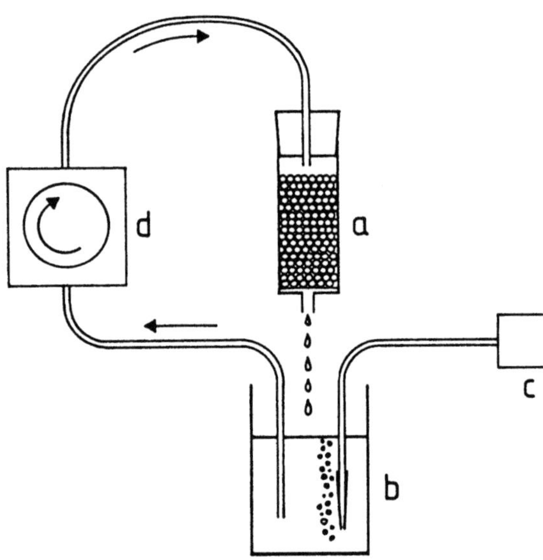

Fig. 2. Continuous flow model system for the biotransformation of β-methyldigitoxin to β-methyl-digoxin by digitoxin 12β-hydroxylase immobilized by entrapment in Ca-alginate. _a_ Column with the alginate beads containing the enzyme; _b_ buffer reservoir; _c_ membrane pump for aerating the buffer; _d_ peristaltic pump

Here, we report on some characteristics of digitoxin 12β-hydroxylase from _Digitalis lanata_ cell cultures and on the immobilization of the enzyme by entrapment of microsomes in a Ca-alginate gel.

MATERIAL AND METHODS

Cell cultures of _Digitalis lanata_ were cultivated as described earlier (Petersen and Seitz 1985).

Preparation of microsomes: 4- to 6-day-old suspension cells of _Digitalis lanata_ were homogenized in a mortar with quartz sand and 1 ml buffer per 1 g fresh weight of cells. The buffer contained 100 mM Tris/HCl pH 7.5, 1 mM DTT, 1 mM EDTA and 0.6 M mannitol. After filtration through Miracloth and centrifugation for 20 min at 8000 g a 1 M $MgCl_2$ solution was added to the supernatant up to a final concentration of 50 mM $MgCl_2$. The preparation was stirred for 20 min on ice, then the membranes were sedimented at 48 000 g for 20 min. The sediment of microsomes was resuspended in buffer up to a final protein concentration of 0.5 mg ml^{-1} for experiments with freely suspended microsomes and 3 mg ml^{-1} for immobilization experiments. The protein concentration was measured according to Bradford (1976).

Immobilization of the microsomes: A microsomal suspension (protein concentration 3 mg ml^{-1}) was diluted 1:5 with a sodium alginate solution in buffer up to a final alginate concentration of 2%. This suspension was dripped into a fivefold volume of ice-cold buffer containing 25 mM $CaCl_2$, which was vigorously stirred by pressing it through a syringe with an injection needle. The resulting beads had a diameter of 2-3 mm. After a hardening time of 2 h the alginate beads were washed in Ca-free buffer three times. 4 g of these alginate beads were used for a digitoxin 12β-hydroxylase assay: The reaction mixture contained 20 μM β-methyldigitoxin (dissolved in 80% methanol) and a NADPH-regenerating system (1 mM $NADP^+$, 5 mM glucose-6-phosphate, 8.4 nkat glucose-6-phosphate dehydrogenase per milliliter) and 2.5-3 mg of microsomal protein or 4 g of alginate beads containing the enzyme in the above mentioned buffer. The total volume was 5 ml for reactions with freely suspended microsomes or 10 ml for experiments with the immobilized enzyme. The reaction proceeded for 4 h at 26°C. Assays with the immobilized enzyme were shaken at 100 rpm to assure the oxygen supply. The cardiac glycosides were extracted into chloroform, washed with petroleum benzine (bp. 40-60°C) and analyzed by HPLC as described previously (Petersen and Seitz 1985).

Continuous flow system: 20 g of alginate beads with the immobilized enzyme were packed into a small column. Then 50 ml of buffer with the appropriate substrate concentrations was pumped continuously through the column and dripped back into the buffer reservoir, which was aerated by a membrame pump. The flow was 50 ml min^{-1}. The system was analyzed for 12β-hydroxylated cardiac glycosides as described above.

RESULTS AND DISCUSSION

Digitoxin 12β-hydroxylase is a cytochrome P-450-dependent enzyme. This was shown by a 40% inhibition of the reaction in a gas phase of 15% O_2, 15% CO and 70% N_2 and the reversal of this inhibition by illumination with 450 nm light. The enzyme needs, besides the substrate (β-methyl-)digitoxin, O_2 and NADPH. NADH could not replace NADPH. The optimal concentration of NADPH either supplied directly or by the regenerating system was 1 mM. The apparent Km value for NADPH was determined to be 26 μM. No synergistic effect of NADP(H) and NADH in the presence or absence of ATP could be observed. This indicates that no transhydrogenases are present in the enzyme preparation and makes a participation of cytochrome b_5 in the electron transport unlikely. β-Methyldigitoxin and digitoxin were hydroxylated at apparent Km values of 7.1 and 10 μM respectively. Cardiac glycosides with the aglycone structure of digitoxigenin (see Fig. 1) were hydroxylated by the enzyme irrespective of their glycosylation degree. A product inhibition by 12β-hydroxylated cardiac glycosides could not be observed. The digitoxin 12β-hydroxylase was stimulated by addition of reducing agents, DTT at a concentration of 1 mM being the best. As inhibitors acted, besides CO, p-hydroxymercuribenzoate, Hg^{2+} and Cu^{2+}. A competitive inhibition was shown for cytochrome c and $NADP^+$. For $NADP^+$ the inhibition constant was 1.1 mM.

The pH optimum was determined to be pH 7.5. Digitoxin 12β-hydroxylase is a membrane-bound enzyme, which is found in the microsomal fraction after $MgCl_2$ precipitation. By sucrose density gradient fractionation and comparison of the distribution of the hydroxylase and marker enzymes, it could be shown that the enzyme is present in the endoplasmic reticulum.

Immobilization of the digitoxin 12β-hydroxylase was tried by several different methods: adsorption to poly-L-lysine-coated glass beads, entrapment in polyacrylamide hydrazide according to Yawetz et al. (1984) and entrapment in k-carrageenan and Ca-alginate (Ibrahim et al. 1986). The best procedure was the entrapment of microsomes in Ca-alginate beads. After the immobilization procedure the enzyme retained 60-70% of its original activity. This is a relatively good value, since often immobilized enzymes lose about 50% of their original activity or the enzymes are less stable after immobilization (Ibrahim et al. 1986). Since the digitoxin 12β-hydroxylase is an extremely stable enzyme, the hydroxylation reaction of freely suspended microsomes or the immobilized enzyme could be observed for more than 20 h without a stubstantial decrease in the reaction rate.

Table 1: Some characteristics of digitoxin 12β-hydroxylase in freely suspended or immobilized microsomes.

	Freely suspended microsomes	Immobilized enzyme
pH-optimum	pH 7.5	pH 7.5
Optimal concentrations of		
- β-methyldigitoxin	20 μM	20 μM
- NADPH	1 mM	1 mM
Stability of hydroxylation reaction at 26°C	>20 h	>20 h
Reaction yield	100%	60-70%

Therefore, the enzyme is suitable for long time biotransformation reactions. The kinetic parameters of digitoxin 12β-hydroxylase were not altered after the immobilization procedure as can be seen in Table 1. This shows, that the enzyme is not damaged during the immobilization procedure and that all carboxylic groups of the alginate are saturated by Ca^{2+}. No loss of protein or enzyme activity out of the Ca-alginate gel could be observed showing that the gel is dense enough to retain the plant microsomes.

A continuous flow system was established as a model system for the use of the immobilized enzyme for a biotransformation reaction (Fig. 2). In this continuous flow system the same reaction rates could be observed as with the immobilized microsomes in a shake incubation. The only limiting factor for the immobilized enzyme was a sufficient osygen supply, which could be assured by aerating the buffer reservoir by a membrane pump.

Since the biotransformation rate by the digitoxin 12β-hydroxylase in freely suspended microsomes or immobilized in alginate beads is about five times lower than by Digitalis lanata cells (Alfermann et al 1974), this reaction is not too useful for a pharmaceutical application, but our model system shows the possibility to immobilize even such fragile enzymes like cytochrome P-450-dependent monooxygenases. The method of immobilizing membrane-bound enzymes by entrapment in Ca-alginate beads could serve as a model system for other plant enzymes.

REFERENCES

Alfermann AW, Bergmann W, Figur C, Helmbold U, Schwantag D, Schuller I and Reinhard E (1974) Biotransformation of β-methyldigitoxin to β-methyldigoxin by cell cultures of Digitalis lanata. In: Mantell SH and Smith H (eds). Plant biotechnology, Cambridge University Press, Cambridge:67-74

Benveniste I, Salaün JP and Durst F (1977) Wounding-induced cinnamic acid hydroxylase in Jerusalem artichoke tuber. Phytochemistry 16:69-73

Bradford MM (1976) A rapid and sensitive method for the quantitation of microgram quantities of protein utilizing the principle of protein-dye binding. Anal. Biochem 72:248-254

Brodelius P, Deus B, Mosbach K, Zenk MH (1979) Immobilized plant cells for the production and transformation of natural products. FEBS Lett 103:93-97

Forkmann G, Heller W and Grisebach H (1980) Anthocyanin biosynthesis in flowers of Matthiola incana - flavanone 3 - and flavonoid 3'-hydroxylases. Z Naturforsch 35c:691-695

Ibrahim M, Decolin M, Batt AM, Dellacherie E and Siest G (1986) Immobilization of pig liver microsomes. Stability of cytochrome P-450-dependent monooxygenase activities. Appl Biochem Biotechnol 12:199-213

Larson RL and Busard JB (1986) Microsomal flavonoid 3'-hydroxylase from maize seedlings. Plant Physiol 80:483-486

Madyastha KM, Meehan TD and Coscia CJ (1976) Characterization of a cytochrome P-450-dependent monoterpene hydroxylase from the higher plant Vinca rosea. Biochemistry 15:1097-1102

Meehan TD and Coscia CJ (1973) Hydroxylation of geraniol and nerol by a monooxygenase from Vinca rosea. Biochem Biophys Res Comm 53:1043-1048

Petersen M, Alfermann AW, Reinhard E, Seitz HU (1987) Immobilization of digitoxin 12β-hydroxylase, a cytochrome P-450-dependent enzyme from cell cultures of Digitalis lanata Ehrh. Plant Cell Reports 6:200-203

Petersen M, Seitz HU (1985) Cytochrome P-450-dependent digitoxin 12β-hydroxylase from cell cultures of Digitalis lanata. FEBS Lett 188:11-14

Reinhard E (1974) Biotransformations of plant tissue cultures. In: Street HE (ed). Tissue culture and plant science 1974, Academic Press London New York San Francisco:433-459

Russell DW (1971) The metabolism of aromatic compounds in higher plants X. Properties of the cinnamic acid 4-hydroxylase of pea seedlings and some aspects of its metabolic and developmental control. J Biol Chem 246:3870-3878

Yawetz A, Perry AS, Freeman A and Katchalski-Katzir E (1984) Monooxygenase activity of rat liver microsomes immobilized by entrapment in a cross-linked prepolymerized polyacrylamide hydrazide. Biochem Biophys Acta 788:204-209

THE EFFECT OF OXYGEN STRESS ON SECONDARY METABOLITE PRODUCTION BY IMMOBILISED PLANT CELLS IN BIOREACTORS

A.K. Wilkinson, P.D. Williams and F. Mavituna
Department of Chemical Engineering
University of Manchester Institute of Science & Technology
P.O. Box 88
Sackville Street
Manchester M60 1QD

INTRODUCTION

A novel bioreactor has been designed to produce secondary metabolites from immobilised plant cell cultures (Mavituna et al, 1987). We have used the production of capsaicin, the hot chilli flavour, by *Capsicum frutescens* as a model system. This is a particularly suitable system for an immobilised cell process since capsaicin is excreted into the liquid medium.

MATERIALS AND METHODS

Callus, freely suspended and immobilised cultures of *Capsicum frutescens* were grown in Schenk and Hildebrant medium at pH 5.8 and 25°C. This medium was used throughout the bioreactor operation reported in Figure 1. It was also used for the first 75 days of bioreactor operation shown in Figure 2 followed by the replacement of this with a modified Schenk and Hildebrandt medium free of nitrogen and hormones.

The reticulated polyurethane foam material used for immobilisation was obtained from Declon (Corby, Northants, UK). Capsaicin was analysed by HPLC and the pure form used as standard was supplied by Sigma.

The cells were immobilised in cubic polyurethane foam particles in situ in the bioreactor. The foam particles were held stationary as a packed bed during immobilisation and later circulated by off-centre air sparging.

EFFECT OF OXYGEN ON GROWTH

With both suspension and immobilised cultures, the concentration of dissolved oxygen affects both the maximum specific growth rate and the biomass productivity. The optimum range of

NATO ASI Series, Vol. H18
Plant Cell Biotechnology. Edited by M.S.S. Pais et al.

the mass transfer coefficient for oxygen, k_La, was found to be 10-15 h^{-1} to give the maximum values of specific growth rate (Wilkinson, 1985). In addition the specific oxygen uptake rate was found to be typically 2.0 x $10^{-3}h^{-1}$ and 1.0 x $10^{-3}h^{-1}$ for suspension and immobilised cultures, respectively.

EFFECT OF OXYGEN ON CAPSAICIN PRODUCTION

Dissolved oxygen concentration affects the appearance of capsaicin in the medium. In the presence of dissolved oxygen at typically 60% saturation, capsaicin was only found in the medium at low concentrations (less than 1 mg/l) and with the use of metabolic precursors, such as phenylalanine or vanillylamine. However, Figure 1 shows that if the dissolved oxygen concentration is allowed to fall to 0-5% the concentration of capsaicin in the medium rises significantly (eg. 17 mg/l). On the restoration of aeration capsaicin can disappear from the medium within 6 hours.

During air sparging some volatile compounds, such as ethylene, and dissolved gases, such as carbon dioxide, may be stripped from the medium (Maurel and Pareilleux, 1985). These compounds may need to be maintained at some critical concentration to stimulate secondary metabolism. In order to distinguish between the effect of stripping and the effect of oxygen concentration, in a separate run the bioreactor was sparged with nitrogen. As can be seen in Figure 2, capsaicin again appeared in the medium during nitrogen sparging. This indicates that it is the reduced dissolved oxygen concentration which stimulates capsaicin production.

Capsaicin disappears from the medium when aeration is restored. In order to investigate the cause of this phenomenon a sample of the medium from the bioreactor which contained capsaicin produced during a non-aerated period was subjected to the following conditions:

- aerated for 24 hours,
- aerated for 24 hours with freely suspended cells of *C.frutescens*,

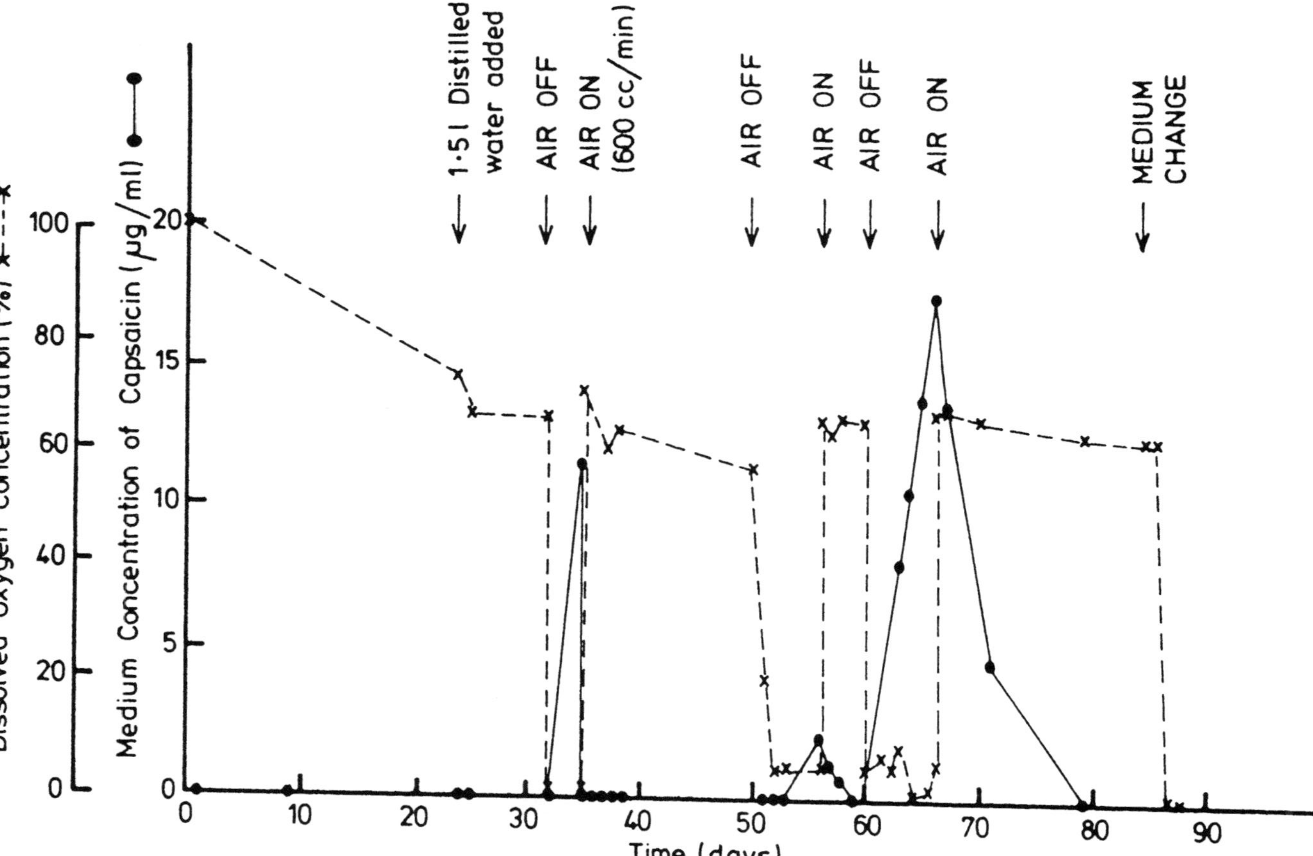

Fig.1. The effect of dissolved oxygen concentration on the production of capsaicin by immobilised <u>C.frutescens</u> in 6 litre bioreactor

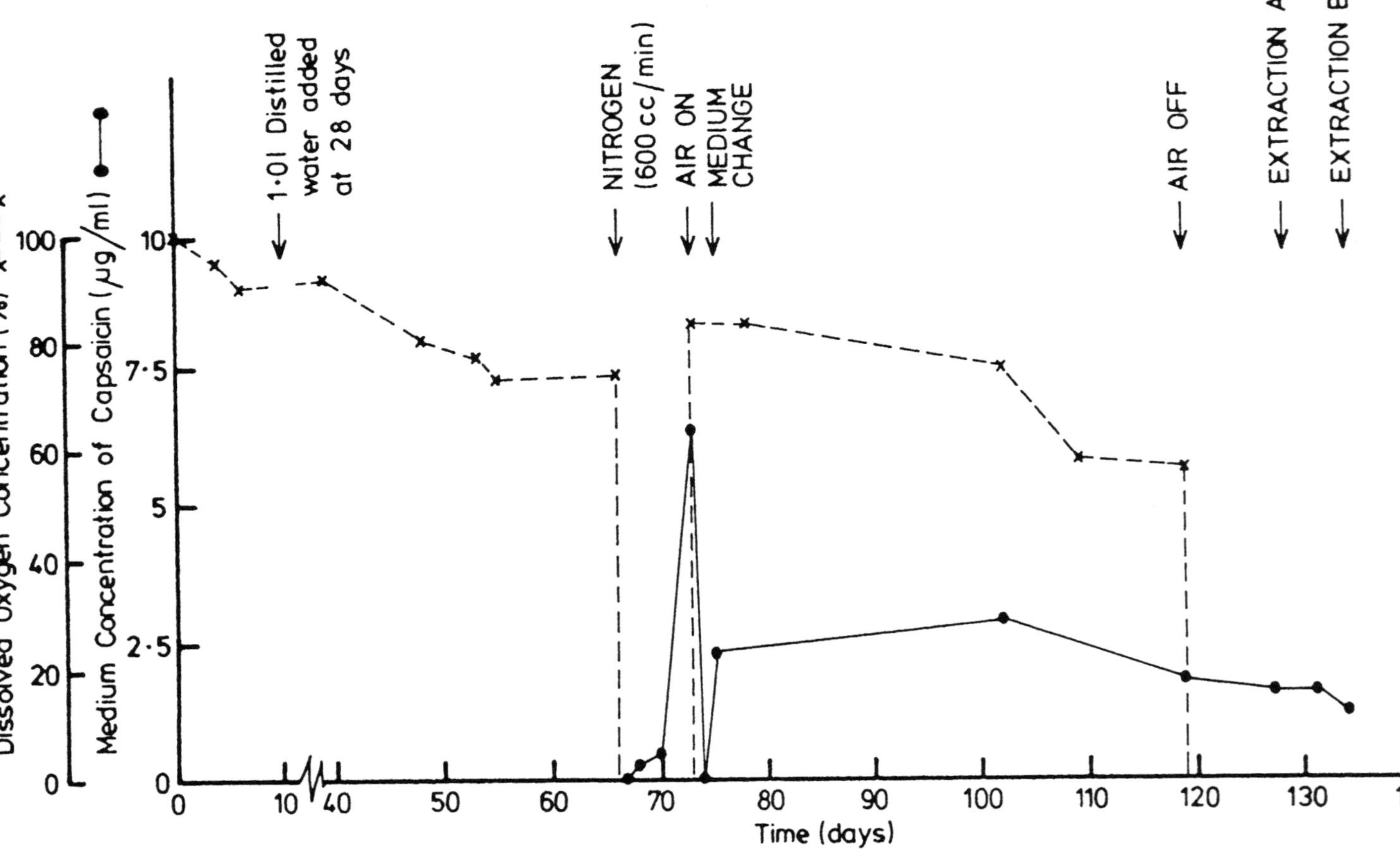

Fig.2. The effect of replacing sparged air with nitrogen on the production of capsaicin by immobilised <u>C.frutescens</u> in a 4 litre bioreactor

- contacted with freely suspended cells of C.frutescens in the absence of aeration.

Analysis of these samples revealed that capsaicin, which was present in the original medium, only disappeared when the medium was aerated in the presence of cells.

CONCLUSION

Relatively high concentrations of dissolved oxygen are required for the growth of plant cells. However, in our model system low concentrations of dissolved oxygen seem to be necessary for capsaicin production. The effect appears not to be associated with other dissolved gases such as ethylene and carbon dioxide. After production, if the dissolved oxygen concentration is increased capsaicin disappears from the medium. This only happens in the presence of cells. These results can be considered as an example of stress induced secondary metabolite production in plant cell cultures.

ACKNOWLEDGEMENTS

The authors would like to acknowledge the Biotechnology Directorate of the SERC and Albright & Wilson Ltd. for their financial support.

REFERENCES

Maurel B and Pareilleux A (1985) Effect of carbon dioxide on the growth of cell suspensions of Catharanthus roseus. Biotech Letts 7 313-318

Mavituna F, Park JM, Wilkinson AK and Williams PD (1987) Characteristics of immobilised plant cell bioreactors. In: Webb C and Mavituna F (eds) Plant and Animal Cells: Process Possibilities. Ellis Horwood Chichester, p92

Wilkinson, AK (1985) Studies on batch growth of suspension cultures of Capsicum frutescens Mill. c.v. annuum. MSc Thesis, UMIST, Manchester, UK.

CONTINUOUS COAGULATION OF MILK USING IMMOBILIZED CELLS OF CYNARA CARDUNCULUS

M.G. Esquîvel[1], M.M.R. Fonseca[2], J.M. Novais[2], J.M.S. Cabral[2] and M.S.S. Pais[3]

INTRODUCTION

Plant rennets are used in Portugal for the small-scale manufacture of certain varieties of cheese, namely "Serra" and "Serpa". In the traditional preparation of these types of cheese the clotting stage is induced by immersing the flowers of the native Compositae *Cynara cardunculus* or *Silybum marianum* in the milk vats. However, the flowers are, by themselves, a potential source of microbial contamination of the product. This problem could be overcome by plant cell culture. Previous work demonstrated that *in vitro* cultivated cells of both species present clotting activity (Fevereiro et al. 1986, Esquîvel et al. 1986), no clotting activity being found in the supernatant (Fonseca et al. 1987).

The caseins are found suspended in milk in the form of micelles. These colloidal aggregates, with an average diameter of 50-100 nm, contain four distinct insoluble proteins (α_{s1}, α_{s2}, β and κ) in a network of calcium-magnesium phosphate (Morr 1967). The main action of chymosin and of the majority of the other rennets is the hydrolysis of the k-casein (at the phenylalanine/methionine bond) in two segments, one hydrophobic and the other hydrophilic. The former stays in the micelle while the second

[1] Instituto Superior de Agronomia, Tapada da Ajuda, 1399 Lisboa Codex, Portugal

[2] Laboratório de Engenharia Bioquîmica, Instituto Superior Técnico, 1000 Lisboa, Portugal

[3] Dept. Biologia Vegetal, Fac. Ciências, 1294 Lisboa Codex, Portugal

NATO ASI Series, Vol. H18
Plant Cell Biotechnology. Edited by M.S.S. Pais et al.

segment goes into solution. The zeta potential of the micelle decreases as a result of the hydrolysis of the k-casein. This leads to micelle destabilization and subsequent aggregation (Darling and Dickson 1979). With the progress of flocculation the micelles interact to form chains which eventually cross-link to form a gel.

Rennets of animal and microbial origin rank second amongst industrial enzymes in terms of sales. At industrial scale they are added in soluble form to milk, 70-90% of the active enzyme being discharged from the cheese plant with whey (Holmes and Ernstrom 1973). In this context the coagulation of milk seems, at first glance, an ideal field for the application of immobilized enzymes (Carlson 1984).

In current cheese manufacture the conditions are such that the two phases of milk clotting (enzymatic and non-enzymatic) overlap considerably in time (Burgess and Shaw 1983). Continuous coagulation of milk using an immobilized enzyme source can only be achieved if (1) the structure of the enzyme matrix is such that contact between the k-casein molecules and the enzyme is possible and (2) separation between the enzymatic and non-enzymatic stages of milk clotting can be obtained in order to avoid clogging of the reactor by flocs.

However, the distinct kinetic characteristics presented by the two stages allow their separation. Most of the published work on continuous coagulation of milk (Carlson et al. 1986) has been based on the fact that the second phase presents a much higher activation energy than the first. Therefore, milk contacting with an immobilized clotting enzyme in refrigerated (temperature $<10^{o}C$) columns allows the hydrolysis to proceed at a reasonable rate while clotting is avoided (Cheryan et al. 1975; Ferrier et al. 1972; Hicks et al. 1975; Taylor et al. 1979). Coagulation takes place by subsequent warming of the column effluent. The major drawbacks encountered with the use of immobilized rennets include (1) short half-life of the catalyst, (2) leakage of the catalyst from the support and (3) difficulties with sanitation during long runs (Fox 1982; Carlson 1984; Carlson et al. 1986). The extent of hydrolysis required for micelle agglomeration to take place decreases with temperature. However, complete or nearly complete k-casein hydrolysis is required for the gel to acquire the proper firmness for cutting (Carlson 1984).

The continuous hydrolysis of milk casein at low temperature is described in this work using calcium-alginate entrapped cells of C. cardunculus in a fixed-bed bioreactor. Milk clotting is induced by raising the temperature of the column effluent.

MATERIAL AND METHODS

Materials: Sodium alginate from BDH, U.K. was used as the matrix for cell gel entrapment; other chemicals were purchased from BDH or Merck. Pasteurized cow milk (pH 6.8) was used as the substrate to test the clotting activity of the cells. The microbial rennet "Rennilase 150L" was a gift from Novo Industri, Denmark.

Cell suspension cultures: Cynara cardunculus suspension cultures were started from friable calli grown from young leaf slices and cultivated batchwise at pH 5.3 in the medium of Tuleike (1966) supplemented with 2,4-dichlorophenoxyacetic acid (1 mg l^{-1}) and kinetin (0.1 mg l^{-1}), on a rotary shaker at 25°C in the dark.

Cell immobilization and continuous coagulation system: The immobilization of C. cardunculus cells was carried out under sterile conditions according to the following method: a concentrated suspension of plant cells was mixed (1:1) with a solution of sodium alginate to give a final alginate concentration of 2.5% (w/v); this mixture was added dropwise to a 50 mM $CaCl_2$ solution; the alginate beads formed were left for 1 h in this solution for hardening and then washed in several batches of sterile water.

The flow system (Fig. 1) was autoclaved prior to filling with the alginate beads. A known dry weight of immobilized cells was loaded into the column (aspect ratio= 10) which was refrigerated at +8°C. The milk, after one passage through the column, was collected in a receiver maintained at +8°C and eventually transferred to a bath at 37°C to induce milk clotting.

The occurrence of cell leakage from the support was checked through the periodic observation of a sample of the column effluent under the microscope. No plant cells were ever detected.

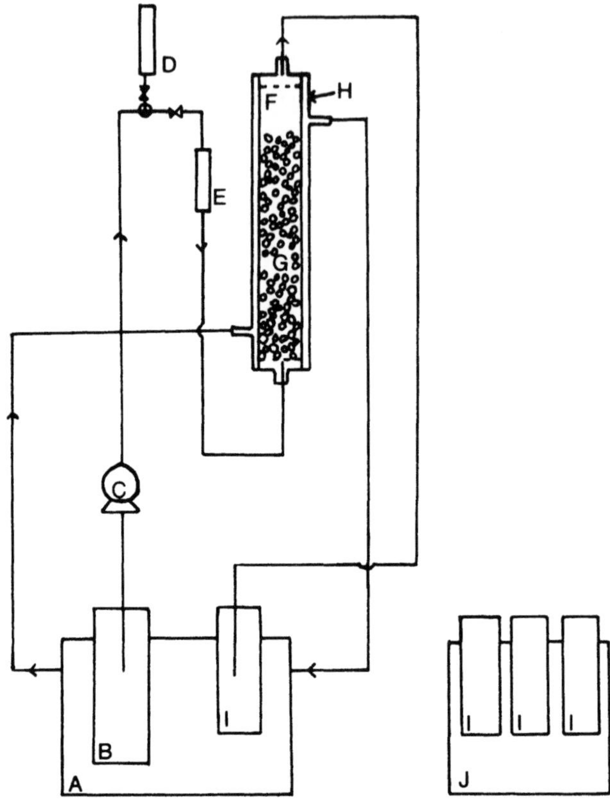

Fig. 1. Flow diagram. A water bath at 8°C; B milk reservoir; C peristaltic pump; D flow meter; E drip-feed; E drip-feed; F glass column (H/D=10); G fixed bed of Ca-alginate beads; H water jacket; I receiver; J water bath at 37°C.

In the blank experiments the column was filled with a similar number of alginate beads without cells, prepared according to the same method.

In order to test the performance of the continuous system with a commercial clotting enzyme, the system was operated with calcium alginate beads containing a rennet of fungal origin ("Rennilase 150L"). The preparation of the immobilized enzyme was performed as follows: the soluble commercial enzyme was diluted $1:10^4$ and mixed (1:1) with a solution of sodium alginate to give a final alginate concentration of 3.5% (w/v); the immobilization was continued as mentioned above for the case of plant cells.

<u>Assays:</u> Biomass was assayed through dry weight determination. Samples were collected on preweighed membrane filters (Whatman, 5-μm pore size) and dried at 80°C for 48 h.

The clotting activity was expressed as the time (h) for milk to clot at 37°C after one passage through the column. Values are averages based on three samples. The clotting time is inversely related to the enzymatic activity (Hicks et al. 1975). Sampling was always followed by the collection of a blank, i.e. a milk sample collected directly from the milk reservoir.

RESULTS AND DISCUSSION

Clotting times between 10.5 and 13 h were obtained in different runs on the second day of operation depending on the cell load of the beads and on the flow rate used (Table 1). This range of clotting times is likely to reflect the true activity of the immobilized plant cell reactor since (1) no clotting activity could be detected in the blank experiments before the fourth day of column operation (see below) and (2) clotting times of the same order of magnitude (10-12 h) were observed in batch coagulation experiments carried out with <u>C. cardunculus</u> cells immobilized under similar conditions.

These results demonstrate the feasibility of using immobilized plant cells with proteolytic activity in a fixed-bed reactor for the continuous hydrolysis of k-casein.

Table 1. The effect of immobilized cell load and milk residence time upon clotting time on the second day of operation of the column

Plant cell concentration in the reactor (mg dry wt. ml^{-1} milk)	Residence time (h)	Clotting time (h)
0.40	4.0	12.0
0.45	1.9	13.0
0.50	2.2	12.0
1.26	3.0	10.5

Separation between the enzymatic and non-enzymatic stages of milk clotting could be accomplished through proper temperature control. In all experiments the blank samples (see Material and Methods) showed a clotting time several hours higher.

The results suggest that a low concentration calcium alginate gel placed in a refrigerated column constitutes an appropriate immobilization matrix for the use of cells with clotting activity. Either the gel pores have a sufficient size to accomodate the large casein micelles or the concentration of free k-casein is considerably enhanced at low operational temperatures, as proposed by Rose (1968).

The coagulation time was found to be inversely proportional to the residence time of milk in the reactor for a fixed cell load (Fig. 2). Other workers (Cheryan et al. 1975; Hicks et al. 1975; Taylor et al. 1979) also found a decrease of the coagulation time as the milk contact time with the immobilized enzyme was increased.

Results in Fig. 2 are averages of three experiments spaced in time. It was noticed that the clotting time decreased with the time of operation of the column for a fixed residence time. In order to investigate the reason for this decrease, blank experiments were set up. Figure 3 compares coagulation times for a fixed residence time obtained with empty beads and with beads containing plant cells. In the blank experiments, the clotting times obtained for the column effluent and for the samples collected from the milk reservoir were similar (>12 h) during the first 3 days of operation. However, from the fourth day onwards, there was a significant contribution of the initially "empty" gel particles which, at that stage, had probably been invaded by the microorganisms of pasteurized milk. The progressive role played by the contaminants is seen in Fig. 3. It can be anticipated that if the enzyme extract were immobilized instead of the whole cells (or if cells with higher specific clotting activity could be obtained), the increase of the activity of the bed would make the contribution of the contaminants negligible (yet uncontrollable). An overnight washing of the column with sterilized distilled water allowed the removal of a substantial fraction of the contaminants. This suggests that at the initial stage of contamination of the bed the contaminants

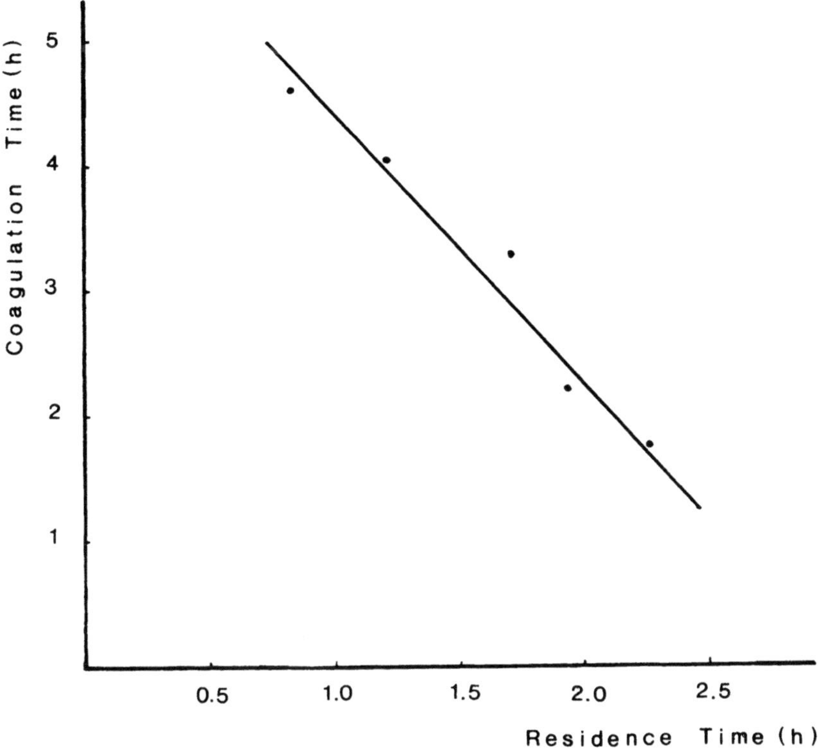

Fig. 2. Relationship between the residence time of milk in the column and the coagulation time using 20 mg (dry wt.) of immobilized cells (duration of experiment 12 days: 12 days)

develop predominantly on the surface of the beads. Thus, one can envisage column operation alternating with "disinfection" cycles.

For the sake of comparison, the performance of the reactor using immobilized C. cardunculus whole cells and an immobilized microbial rennet is shown in Table 2.

This work contributes to the elucidation of the potential use of immobilized biocatalysts in the continuous coagulation of milk.

Acknowledgements. We thank the Junta Nacional de Investigação Científica e Tecnológica (JNICT) for the financial support under Research Contract No. 520.83.47.

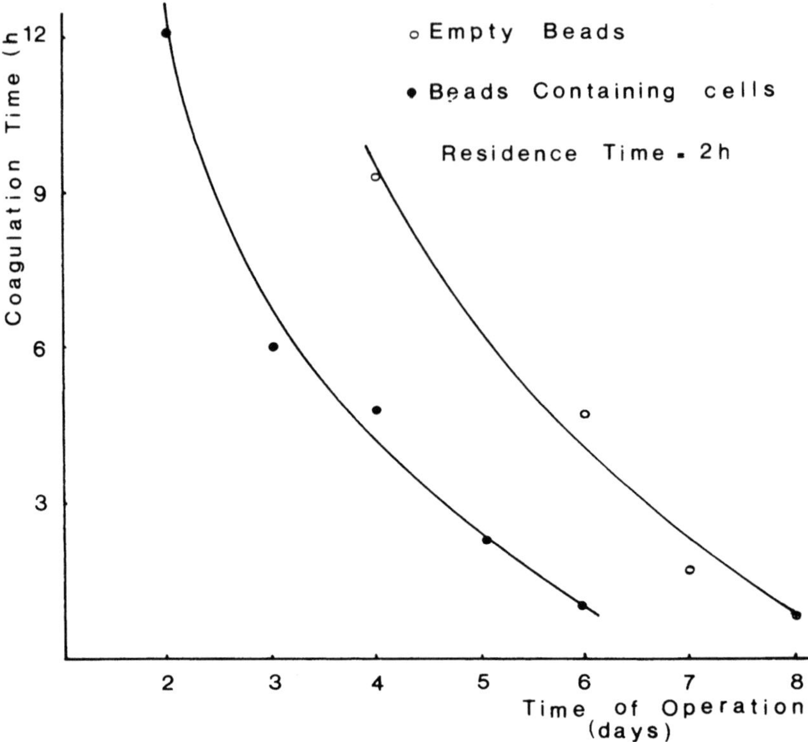

Fig. 3. Influence of the time of operation of the fixed bed of calcium alginate beads on the coagulation time

Table 2. Clotting activity of the column loaded with _C. cardunculus_ cells and with a commercial rennet

Catalyst	Residence time (h)	Clotting time (h)
_C. carduncul_us cells	0.85	4.75
"Rennilase"	0.13	0.50

REFERENCES

Burgess K and Shaw M (1983) The application of enzymes in industry. In Godfrey T and Reichelt J (eds) Dairy industrial enzymology. The Nature Press New York, p 260-283

Carlson A (1984) Coagulation of milk with immobilized enzymes. Enzyme Microb Technol 6:46-47

Carlson A, Hill GC and Olson NF (1986) The coagulation of milk with immobilized enzymes: critical review. Enzyme Microb Technol 8:642-650

Cheryan M, Van Wyk PJ, Olson NF and Richardson T (1975) Continuous coagulation of milk using immobilized enzymes in a fluidized bed reactor. Biotechnol Bioeng 17:585-598

Darling DF and Dickson J (1979) Electrophoretic mobility of casein micelles. J Dairy Res 46:441

Esquîvel MG, Fonseca MMR, Cabral JMS, Novais JM, Fevereiro P and Pais MSS (1986) Coagulaçao do leite utilizando células de Cynara cardunculus. Proc. 3º Encontro Nac. Biotechnol. Lisbon p. 55

Ferrier KL, Richardson T, Olson NF and Hicks CL (1972) Characteristics of insoluble pepsin used in a continuous milk clotting system. J Dairy Sci 55:726-734

Fevereiro P, Cabral JMS, Fonseca MMR, Novais JM and Pais MSS (1986) Callus and suspension culture of Silybum marianum. Biosynthesis of proteins with clotting activity. Biotechnol Lett 8:19-24

Fonseca MMR, Cabral JMS, Fevereiro P, Pais MSS and Novais JM (1987) Milk clotting with free and immobilized plant cells of Silybum marianum Ann N Y Acad Sci 501:358-361

Fox PF (1982) Exogenous enzymes in dairy technology. In Dupuy P (ed) Use of enzymes in food technology. Tec and Doc Lavoisier, Paris pp 135-158

Hicks CL, Ferrier LK, Olson NF and Richardson T (1975) Immobilized pepsin treatment of skim milk and skim milk fractions. J Dairy Sci 58:19-24

Holmes DG and Ernstrom CA (1973) Distribution of milk-clotting enzymes between curd and whey and their survival during cheddar cheese making. J Dairy Sci 56:622

Morr CV (1967) Effect of oxalate and urea upon ultracentrifugation properties of raw and heated skim milk casein micelles. J Dairy Sci 50:1744-1751

Rose D (1968) Relation between micellar and sorum casein in bovine milk. J Dairy Sci:1897-1902

Taylor MJ, Olson NF and Richardson T (1979) Coagulation of skim milk with immobilized proteases. Proc Biochem (Feb.):10-26

Tuleike W (1966) Continuous cultures of higher plant cells in liquid medium. Ann N Y Acad Sci 139:162-175

ENGINEERING ASPECTS OF PLANT CELL CULTURE

M.M.R. Fonseca[1], F. Mavituna[2] and P. Brodelius[3]

Summary: Some of the ideas presented at the panel discussion on engineering aspects of plant cell culture are brought together, with special emphasis on mass transfer problems and on the release of intracellular products by permeabilized cells.

Key words: Aeration, electroporation, immobilized plant cells, permeabilization, ultrasonication.

INTRODUCTION (M.M.R. Fonseca)

Medium- to large-scale plant cell culture is prone to pose problems to the process engineer which differ, either in character or intensity, from those generally encountered in microbial processes. At the moment very few processes using *in vitro* cultivated plant cells have reached even a pilot plant scale. Thus, feedback information is scarce. Experience gathered at small scale permits, nevertheless, to foresee various engineering aspects which will be pertinent on scaling-up.

In contrast with microbial cultures, plant cell cultures present (1) higher instability, (2) lower growth rates, (3) higher sensitivity to hydrodynamic stress, (4) a more pronounced trend to form aggregates, (5) lower oxygen demands, (6) the tendency

[1]Laboratório de Engenharia Bioquímica, Instituto Superior Técnico, 1096 Lisboa Codex, Portugal

[2]Chemical Engineering Department, UMIST, P.O. Box 88, Manchester M60 1QD, U.K.

[3]Institute für Biotechnologie, ETH, Hönggerberg, CH-8093, Zürich, Switzerland

NATO ASI Series, Vol. H18
Plant Cell Biotechnology. Edited by M. S. S. Pais et al.

for differentiation and (7) reduced rates of excretion of secondary metabolites.

In both secondary metabolite production and biotransformation processes, the early stages of inoculum preparation and cell mass production are followed by the exploitation of the biosynthetic/ biotransformation potential of the biomass in appropriate production media, and finally by product recovery.

The general aim of inoculum preparation is to produce a finely dispersed and actively growing suspension culture. This can be achieved through the propagation of a preinoculum suspension or, as recently proposed by Mavituna et al. (1987a), through callus disintegration, obtained by submitting calli suspended in liquid media to high shear for a few seconds. The resulting fine suspensions present low viability but can yield highly viable growing cultures after approximately 10-15 days. Since the residence time in the biomass-producing reactor is dictated both by the growth rate and the percentage of inoculum, a need to adjust the latter can arise in order to compensate for slow growth rates.

Slow growth rates impose strict sterility requirements. The increase of the sterility levels of liquid media, equipment and gaseous streams and the use of antifungal and/or antibacterial agents are likely to be rewarding practices.

Plant cell transport involves the risk of both contamination and shear damage. The latter is reduced if transport is achieved by gravity, pneumatic means or the use of sterilizable low shear pumps.

Until recently the airlift reactor was the most advocated reactor configuration for the culturing of plant cells. Conventionally stirred tank reactors were generally considered to be disadvantageous due to impeller shear damage. However, Tanaka (1981), taking as a model a system of highly sensitive cells, found that, for culturing plant cells at high density, a fermenter with a large modified paddle-type impeller performed better than an airlift. Also Ulbrich et al. (1985) reported an increase in rosmarinic acid productivity by Coleus blumei when they shifted from an airlift to a mechanically stirred reactor equipped with a specially designed, slow running, helical agitator. The aspect ratio of the reactors was the same (height:diameter = 3:1), as was the aeration rate.

The recovery of intracellular products is usually incompatible with the maintenance of the integrity of the cell (cf. Permeabilization of Plant Cells). Thus, the normal approach is to carry out production of intracellular metabolites with freely suspended cells.

Whenever plant cells can be reused, i.e. when the product is naturally excreted or when excretion can be induced without causing a large extent of damage, the logical reactor choice for the production stage is an immobilized cell reactor. The use of an immobilized cell reactor saves growth costs which are usually considerable for eukaryotic cells. Such a benefit is obviously dependent on the operational stability of the reactor, a parameter which is often overlooked during preliminary evaluations of reactor performance. The possibility of continuous or cyclic cell revival is worthy of consideration.

The immobilization of plant cells requires mild methods. So far, gel entrapment has been the immobilization method most frequently adopted for plant cells (Brodelius 1983; Rosevear and Lambe 1985; Novais 1987). However, natural immobilization using biomass support particles (Lindsey et al. 1983; Mavituna and Park 1985; Mavituna et al. 1987b) or hollow fibres (Shuler 1981; Shuler et al. 1983; José et al. 1983; Prenosil and Pedersen 1983) is likely to take the lead since it allows immobilization *in situ*. This avoids biomass transfer and manipulation which, in turn, make the maintenance of aseptic conditions easier; scale-up is also facilitated.

Two further aspects of plant cell culture are addressed in detail in the following sections.

MASS TRANSFER (F. Mavituna)

The transfer rate of substrates and products between the bulk fluid and the plant cells can play an important role in the physiology of cells and the overall productivity of the system. For mass transfer considerations, two types of culture systems can be identified: homogeneous systems which consist of very fine plant cell suspension and protoplast cultures; and heterogeneous systems which include organ, callus, clumpy suspension cultures and immobilized plant cells. In the design of bioreactors for plant organ, tissue and cell cultures, it is important to know whether mass transfer would be the rate-limiting step and if so, whether it can be avoided by the careful choice of certain design and process parameters.

The factors which affect the extent of mass transfer limitations are the solubilities, the effective diffusion coefficients, the intrinsic metabolic rates of the cultures, the size and shape of the cell aggregates, immobilized cells or plant organs, concentrations and mixing conditions of the bulk fluid, and biomass concentration. Perhaps an example which would illustrate best the effect of these factors is the transfer of oxygen from the gaseous phase to a plant cell culture. The task of the process engineer will be to supply enough oxygen in order to meet the biological demand.

The biological demand or oxygen uptake rate can be quantified by:

$$r_o = q_o \cdot x ,$$

where r_o is the volumetric oxygen uptake rate (kg oxygen m^{-3} medium h^{-1}), q_o is the specific oxygen uptake rate (kg oxygen kg^{-1} dry biomass h^{-1}) and x is the biomass concentration (kg dry biomass m^{-3} medium). The specific oxygen uptake rate, q_o, can be determined experimentally (Sinclair and Mavituna 1983) and it may be possible to express it in terms of other system parameters. For example, for a freely suspended microbial system, q_o would generally be expressed by:

$$q_o = q_{o,max} \frac{C_o}{K_o + C_o} ,$$

where $q_{o,max}$ is the maximum specific rate of oxygen uptake (kg oxygen kg^{-1} dry biomass h^{-1}), K_o is the saturation constant for the biological system (kg oxygen m^{-3} medium) and C_o is the concentration of oxygen available to cells in the medium (kg oxygen m^{-3} medium). Similar quantification of the specific oxygen uptake rate, q_o, for plant cell cultures should be possible.

This biological demand should be met by the physical supply or mass transfer rate; for a dynamic equilibrium at least they should be equal. The oxygen transfer rate from the gaseous phase to the liquid medium can be described by:

$$N = k_La\ (C^*_L - C_L),$$

where N is the oxygen transfer rate (kg oxygen m^{-3} medium h^{-1}), k_L is the overall mass transfer coefficient for oxygen (m h^{-1}), a is the specific interfacial area between the gaseous and the liquid phase (m^2 interfacial area m^{-3} medium), C_L is the concentration of oxygen in the liquid (kg oxygen m^{-3} medium) and C^*_L is the hypothetical oxygen concentration in the liquid which would be in equilibrium with the gaseous phase (kg oxygen m^{-3} medium).

When the physical transfer rate is equal to the biological demand, then:

$$N = r_o$$
$$k_La\ (C_L^* - C_L) = q_o\ .\ x.$$

From this equality, it is already obvious how the numerical value of various factors listed previously will determine whether the mass transfer is the limiting step for the metabolic activity of the cells (Mavituna and Sinclair 1985). The metabolic rate, q_o, for plant cell cultures is usually lower than those found in microbial cultures. A typical figure for *Capsicum frutescens* cultures would be 2 x 10^{-3} kg oxygen kg^{-1} dry biomass h^{-1}. Therefore, oxygen supply should be able to meet the biological demand so long as the biomass concentration, x, is low. However, if the plant cell cultures become too concentrated, the overall volumetric oxygen uptake rate may become too large and the oxygen transfer rate may not be large enough to meet the demand.

The solubility of oxygen in aqueous media is low which affects both C^*_L and C_L. For example, the solubility of oxygen in pure water in equilibrium with air at a pressure of 1 atm, at 25°C is about 8 x 10^{-3} kg oxygen m^{-3} (Sinclair and Mavituna 1983). It is lower in a medium which contains sugar, various ions and other compounds. Increasing the concentration of oxygen in the gaseous phase (supplying pure oxygen as an ultimate resort) and/or increasing the pressure both increase the solubility of oxygen in the liquid phase.

The factors which affect k_L and a are the mixing conditions in the bulk liquid, the diffusion coefficient, the viscosity and the surface tension of the medium, air-flow rate, gas hold-up and the bubble size. A typical value for k_L is 1.44 m h^{-1}. The specific interfacial area a is difficult to measure, therefore, for mass transfer considerations, the two parameters are combined and referred to as the volumetric mass transfer coefficient, k_La. Since plant cell cultures can be sensitive to shear sustained over long periods of bioreactor operation, the typical k_La values achieved in bioreactors suitable for plant cultures are 10-100 h^{-1} (Atkinson and Mavituna 1985).

For a heterogeneous plant cell culture, such as immobilized plant cells, oxygen will have to diffuse through the immobilized cell matrix to reach the innermost cells. The plasmalemma, i.e. the cell membrane surrounding the cytoplasm, presents the major barrier to the diffusion of solutes into and out of plant cells. However, the molecular diffusion as described by Fick's law can still be used with the incorporation of a partition coefficient which accounts for the permeability of the membrane. Because of the high lipid content of the membranes, the relative ease of penetration of a membrane depends on the lipid solubility of the transferring substance. The diffusion of solutes across a cell wall, which surrounds the plasmalemma, occurs mainly in the water located in the numerous interstices, which are often about 10 nm in diameter. Thus, the mass transfer from the external solution up to the plasmalemma can be in aqueous channels through the cell wall and hence again can be described by molecular diffusion theory. The transport of solutes between plant cells is mostly through the plasmodesmata. They usually occupy about 0.1-0.5% of the surface area of the cell. Because of the high water content of the cytoplasm, diffusion coefficients for solutes are similar to their values in water, and the molecular diffusion theory in the form of Fick's law can be applied to describe the mass transfer from one cell to another through the plasmodesmata.

There are of course exceptions to this simplified approach. An example is the active transport across the membranes and the other is mixing in the cell due to cytoplasmic streaming. However, considering that diffusion through the plasmodesmata is more than 1000-fold greater than the simultaneously occurring mass transfer across the plasmalemmas (Nobel 1974), and that the concern from a process engineering point of view is for mass transfer on a macroscale, there is sufficient justification for applying molecular diffusion theory to mass transfer through the plant cell aggregates or tissue. Discrepancies due to other means of transport can be lumped together into the "effective" or "apparent" diffusion coefficient.

Assuming molecular diffusion for mass transfer through the plant cell aggregates immobilized in reticulate foam matrices and through the callus tissue of Capsicum frutescens, the effective diffusion coefficients for glucose have been experimentally determined (Mavituna et al. 1987c). For the callus tissue it was found to lie in the range of 0.01×10^{-5} - 0.1×10^{-5} m^2 h^{-1}, and for the immobilized cells in the range of 0.5×10^{-5} - 5×10^{-5} m^2 h^{-1} compared with the diffusion coefficient of 0.25×10^{-5} m^2 h^{-1} for glucose in water. If the plant cells are immobilized in various gels, then the solutes will have to diffuse through the gel layer surrounding individual cell aggregates which will reduce the mass transfer rates.

In addition to oxygen, several other substrates and products encountered in plant cell cultures have low solubilities in aqueous media. In such cases an organic solvent which may be immiscible with the aqueous phase may be used as a source for the substrates or as sink for the products. Then, liquid-liquid mass transfer rates will have to be considered. For example, capsaicin has low solubility in aqueous media at about 0.04 kg m^{-3} at 25°C and it exhibits feedback inhibition on its own production. Therefore, it has been extracted into sunflower oil successfully during its production (Wilkinson et al. 1987a).

Another mass transfer problem may arise from the loss of some volatile compounds from the medium due to the stripping effect of aeration. If these compounds are necessary for the cell metabolism, they may have to be supplied with the incoming air stream or separately, to reduce or prevent their loss.

Both for oxygen and such volatile compounds, making a material balance for the gaseous phase should be a very useful exercise if the components of the gaseous phase can be analyzed accurately and the changes in the composition of the inlet and outlet streams are large enough to be detected.

An interesting aspect of aeration which has recently arisen from the experimental results of various research groups (e.g. Wilkinson et al. 1987b) is the effect of dissolved oxygen concentration on the secondary metabolite production by plant cell cultures. It appears that for the production of certain secondary metabolites it is necessary to reduce the concentration of the dissolved oxygen in the medium. However, at this stage it is not clear as yet, whether this production is due to the induced chemical stress or to the prevention of oxidation of the product by enzymes present in the medium, such as phenoloxidase. In any case, this effect should be investigated with other products.

It may be necessary to reduce aeration for the production of secondary metabolites, yet it is also necessary to supply enough oxygen during the growth of the cultures. Furthermore, it has been found by various research groups that plant cell suspension cultures of several species have optimum dissolved oxygen concentrations for the maximum specific growth rates. Therefore, it is necessary to obtain and publish the aeration data accurately and adequately. Reporting the aeration level in terms of vvm (volume of gas flow per volume of liquid per minute) alone is far from adequate. It is necessary to measure and report the volumetric mass transfer coefficient for oxygen, k_La, the dissolved oxygen concentration, the biomass concentration and the specific oxygen uptake rate for the data to be meaningful for process engineers.

PERMEABILIZATION OF PLANT CELLS (P. Brodelius)

Cultivated plant cells often store secondary products within the vacuoles. The transport mechanism responsible for the translocation of substances over the tonoplast into the vacuoles has not been fully evaluated. Active transport (requiring ATP) has been suggested. Another possible transport system is based on the so-called ion trap mechanism. In this case the transport over the membrane is driven by a pH gradient over the tonoplast. Through an ion complex formation the equilibrium is shifted towards the deposit of the substance within the vacuole.

The biotechnological utilization of plant cell cultures (especially immobilized cells) is hampered by the fact that most substances are intracellular. Attempts are now being made to induce product release from plant cells. Here, the techniques involving permeabilization of membranes will be briefly reviewed. The ion trap mechanism may be employed to release products from cells by changing the extracellular pH. This does not involve a change in membrane permeability but merely a change of the equilibrium of product distribution and therefore this approach will not be discussed further.

Various techniques have been employed to make membranes within plant cells permeable to various substances. In order to release secondary products from vacuoles of a cell two membrane barriers have to be penetrated (i.e. the plasma membrane and the tonoplast). Chemical permeabilization, electroporation and ultrasonication are examples of methods employed to permeabilize these membranes.

A. Chemical Permeabilization

Various chemical substances may be employed to permeabilize plant cells (Felix et al. 1981). DMSO has been used to release indole alkaloids from immobilized cells of Catharanthus roseus (Brodelius and Nilson 1983). After quantitative release of the product the cells appeared to be viable. A repeated permeabilization could be carried out and a cyclic process involving product formation, permeabilization and cell recovery was suggested. This procedure would allow the reutilization of the expensive biomass. However, more extensive studies have shown that this study was carried out on an exceptional model system (Brodelius 1987). The amount of permeabilizing agent required for quantitative release of intracellular products may be determined for any cell culture. The concentration of some permeabilizing agents required to release 50 and 90% of the product from three different cell cultures are listed in Table 1. The viability of the cells was very low after the permeabilization. This is probably not due to the toxicity of the permeabilizing agents but to the destruction of cell compartmentation and the liberation of intracellular toxic substances (e.g. proteases and phenolics).

Table 1. Concentration of various permeabilizing agents required for the release of 50 and 90% of intracellularly stored products from cultivated plant cells (Brodelius 1987)

Permeabilizing agent	Catharanthus roseus		Chenopodium rubrum		Thalictrum rugosum	
	50%	90%	50%	90%	50%	90%
DMSO (% v/v)	3	7	10	35	13	30
Phenethylalcohol (% v/v)	1.04	1.16	0.86	0.98	0.60	0.80
Chloroform (%) saturation)	53	66	54	64	50	67
Triton X-100(ppm)	140	185	185	230	140	210
Hexadecyltrimethylammonium bromide (ppm)	44	72	22	84	24	60

B. Electroporation

Electroporation is widely used to improve the transfer of genes into plant protoplasts (Shillito et al. 1985). We have been involved in studies on the use of electroporation for the release of secondary products from cultivated plant cells (Brodelius et al. 1987). The technique has been used to release betanin and berberine from cells of C. rubrum and T. rugosum respectively. These two cell lines show a different response. Freely suspended cells of T. rugosum are completely permeabilized at a voltage of 5 kV cm^{-1} while around 10 kV cm^{-1} is required for the complete release of betanin from C. rubrum cells. The viability of the treated cells was determined by a plating procedure and the viability curve is essentially the reverse of the release curve. It is clear that electroporation resulting in product release also leads to a decrease in cell viability. The release of berberine from alginate or agarose entrapped cells of T. rugosum has also been studied. Cells entrapped in agarose respond to electroporation in the same manner as freely suspended cells, while alginate-entrapped cells show a somewhat different response. At lower voltages (e.g. 0.6 and 1.2 kV cm^{-1}) only the alginate-entrapped cells released significant amounts of product. The differences observed were ascribed to the different ion character of the two polymers.

C. Ultrasonication

Ultrasonication has been used to release betanin cell suspension cultures of _Beta vulgaris_ (Kilby and Hunter 1986). The cells were treated with continuous ultrasound (1.02 MHz) at 3 W cm^{-2} for periods ranging from 10 to 60 s. Cells sonicated for 10 s did not release vacuolar-stored betanin, whereas those sonicated for a longer period of time released betanin into the medium during the initial 30 min of the post-sonication incubation. The quantity and rate of betanin release increased in proportion to the sonication period. The investigators concluded that ultrasound stimulated the release of betanin from _B. vulgaris_ cells _in vitro_ by a non-thermal, cavitation-mediated phenomenon. The effect of sonication on cell viability was not reported but experimental data indicate that the mechanism of betanin efflux involved a sonically-induced, reversible alteration in cell membrane permeability.

The results so far obtained with various permeabilization methods are relatively discouraging. It appears very difficult, if not impossible, to release vacuolar substances into the surrounding medium by permeabilization of the membrane without killing the cells. Other approaches to release intracellular products must be evaluated.

REFERENCES

Atkinson B and Mavituna F (1985) Biochemical engineering and biotechnology handbook, p. 772, 2nd printing, MacMillan Publishers Ltd.

Brodelius P (1983) in Immobilized cells and organelles, vol. 1, p. 27, B. Mattiasson (Ed.), CRC Press, Florida

Brodelius P and Nilson K (1983) Permeabilization of immobilized plant cells, resulting in release of intracellularly stored products with preserved viability. Eur J Appl Microbiol Biotechnol 17:275-280

Brodelius P, Funk C and Shillito RD (1987) Permeabilization of cultivated plant cells by electroporation for release of intracellularly stored secondary products. Plant Cell Rep, in press

Brodelius P (1987) Permeabilization of plant cells for release of intracellularly stored products: viability studies. J Appl Microbiol Biotechnol, in press

Felix H, Brodelius P and Mosbach K (1981) Enzyme activities of the primary and secondary metabolism of simultaneously permeabilized and immobilized plant cells. Anal Biochem 116: 462-470

José W, Pedersen H and Chin CK (1983) Immobilization of plant cells in a hollow-fiber reactor. Ann N Y Acad Sci 413:409-412

Kilby NJ and Hunter CS (1986) Ultrasonic stimulation of betanin release from Beta vulgaris cells in vitro: a non-thermal, cavitation-mediated effect. Poster No. 375 presented at VI International Congress of Plant Tissue and Cell Culture, August 3-8, 1986, Minneapolis. Minn. USA

Lindsey K, Yeoman MM, Black GM and Mavituna F (1983) A novel method for the immobilization and culture of plant cells. FEBS Letts. 155(1):143-149

Mavituna F and Park JM (1985) Growth of immobilized plant cells in reticulated polyurethane foam matrices. Biotech Letts 7: 637-640

Mavituna F and Sinclair CG (1985) A graphical method for the determination of critical biomass concentration for non-oxygen-limited growth. Biotech Letts 7(2):69-74

Mavituna F, Wilkinson AK and Williams PD (1987a) in Plant and animal cells, C. Webb and F. Mavituna (Eds.). Ellis Horwood Publ

Mavituna F, Park JM, Wilkinson AK and Williams PD (1987b) in Plant and animal cells, C. Webb and F. Mavituna (Eds.). Ellis Horwood Publ

Mavituna F, Park JM and Gardner D (1987c) Determination of the effective diffusion coefficient of glucose in callus tissue. The Biochem Eng J 34:B1-B5

Nobel PS (1974) Introduction to biophysical plant physiology, W.H. Freemann (Ed.). San Francisco, CA

Novais JM (1987) Methods of immobilization of plant cells, Proc. of NATO ASI on Plant Cell Biotech., Portugal, March 29-April 11

Prenosil JE and Pedersen H (1983) Immobilized plant cell reactors, Enzyme Microb Technol 5:323-331

Rosevear A and Lambe CA (1985) Immobilized plant cells, in Adv. in Biochem. Eng., vol. 31, p. 37, A. Fiechter (Ed.), Springer, Berlin

Shillito R, Saul MW, Paszkowski J, Mueller M and Potrykus I (1985) High efficiency direct gene transfer to plants. Bio/Technology 3:1099-1103

Shuler ML (1981) Production of secondary metabolites from plant tissue culture: problems and prospects. Ann N Y Acad Sci 369: 65-79

Shuler ML, Sahai OP and Hallsby GA (1983) Entrapped plant cell tissue cultures. Ann N Y Acad Sci 413:373-382

Sinclair CG and Mavituna F (1983) in Filamentous fungi, vol. 4, p. 20, J.E. Smith, D.R. Berry and B. Kristiansen (Eds.). Edward Arnold Publ Ltd

Tanaka H (1981) Technological problems in the cultivation of plant cells at high density. Biotech and Bioeng 23:1203-1218

Ulbrich B, Wiesner W and Arens H (1985) in Primary and secondary metabolism of plant cell cultures, K.H. Neumann et al. (Eds.), Springer, Berlin

Wilkinson AK, Williams PD and Mavituna F (1987a) Effect of continuous extraction on production of a secondary metabolite by immobilized plant cells, Proc. of the 4th Eur. Cong. on Biotechnol., vol. 2, pg. 574, Amsterdam, June 14-20

Wilkinson AK, Williams PD and Mavituna F (1987b) The effect of oxygen stress on secondary metabolite production by immobilized plant cells in bioreactors, Proc. of NATO ASI on Plant Cell Biotechnology, Portugal, March 29 - April 11

CLONING AND CELL SORTER

Annie Bariaud-Fontanel[1], Marc Julien[2], Pierre Coutos-Thevenot[2], Spencer Brown[2], Didier Courtois[3] and Vincent Petiard[3]

[1]Faculty of Pharmaceutical Sciences, Dept. of Pharmacognosy, Kyoto University, Kyoto, Japan

[2]Service de Cytomêtrie et Physiologie Cellulaire Vêgêtale, CNRS, BP 1, 91190 - Gif sur Yvette, France

[3]Francereco, BP 0166, 37001 Tours Cedex, France (To whom correspondence should be sent)

KEYWORDS: Alkaloid, berberine, _Catharanthus roseus_, cell sorter, cloning, flow cytometry, serpentine, _Thalictrum minus_ .

INTRODUCTION: CELL CLONING AND VARIABILITY

The ability of plant cell cultures to produce useful compounds (pharmaceutical, food additives, cosmetics) with large-scale production has been investigated by numerous authors from scientific and economical points of view. (For reviews see: Zenk and Deus 1982, Kurz and Constabel 1983, Rosevear 1984. Yamada and Hashimoto 1984, Brodelius 1985, Dougall 1985, Sahai and Knuth 1985, Collinge 1986, Fowler 1986a, Fujita and Tabata 1986, Mac Laren 1986.)

The typical method to obtain cell lines consists of a mass selection of plants, initiation of calli from explants of high-producing plants under various culture conditions. After the selection of a high-producing strain (Zenk et al. 1977, Tabata et al. 1978, Berlin and Sasse 1985, Berlin et al. 1985, Heinstein 1985), the culture conditions must be optimized with regards to the scaling-up for product formation (Misawa 1985, Tabata and Fujita 1985, Shuler and Hallsby 1985, Fowler 1986b, Lee and An 1986).

It is well known that the derived strains, even issued from a single genotype, are phenotypically different.

Moreover, although it is possible to obtain stable strains by such a method, the selection efficiency is relatively low and productivity is frequently unstable. Besides this somaclonal variation (Evans et al. 1984, Larkin et al. 1985) between

NATO ASI Series, Vol. H18
Plant Cell Biotechnology. Edited by M.S.S. Pais et al.

strains (intercultural variability), an intracultural heterogeneity occurs in cell cultures (cell-to-cell variation). To suppress this heterogeneity and to select stable high-producing strains, cloning with protoplasts, single cells, or small cell aggregates has been used (Ogino et al. 1978, Constabel 1983, Fujita et al. 1985, Sato and Yamada 1984).

In fact, because of the contradictory results reported, the determination of the most efficient method is very difficult. For example, Yamamoto et al. (1982) succeeded in obtaining a high anthocyanin-producing strain of Euphorbia millii by cell-aggregate cloning, but not by single-cell cloning. In contrast, Fujita et al. (1985) reported that high shikonin-producing, stable cell lines of Lithospermum erythrorhizon have been obtained by protoplast selection, while cell lines selected by cell-aggregate cloning showed unstable productivity. They suggested that the stability of cell lines isolated by protoplast cloning could be due to the uniformity of the protoplast-derived cells. However, this finding is in conflict with the results observed in Daucus carota by Dougall et al. (1980), Anchusa officinalis by Ellis (1985), Catharanthus roseus by Petiard et al. (1985), Hyoscyamus muticus by Oksman-Caldentey and Strauss (1986).

These authors showed that the cloning processes (isolation of single cells or protoplasts and subcultures) induce a destabilization effect on the phenotype and the genotype of the cells leading to unstable clones. However, in some cases, this problem can be resolved by repetitive selection until a stable productivity is reached (Ogino et al. 1978, Sato and Yamada 1984).

From these observations, two distinct levels of studies are warranted:

1. Determination of the nature and origin of the variability, especially the cell-to-cell variation observed in cell culture and the somaclonal variation of clones derived therefrom. This variation could be due to:

a) A genetic phenomenon defined as a stable modification (showing or not showing Mendelian inheritance) which is reversible with a very low frequency.

b) An epigenetic phenomenon defined as a modification in structure, organization or regulation of the genotype, having a large inheritance under stable culture conditions but more or less easily reversed by any modification of these conditions.

Another explanation for the cell heterogeneity within a strain is the existence of differences in kinetics, defined as an asynchroneity in the differentiation process, at any culture time. But, as suggested by Hall and Yeoman (1986), it could be possible that a high-producing subpopulation in a cell culture is more or less fixed by the "spatial heterogeneity" (i.e. due to variations in the culture environment around the cells). This hypothesis relates to the observation that metabolite production is often dependent on the degree of differentiation (for example, the size of clumps in cell culture).

2. Enhancement of the selection efficiency:
From a fundamental and practical point of view, the emergence of the flow cytometric and cell-sorter techniques in plant cell biology could improve the exploitation of plant cell cultures (for review see Brown 1984). On the one hand, it could allow the assessment of a large number of cells according to various parameters (cell size or structure, fluorescent markers), an alternative approach to studying the variability within a cell culture by single-cell or protoplast cloning.

On the other hand, flow cytometry should enable the sorting of homogeneous subpopulations of high-producing cells, and may thereby avoid the putative induced variation ascribed to the "singleness" of isolated cells (Yamada and Mino 1986, Ellis 1985, Petiard et al. 1985).

This work describes studies on the heterogeneity within cell cultures manifest through different cell-cloning processes, namely, as cell aggregates with <u>Thalictrum minus</u> cells, as protoplasts with the CR_2 cell line of <u>Catharanthus roseus</u> and as protoplasts processed with a cell sorter with the C_{20} cell line of <u>Catharanthus roseus</u>. Possible sources of variation in the derived clones are discussed.

MATERIAL AND METHODS

Primary cloning of Thalictrum minus cells:

Suspension cultures of Thalictrum minus were established (Nakagawa et al. 1984) and subcultured every 2 weeks in the dark at 25°C in liquid basal medium (Linsmaier-Skoog 1965) containing 1 μm 2,4-D.

Eight day-old cultures were filtered through nylon mesh filters. The filtered suspensions, containing 90% single cells and 10% aggregates of two cells, were cultured in liquid medium (1 ml in 3-cm petri dishes at a density of 8000 cells ml^{-1}). After 2 weeks, the initial divisions occurred and cultures were mixed with melted agarose medium and 2 ml were plated in 6-cm petri dishes at a density of 350 aggregates ml^{-1} (about 1000 cells ml^{-1}). After 1-2 months, the colonies were transferred onto LS agar medium (1%) and subcultured at 25°C in the dark at intervals of 1 month. After 4-6 months, 0.3 g of cells was transferred to 10 ml LS liquid medium containing 1 μM 2,4-D. Cell suspension cultures were subcultured every 2 weeks under the conditions described above.

Secondary cloning:

After 12 subcultures (6 months) in LS medium with 1 μM 2,4-D, the cloning process was repeated on clones 1, 7 and 51. After 4 months of subculture on solid medium, the subclones (0.3 g in 50-ml flasks) were cultured for 14 days in 10 ml LS liquid medium with 10 μM BA and 50 μM NAA to analyze their berberine production and their growth.

Small cell-aggregate cloning:

Aggregates of 1-20 cells were isolated by filtration, mixed with melted agarose and cultured at low density (300 aggregates ml^{-1}). After 1 month, the colonies were transferred onto agar medium and subcultured under the same culture conditions as those described above.

Analysis of berberine production:

For analysis, 0.3 g of cells were inoculated into 10 ml LS medium with 10 μM BA and 50 μM NAA and cultured for 14 days. Cells and berberine crystals were collected by filtration on Miracloth and extracted at 75°C for 5 h. The medium was passed through SEP-PAK C_{18} cartridges and the alkaloids were eluted with MeOH. The MeOH extracts were subjected to HPLC as described by Nakagawa et al. (1984).

Cloning of _Catharanthus roseus_ strains:

All cultures were performed on solid media. Culture conditions, cloning procedure and subculture, alkaloid extraction and analysis of the strain CR_2 have been described earlier (Petiard et al. 1982, Petiard and Courtois 1983, Gueritte et al. 1983, Petiard et al. 1985). Flow cytometric techniques used with the strain C_{20} have been described by Brown et al. (1986); protoplast isolation and analytical HPLC of serpentine content were according to Brown et al. (1984) and Renaudin et al. (1986).

RESULTS AND DISCUSSION

Cloning of _Thalictrum minus_ cells:

Primary and secondary clonal cell lines derived from cell suspensions of _Thalictrum minus_ were established using the methodology described in Fig. 1. The berberine production was determined on the basis of the two-step culture method using LS medium, 1 μM, 2,4-D for cell growth and LS medium, 10 μM BA, 50 μM NAA for alkaloid production.

In all the clones, about 90% of the berberine was released into the medium and HPLC analysis showed that the berberine represented between 95-99% of the alkaloids produced by the cells.

While no differences in the secretion ability and in the qualitative pattern of the alkaloids produced could be observed, the level of berberine production differed greatly among the clonal lines (Fig. 2A).

Thalictrum minus clones also differed with respect to production stability. As shown in Fig. 3, during extended cultures, 28% of the clones remained stable while the production level of other lines increased or decreased with passage number. This instability indicates that the variations observed among the clones were not due to the isolation of different genotypes existing in the original culture, but rather suggests that the cloning process (isolation and subculture) induced a destabilization of the phenotype or the genotype of the cells.

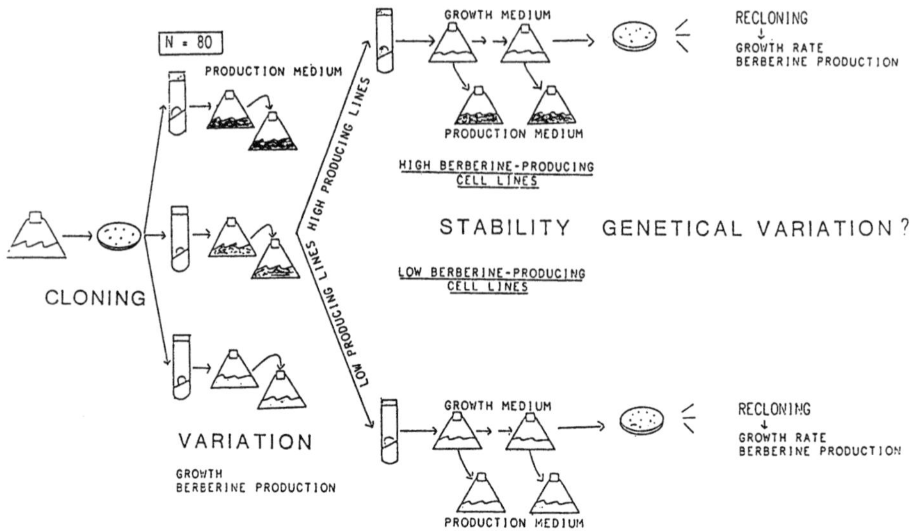

Fig. 1. Procedure for the study of somaclonal variation in cell suspension of Thalictrum minus

The data of Ellis (1985) on the production of rosmarinic acid by nine clonal lines of Anchusa officinalis led to the same conclusion.

This hypothesis is also supported by the results observed after the second cloning cycle. Secondary clonal lines were established from three primary clones. The subclones selected from a high- and a medium-producing line gave a bimodal distribution (Fig. 2B,C).

These results indicate that these two selected clones had developed as a mixture of two populations, one with a low growth rate and low berberine-producing cells and the other with a high growth rate and high-producing cells (data not shown). With regards to the subclones selected from a low-producing clone (Fig. 2D), although there was no appearance of any high-producing lines, 45% of the subclones produced more berberine than the original clone, showing some reversibility for this character. Such changes support the conclusion that the origin of variation is mainly epigenetic.

Concerning the practical interest of cloning methods, examination of berberine production after 8 months of subculture of cell lines isolated by cloning single cells and small cell

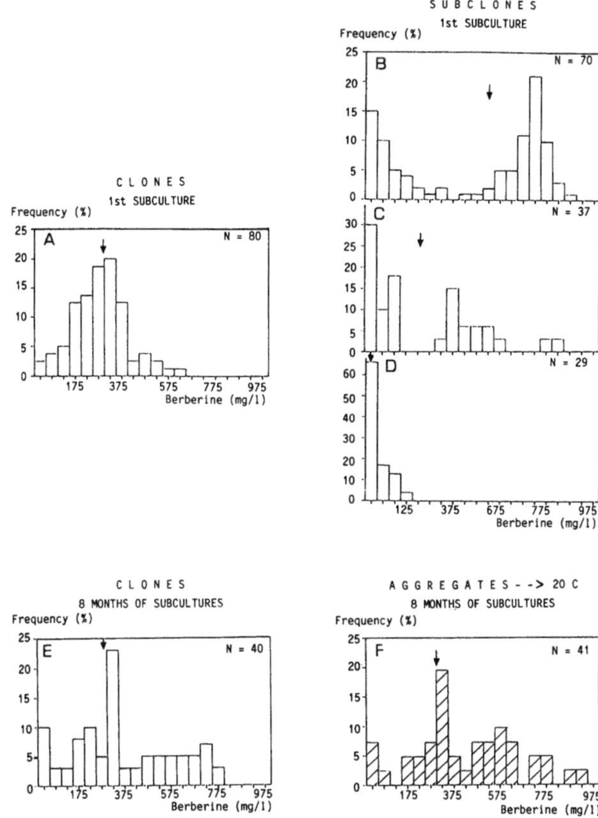

Fig. 2. Distribution of berberine production of clones and subclones. A clones, 1st subculture; B,C,D subclones, 1st subculture; E and F comparison between clones and aggregates, 8-month subculture. N Number of cell lines examined. Arrows indicate the berberine production of the original strain or clone

aggregates shows that the latter is an efficient methodology to select some high berberine-producing cell lines (Fig. 2 E,F).

As in the primary cloning, berberine production is highly disperse. But, when small cell aggregate cloning was used, the mean berberine production was higher (0.429 g l^{-1} instead of 0.367 g l^{-1}) and the best cell line produced more berberine after 14 days of culture (0.91 g l^{-1} instead of 0.76 g l^{-1}).

In both cases, the berberine production was faster and higher than that of the original strain (0.63 g l^{-1} in 20 days).

Cloning of the Catharanthus roseus strain CR_2:

This strain, established in 1975, contains a wide diversity of alkaloids, of which 13 were isolated (Petiard et al. 1982,

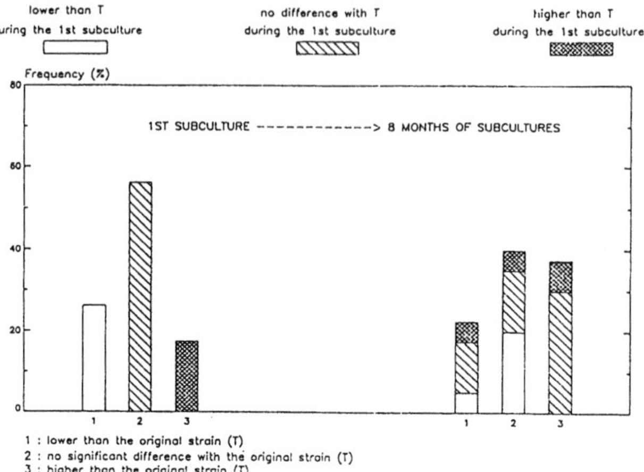

Fig. 3. Distribution of berberine production of clones during the first subculture and after 8 months of subcultures

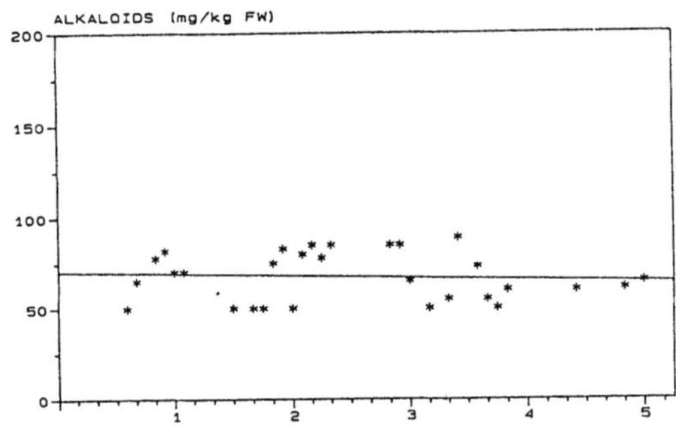

Fig. 4. Crude alkaloid content of CR_2 strain variations over 5 years of subcultures

Gueritte et al. 1983). Figure 4 shows that this strain has been relatively stable with respect to total alkaloid content for several years. Nineteen clones were obtained after subculture of protoplasts at low density (80 protoplasts ml^{-1}). Subclones were obtained from clones of 18 months. All clones and subclones were cultivated on the same medium as the mother strain.

Petiard et al. (1985) have shown that all the clones are distinctive for the studied parameters (chlorophyll and alkaloid content, alkaloid diversity).

These clones were unstable and the variability decreased with time as shown in Fig. 5A, B, C for the diversity of alkaloids (number of alkaloids detected on TLC with CAS reagent). No clone appears more interesting than the mother strain. It is improbable that, to account for such results, at least 19 different chemotypes existed within the mother strain.

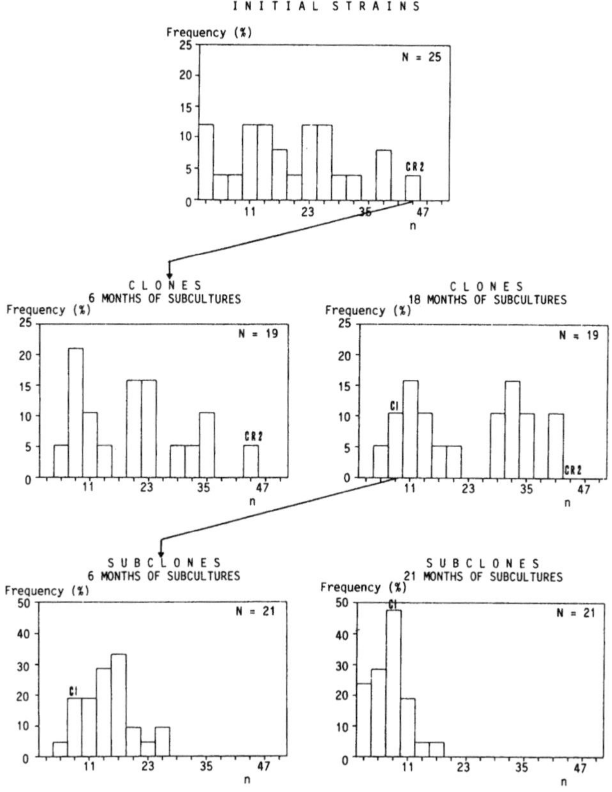

Fig. 5. Distribution of the number of alkaloids (N) detected in initial strain collection (A), clones from CR_2 strain (B, C 6 and 18 months of subculture) and subclones from one low-producing clone (D, E 6 and 21 months of subculture). The detection was made on TLC with CAS reagent

Nevertheless, to check this hypothesis, a secondary cloning was performed with a high- and a low-producing clone. From the low-producing clone, an unstable variation was again observed among the subclones between 6 and 21 months (Fig. 5D, E shows the results obtained on the number of alkaloids). This result supports the idea that the cloning procedure itself induced a new variability. Moreover, even after trying various culture conditions (media, feeder layers, nurse cultures), it was impossible to obtain any subclone from the high-producing clone (although the clone itself was derived from cells obtained by the same cloning procedure). This could explain the fact that no clone more interesting than the mother strain was obtained with the first cloning, due to a possible counter selection.

These results show that a destabilization occurred during the cloning and subcloning, followed by a new genetic or epigenetic stable state.

Cell sorting of Catharanthus roseus C_{20} protoplasts:

The natural blue fluorescence of such protoplasts has often been attributed to indole alkaloid serpentine. Figure 6 shows that, indeed, the fluorescence intensity was highly related to the serpentine content.

From protoplasts of an exponentially growing culture, protoplasts were sorted on "modal fluorescence" (the modal decade of the population, eliminating notably those with very low and very high fluorescence), and "rare" fluorescence (the most highly fluorescent protoplasts per thousand).

Subculture of the low fluorescent protoplasts failed. As a control, the subculture of protoplasts without sorting (stock) was performed, as well as an analysis of protoplasts without subculture (C_{20} analysis).

Figure 7 shows the results obtained on some derived strains. As for Catharanthus roseus CR_2 and Thalictrum minus cloning, a variability was observed after 2 months of subculture (four strains issued from modal and stock, three from the rare subpopulation of the mother strain). Moreover, these strains are unstable: e.g., when a "rare" population was subcultured for 7 months (Fig. 8), its fluorescence intensity became very similar to the control (stock) and mode.

Bi-parametric analysis:

9-Amino-acridine (9-AA) has been used to study the indivividual value for vacuolar pH in protoplasts (Brown et al. 1984). A correlation was shown between this vacuolar pH and serpentine content. Figure 9 shows that, in comparison with the control culture, in a "rare" culture (i.e. protoplasts sorted at the most fluorescence intensity), there was an additional subpopulation with a high serpentine content but low 9-AA accumulation. While the main population reflects a model of serpentine accu-

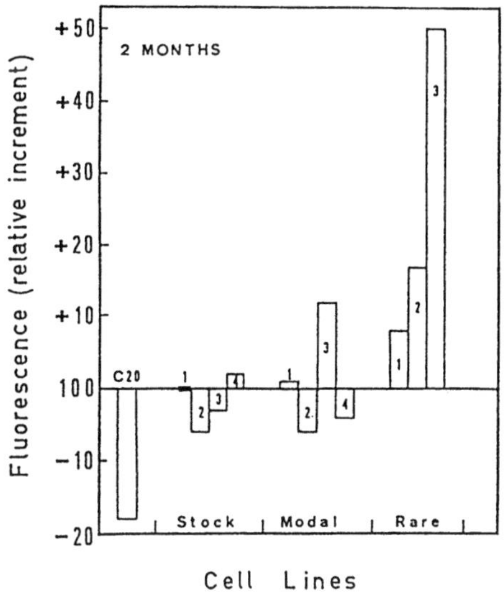

Fig. 6.

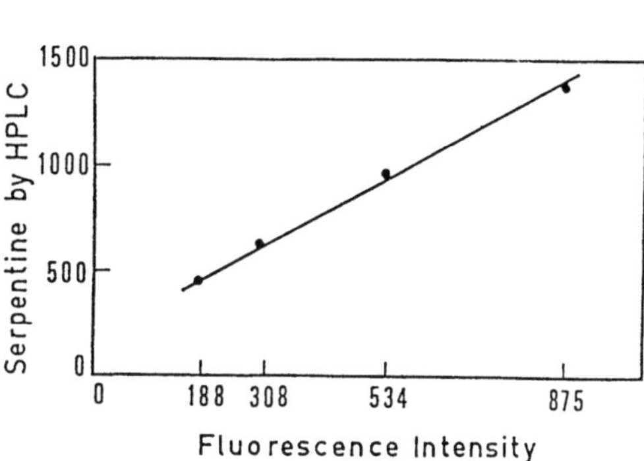

Fig. 7

Fig. 6. Relation between the fluorescence intensity by cytometry and the cellular serpentine concentration. Four classes of <u>Catharanthus roseus</u> protoplasts, having mean relative fluorescence as indicated (X-axis, 1024 channel analyzer), were sorted and subsequently analyzed by HPLC for serpentine (Y-axis, $\mu g\ l^{-1}$ after correction for differing cell volumes). $y = 1.38\ x + 205.5$; $r^2 = 0.998$

Fig. 7. Histograms of fluorescence means relative to stock 1, obtained by cytometric analysis for the different cell lines at 2 months. Measurements were made on protoplasts obtained from the <u>Catharanthus roseus</u> cell suspensions 5 days after their transfer (exponential phase of growth). Cultures were derived from protoplasts without sorting (<u>stock</u>), or sorted on modal fluorescence (<u>modal</u>), or sorted as the most intense fluorescence (<u>rare</u>). The initial cell line, C_{20}, was also maintained

mulation related to the pH gradients (Renaudin et al. 1985), a few protoplasts do not conform to this model. The significance of these results needs further investigations: apparently, this subpopulation does not accumulate serpentine at a high level but has a high biosynthetic capability. Moreover, the relation between fluorescence intensity and serpentine content must be verified for this particular subpopulation to check whether the increase of natural blue fluorescence is really due to serpentine as in the mother strain.

On the one hand, flow cytometry is pertinent for the selection of high-producing strains, allowing the processing of a

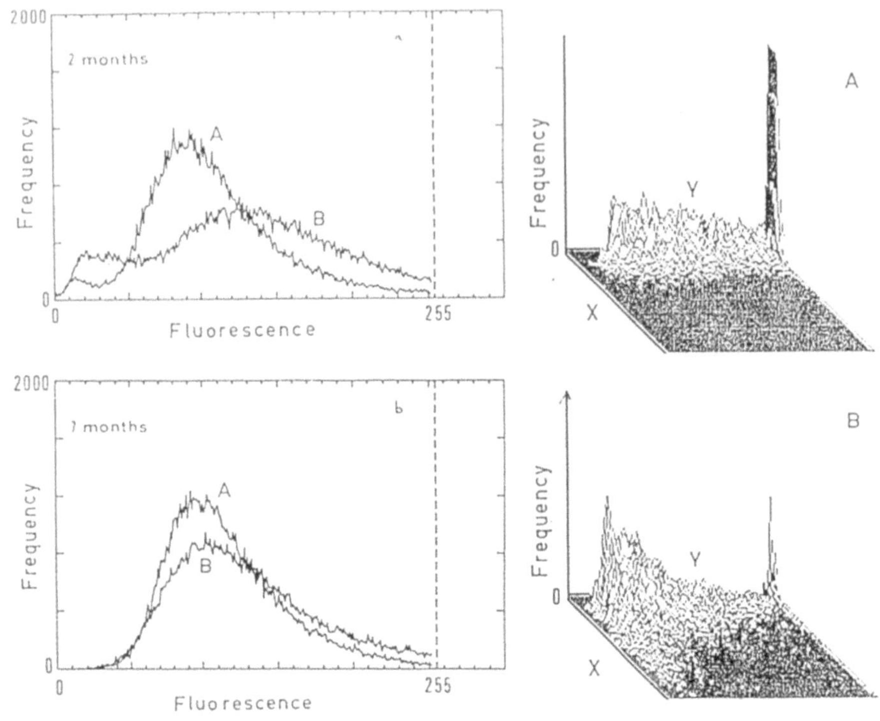

Fig. 8 Fig. 9

Fig. 8. Histograms of blue fluorescence in 50 000 protoplasts from exponentially growing cells (5 days after transfer) from lines established by protoplast culture. a After 2 months of culture from protoplasts without sorting (A) or sorted as the most fluorescent (B). A subpopulation (7%) occurs in the highest class (channel 255) for line B. b After 7 months of culture from protoplasts sorted on mode (A) or sorted as the most fluorescent (B)

Fig. 9. Three-dimensional cytogram of serpentine fluorescence (X) as a function of 9-AA accumulation (Y) for 10 000 protoplasts showing two different tendencies. A Control C_{20} culture; B rare culture where there is another category characterized by high serpentine and low 9-AA accumulation

large number of cells and the subsequent enrichment in high-producing cells. Particularly, further investigations are needed to check whether, as with small cell aggregate cloning of *Thalictrum minus*, repeated cell sorting of the "rare" population at an early stage could lead to a stable high-producing strain.

On the other hand, the analysis of the heterogeneity within a culture becomes quantitative at the cellular level and several parameters should be analyzed. This possibility of a multi-parametric analysis is one of the most promising perspectives (Brown et al. 1986).

CONCLUSION

Results obtained on three different cell lines with three cloning procedures show that the cloning procedure itself induces a variation in derived clones. This variation rather suggests an epigenetic phenomenon, due to its relative reversibility and its long-term nature (several months for restabilization). In accord with other authors (Ellis 1985, Hall and Yeoman 1986, Yamada and mino 1986), the heterogeneity observed within a cell culture is apparently not due to the presence of numerous different chemotypes.

The fact that no clone seems to be more interesting than the mother strain in the two Catharanthus roseus cell lines could be due to the asynchroneity of cells within the mother strains and also to a counterselection related to the induced variation. For example, no subclone has been obtained from high-producing clones of CR_2.

As was suggested by Yamada and Mino (1986), the instability of derived clones could be due to the "singleness" of cells isolated from aggregates in which the cells interact. Hall and Yeoman (1986) consider that producing cells within a culture are controlled by spatial and temporal heterogeneity, i.e. the proportion of producing cells is limited by culture conditions. If this is the case, cell sorting does not seem to be a useful method for obtaining a high-producing strain (our "rare" populations are not stable). But, in some cases, as with clones of Thalictrum minus, it is possible to obtain a higher producing strain than the mother strain. Our opinion is that, by repeated cloning, it is possible to exploit the induced variation to enhance the proportion of producing cells if these clonings occur before the restabilization of clones, i.e. if a constant selection pressure is applied during several cloning cycles. This requires a tool for the early detection of the interesting subpopulations among the clones and the processing of a large number of cells, which could involve cell sorting and flow cytometry.

Acknowledgements. The close collaboration of J.P. Renaudin, INRA (Epoisse) in acknowledged, notably for the results appearing in Fig. 6-9.

REFERENCES

Berlin J, Sasse F (1985) Selection and screening techniques for plant cell cultures. In: Fiechter A (ed). Advances in biochemical engineering biotechnology. Springer 1985 Berlin 31: 99-132

Berlin J, Beier F, Fecker L, Forche E, Noe W, Sasse F, Schiel O and Wray V (1985) Conventional and new approaches to increase the alkaloid production of plant cell cultures. In: Neumann KH, Barz W and Reinhard E (eds). Primary and secondary metabolism of plant cell cultures. Springer, 1985 Berlin:272-280

Brodelius P (1985) Utilization of plant cell cultures for production of biochemicals. Hereditas Suppl. 3:73-81

Brown SC (1984) Analysis and sorting of plant material by flow cytometry. Physiol. Veg. 22:341-349

Brown SC, Renaudin JP, Prevot C and Guern J (1984) Flow cytometry and sorting of plant protoplasts: technical problems and physiological results from a study of pH and alkaloids in Catharanthus roseus. Physiol Veg 22:541-554

Brown SC, Jullien M, Coutos-Thevenot P, Muller P and Renaudin JP (1986) Present developments of flow cytometry in plant biology. Biol of the Cell 58:173-178

Collinge M (1986) Ways and means to plant secondary metabolites. Trends in Biotech 4:299-301

Constabel F (1983) Protoplast technology applied to metabolite production. Int Rev Cyt Suppl 16:209-217

Deus-Neumann B and Zenk MH (1984) Instability of indole alkaloid production in Catharanthus roseus cell suspension cultures. Planta Med 50:427-431

Dougall DK (1985) Chemicals from plant cell cultures: yields and variation. In: Zaitlin M, Day P and Hollaender P (eds). Biotechnology in plant science. Academic Press 1985 Orlando: 179-190

Dougall DK, Morris Johnson J and Whitten GH (1980) A clonal analysis of anthocyanin accumulation by cell cultures of wild carrot. Planta 149:292-297

Ellis BE (1985) Characterization of clonal cultures of Anchusa-officinalis derived from single cells of known productivity. J Plant Physiol 119:149-158

Evans DA, Sharp WR and Medina-Filho HP (1984) Somaclonal and gametoclonal variation. Amer J Bot 71:759-774

Fowler MW (1986a) Process strategies for plant cell culture. Trends in Biotech 4:214-219

Fowler MW (1986b) Physiological factors affecting product yield in plant cell cultures. Petic Sci, 17:595-601

Fujita Y and Tabata M (1986) Secondary metabolites from plant cells: pharmaceutical application and progress in commercial production. In: Somers DA, Gengenbach GB, Biesboer DD, Hackett WP and Green CE (eds). VIth International Congress of Plant Tissue and Cell Culture, August 3-8 1986, Abstract p. 2

Fujita Y, Takahashi S and Yamada Y (1985) Selection of cell lines with high productivity of shikonin derivatives by protoplast culture of Lithospermum erythrorhizon cells. Agric Biol Chem 49:1755-1759

Gueritte F, Langlois N and Petiard V (1983) Métabolites secondaires isolés d'une culture de tissus de Catharanthus roseus. J Nat Prod 46:144-148

Hall RD and Yeoman MM (1986) Temporal and spatial heterogeneity in the accumulation of anthocyane in cell cultures of Catharanthus roseus (L) G. Don. J Exp Bot, 37:48-60

Heinstein PF (1985) Future approaches to the formation of secondary natural products in plant cell suspension cultures. J Nat Prod 48:1-9

Kurz WGW and Constabel F (1983) Aspects affecting biosynthesis and biotransformation of secondary metabolites in plant cell cultures. CRC Critical Reviews in Biotechnology 2:105-118

Larkin JP, Brettel RIS, Ryan SA, Davies PA, Pallotta MA and Scowcroft WR (1985) Somaclonal variation: impact on plant biology and breeding strategies. In: Zaitlin M, Day P and Hollaender A (eds). Biotechnology in plant science. Academic Press 1985 Orlando:83-100

Lee JM and An G (1986) Industrial application and genetic engineering of plant cell cultures. Enzyme Microb Technol 8:260-265

Linsmaier EM and Skoog F (1965) Organic growth factor requirements of tobacco tissue cultures. Physiol Plant, 18:100-125

MacLaren JS (1986) Biologically active substances from higher plants: status and future potential. Pestic Sci 17:559-578

Misawa M (1985) Production of useful plant metabolites. In: Fiechter A (ed). Advances in biochemical engineering/biotechnology. Springer 1985 Berlin 31:59-88

Nakagawa K, Konagai A, Fukui H and Tabata M (1984) Release and crystallization of berberine in the liquid medium of Thalictrum minus cell suspension cultures. Pl Cell Rep 3:254-257

Ogino T, Hiraoka N, and Tabata M (1978) Selection of high nicotine producing cell lines of tobacco callus by single-cell cloning. Phytochemistry 17:1907-1910

Oksmann-Caldentey KM and Strauss A (1986) Somaclonal variation of scopolamine content in protoplast-derived cell culture clones of Hyosciamus muticus. Planta Med 1:6-12

Petiard V and Courtois D (1983) Recent advances in research for novel alkaloids in Apocynaceae tissue cultures. Physiol Veg, 21:217-227

Petiard V, Courtois D, Gueritte F, Langlois N and Mompon B (1982) New alkaloids in plant tissue cultures. In: Fujiwara A (ed). Vth International Congress of Plant Tissue and Cell Culture. Maruzen Co. 1982 Tokyo:309-310

Petiard V, Baubault C, Bariaud A, Hutin H and Courtois D (1985) Studies on variability of plant tissue cultures for alkaloid production in Catharanthus roseus and Papaver somniferum callus cultures. In: Neumann KH, Barz W and Reinhard E (eds). Primary and secondary metabolism of plant cell cultures. Springer 1985 Berlin:133-142

Renaudin JP, Brown SC and Guern J (1985) Compartmentation of alkaloids in a cell suspension of Catharanthus roseus: a reappraisial of the role of pH gradients. In: Neumann KH, Barz W and Reinhard E (eds). Primary and secondary metabolism of plant cell cultures. Springer 1985 Berlin:124-132

Renaudin JP, Brown SC, Barbier-Brygoo H and Guern J (1986) Quantitative characterization of prooplasts and vacuoles from suspension-cultured cells of Catharanthus roseus. Physiol Plant 68:695-703

Rosevear A (1984) Putting a bit of colour into the subject. Trends in Biotech 2:6

Sahai O and Knuth M (1985) Commercializing plant tissue culture processes: economics, problems and prospects. Biotech Progress 1:1-9

Sato F and Yamada Y (1984) High berberine producing cultures of Coptis japonica cells. Phytochemistry 23:281-285

Shuler ML and Hallsby GA (1985) Bioreactor considerations for chemical production from plant cell cultures. In: Zaitlin M, Day P and Hollaender A (eds). Biotechnology in plant science. Academic Press 1985 Orlando:191-206

Tabata M and Fujita Y (1985) Production of shikonin by plant cell cultures. In: Zaitlin M, Day P and Hollaender A (eds). Biotechnology in plant science. Academic Press 1975 Orlando: 207-218

Tabata M, Ogino T, Yoshioka K, Yoshikawa N and Hiraoka N (1978) Selection of cell lines with higher yield of secondary products. In: Thorpe TA (ed). Frontiers of plant tissue culture. University of Calgary:213-222

Yamada Y and Hashimoto T (1984) Secondary products in tissue culture. In: Collins GB and Petolino JG (eds). Application of genetic engineering to crop improvement. Martinus Nijhoff 1984 Dordrecht:561-604

Yamada Y and Mino M (1986) Instability of chromosomes and alkaloid content in cell lines derived from single protoplasts of cultured Coptis japonica cells. Cur Top Develop Biol 20: 409-417

Yamamoto Y, Mizuguchi R and Yamada Y (1982) Selection of a high and stable pigment producing strain in cultured _Euphorbia millii_ cells. Theor Appl Genet 61:113-116

Zenk MH and Deus B (1982) Natural product synthesis by plant cell cultures. In: Fujiwara A (ed). Vth International Congress of Plant Tissue and Cell Culture. Maruzen Co. 1982 Tokyo:391-394

Zenk MH, El-Shagi H, Stockigt J, Weiler EW and Deus B (1977) Formation of indole alkaloids serpentine and ajmalicine in cell suspension cultures of _Catharanthus roseus_. In: Barz W, Reinhard E and Zenk MH (eds). Springer 1977 Berlin:27-44

SECRETION OF THIOPHENES BY DIFFERENTIATED CELL CULTURES OF *TAGETES* SPECIES

Johannes P.F.G. Helsper[1], David H. Ketel[1], Anne C. Hulst[2] and Hans Breteler[1]

ABSTRACT

Cell aggregates of *T. patula* synthesize and secrete the thiophenes BBT, BBTOH and BBTOAc (see Fig. 1), when cultured in liquid medium. There is a positive correlation between thiophene synthesis and aggregate diameter in the 1 to 12 mm range. No thiophenes are produced by smaller or larger aggregates. Differentiated, hormone-autotrophic "root" cultures, obtained from *T. patula* tissues after transformation by *Agrobacterium tumefaciens* LBA 8370, show an extended period of stable BBT and BBTOAc production. The results support the hypothesis that morphological differentiation is required for thiophene biosynthesis. They also show the value of differentiated cell cultures of *T. patula* for the production of natural biocides by plant cell biotechnology.

INTRODUCTION

Several species of the genus *Tagetes* (marigolds) and related genera produce thiophenes. Thiophenes are heterocyclic compounds (Fig. 1) with biocidal activity against a wide spectrum of organisms. The quantity and type of thiophenes are not only related to plant species (Bohlmann et al. 1973) but also to plant organ and developmental stage (Sütfeld 1982).

---

[1]NOVAPLANT Cell Biotechnology Group, Research Institute Ital, P.O. Box 48, 6700 AA Wageningen, The Netherlands

[2]Agricultural University, Department of Food Science, Food and Bioengineering Group, De Dreijen 12, 6703 BC Wageningen, The Netherlands

NATO ASI Series, Vol. H18
Plant Cell Biotechnology. Edited by M.S.S. Pais et al.

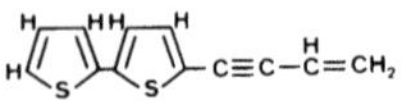

BBT

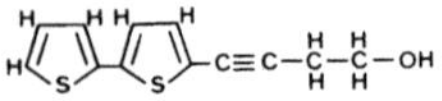

BBTOH

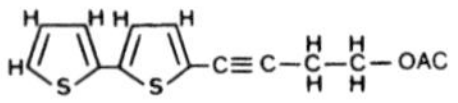

BBTOAc

Fig. 1. The major thiophenes occurring in *Tagetes patula*. The systematic formula names are given in the introduction

Cell cultures of several *Tagetes* species produce considerable amounts of thiophenes (Ketel 1987). The major thiophenes produced by calli and cell suspensions of *Tagetes patula* are 5-(4-hydroxy-1-butinyl)-2,2'-bithiophene (BBTOH, Fig. 1), its acetate ester BBTOAc, and 5-(1-butinyl-3-ene)-2,2'-bithiophene (BBT). Secretion of the relatively polar BBTOH into liquid culture media (Ketel 1987) may be of biotechnological significance.

In this work we describe the synthesis and secretion of thiophenes by cell aggregates of *Tagetes patula* in which different levels of differentiation have been obtained by either transformation with *Agrobacterium tumefaciens* LBA 8370 or by varying the aggregate diameter.

MATERIAL AND METHODS

Plant material

Plants of *Tagetes patula* L. c.v. Nana furia were grown in a greenhouse. Cell cultures were obtained from calli of leaf explants grown on MS medium (pH 5.8), supplemented with naphthylacetic acid (5 mg l^{-1}), benzyladenine (0.5 mg l^{-1}) and polyvinylpyrrolidone-40 (1 g l^{-1}). Growth medium for *Agrobacterium* transformants contained no plant growth regulators, but carbenicillin (0.1 g l^{-1}) was added to inhibit bacterial growth. Batches of cell aggregates of defined diameters were obtained from heterogeneous, untransformed cell cultures by passage through stainless steel sieves. All cell cultures were grown at 24°C under continuous light (2 klx). Each third or fourth day the growth medium was substituted with fresh medium.

Extraction and analysis of thiophenes

Lyophilized cell suspensions and aggregates were extracted with acetone (5 ml g^{-1} fresh wt.). The extract was filtered through glass wool and reduced in volume under a nitrogen stream. Thiophenes were extracted with n-hexane (5 ml g^{-1} fresh wt.) for 24 h in the dark. After evaporation to dryness thiophenes were redissolved in n-hexane and stored at -18°C until analysis. Thiophenes in growth media were extracted with an equal volume of n-hexane and treated further as described above.

High performance liquid chromatography of thiophenes was performed on a Silica Si 100 column (Serva 3 μ, 110 x 4.6 mm), connected to a Waters 6000 A pump system. Hexane-dioxane (95:5, v/v) was used as the mobile phase at 1.5 ml min^{-1}. Thiophenes were quantified with the thiophene α-terthienyl as a standard by measuring the absorbance at 350 nm and identified on the basis of retention time and the UV-absorption spectrum (230-400 nm).

RESULTS AND DISCUSSION

Thiophene production in cell aggregate suspension cultures

Thiophenes were only synthesized in cell aggregates larger than 1 mm (Fig. 2). Thiophene production increased with aggregate diameter up to about 12 mm. In this range there was also an increase in the level of cellular differentiation, which was present in the form of heterogeneity of cell size, and the occurrence of cork tissue and tracheids. These results indicated a positive correlation between cellular differentiation and thiophene production. Aggregates with a diameter larger than 13 mm became hollow and did not produce thiophenes (Fig. 2).

The major thiophenes observed in the cell aggregates of 11 mm were BBTOH (150 nmol g^{-1} dry matter), BBTOAc and BBT (together about 500 nmol g^{-1} dry matter). At 90 g cell material (fresh weight) per liter about 1 g of thiophenes was secreted over 3 weeks with BBTOH as the predominant (75%) thiophene species. The amounts of polar (BBTOH) and non-polar (BBTOAc and BBT) thiophenes in cell aggregates of *Tagetes patula* gradually decreased with more than 90% during the first 60 to 80 days of subculturing in liquid medium.

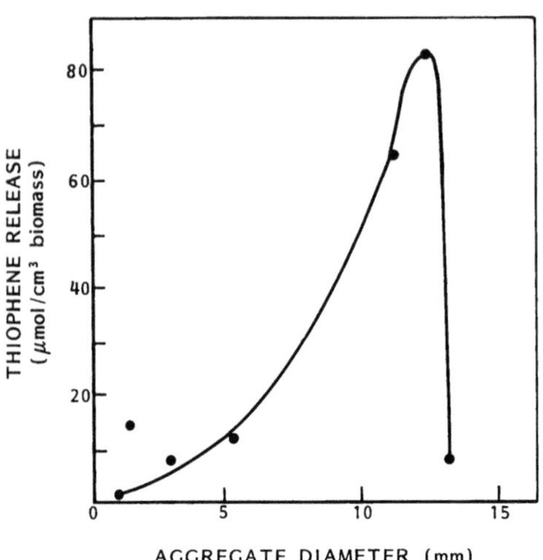

Fig. 2. Thiophene secretion by cell aggregates of Tagetes patula with defined diameters

Thiophene production in differentiated, transformed "root" cultures

Infection of Tagetes patula plants with Agrobacterium tumefaciens LBA 8370 frequently resulted in the formation of rhizoid tumours. These could be subcultured in liquid medium and maintained a rootlike structure.

The content of BBTOH in these "root" cultures was similar to the maximum level observed in untransformed cell aggregates, but the levels of BBTOAc and BBT were 50- to 100-fold higher. Secretion of BBTOH was negligible, but BBTOAc and BBT were secreted weekly to the extent of 0.2-1 mg^{-1}.

In contrast to untransformed cell aggregates of Tagetes patula, both synthesis and secretion rates for thiophenes in "root" cultures remained constant for more than 1 year.

REFERENCES

Bohlmann CF, Burkhardt T, Zdero C (1973) Distribution of acetylenes. In: Bohlmann CF, Burkhardt T, Zdero C (eds) Naturally occurring acetylenes. Academic Press, London New York, pp. 340-505

Ketel DH (1987) Callus and cell cultures of Tagetes species in relation to production of thiophenes, Thesis Agricultural University, Wageningen, The Netherlands

Sütfeld R (1982) Distribution of thiophene derivatives in different organs of Tagetes patula seedlings grown under various conditions. Planta 156:536-540

THE USE OF HAIRY ROOTS TRANSFORMED BY AGROBACTERIUM RHIZOGENES FOR THE PRODUCTION OF PLANT SECONDARY PRODUCTS IN VITRO

John D. Hamill[1], Adrian J. Parr[1], Cathie Martin[2], Nick J. Walton[1], Richard Robins[1] and Mike J.C. Rhodes[1]

ABSTRACT

The study of plant secondary metabolism, with the aim of biotechnological exploitation using tissues grown in vitro, has traditionally tended to use dispersed cell cultures or callus. The cultures usually have a strong tendency to be genetically and biochemically unstable and often synthesize very low levels of secondary products. Recently we have begun to study the potential of "hairy root" cultures, resulting from the infection of dicotyledonous plants with *Agrobacterium rhizogenes*, for the production of secondary products *in vitro*. We have found that such cultures produce the secondary products synthesized in the roots of the plant species in question in qualitative proportions typical of the parent plant and in quantitative levels at least as high as those found in the plant. Such cultures grow very rapidly *in vitro* in simple media devoid of phytohormones. They are biochemically stable and we have evidence that their chromosome number remains stable for at least a year in culture.

Such cultures are amenable to genetic mainipulation, raising the possibility of increasing the secondary product synthesizing capacity of hairy roots grown in vitro. Coupled with the inherent biochemical and genetic stability of such tissues, these methods may facilitate the biotechnological exploitation of hairy roots as a source of valuable secondary products.

---

[1] Plant Biotechnology Group, AFRC Institute of Food Research, Colney Lane, Norwich NR4 7UA, U.K.

[2] John Innes Institute, Norwich, U.K.

NATO ASI Series, Vol. H18
Plant Cell Biotechnology. Edited by M.S.M. Pais et al.

INTRODUCTION

The study of plant secondary metabolism using plant tissues grown in vitro, with the aim of biotechnological exploitation, has traditionally used dispersed cell suspensions or callus. Such cultures often have poor production capacity and in addition are usually genetically unstable (Deus-Neumann and Zenk 1984, D'Amato 1985). Recently, we have begun to study the use of hairy root cultures, derived from dicotyledonous plants after infection with Agrobacterium rhizogenes, for the synthesis of valuable plant secondary products in vitro (Hamill et al. 1986). T-DNA [T_{Left} (T_L) and T_{Right} (T_R)] is transferred to the plant genome from the Ri plasmid of A. rhizogenes (Chilton et al. 1982, Huffmann et al. 1984, De Paolis et al. 1985) and the subsequent hairy root tissues which form may be cultured aseptically in vitro in simple media devoid of phytohormones.

We have found that hairy root tissue, from a number of plant species, grows rapidly in simple media (e.g. B5 salts (Gamborg et al., 1968) and sucrose. Typical fresh weight doubling times of hairy roots of Nicotiana rustica, Beta vulgaris and Datura stramonium are 48-60 h (Hamill et al. 1986, Payne et al. 1987). Such cultures produce the secondary products typical of the plant species. The levels of production, on a per gram tissue basis, were at least as high as were found in the parent plant species. We have analyzed hairy root cultures of seven different species in the Nicotiana genus, each characterized by its own particular synthetic pattern of the four major Nicotiana alkaloids, Nicotine, Nornicotine, Anatabine and Anabasine. It was found that the hairy root cultures synthesized the alkaloids characteristic of each species both in qualitative and quantitative terms. A proportion of the alkaloids were released into the medium (Parr and Hamill 1987). Hairy root tissue can grow from small inocula without the need for medium conditioning, e.g. 0.1 - 0.2 g N. rustica hairy root tissue is typically inoculated into 50 ml fresh medium of each fortnightly subculture. We have successfully grown hairy root tissues in fermenters and monitored the effects of altering the composition of the media upon the production and release of alkaloids in the culture (Rhodes et al. 1986a,b).

STABILITY

While hairy root cultures usually grow faster than untransformed root cultures, and have a much greater degree of branching, they are, nevertheless, organized and differentiated structures which have many similarities with the normal division and maturation sequence of root cells of intact plants. Thus there is a meristematic region at the root tip, a zone of cell elongation and an area of mature tissue. Using Beta vulgaris hairy root cultures, we observed that the red pigment (betacyanin) was synthesized primarily in the zone of mature tissue and not in the meristematic region of the root tip. However, a fast growth rate, coupled with the high degree of lateral root branching characteristic of hairy root lines, contributed to the net accumulation of 5.7 mg betacyanin l^{-1} day^{-1} in growing hairy root lines cultures of B. vulgaris (Hamill et al. 1986). This inherent physiological stability of hairy roots contrasts with the marked degree of heterogeneity within many populations of cell suspensions, both in terms of cell morphology and biosynthetic capacity (Yeoman and Street 1973, Deus and Zenk 1982, Neumann et al. 1983).

We have found hairy root cultures of Nicotiana rustica, Datura stramonium and Beta vulgaris to be stable in terms of their biosynthetic capacity over a period of more than 1 year in culture. All hairy root cultures which we have examined possess the euploid chromosome number typical of the species (Aird et al. 1988). This is in marked contrast to cell suspension cultures which usually show considerable chromosomal drift from the euploid number (D'Amato 1985). In a comparative study Aird et al. (1987) have shown that a hairy root line of N. rustica ($2n = 4x = 48$) which had been cultured for 17 months, possessed a chromosome number of 48, while a cell suspension of the same variety, and initiated at the same time, possessed few if any euploid cells. The majority of cells were aneuploid with $2n = 16-72$. Therefore the major advantage of hairy root cultures, over cell suspension cultures, is their inherent genetic stability which is reflected in their biosynthetic capacity.

PROSPECTS FOR IMPROVEMENT

Initial studies suggest that methods which are applicable to the genetic improvement of plants are applicable to the improvement of the biosynthetic capacity of hairy roots. Highly productive varieties of plants may be developed and used as starting material from which hairy roots can be developed (Parr and Hamill 1987). Regeneration of hairy roots from disorganized cell suspensions may facilitate recovery of biochemical mutants (Robins et al. 1987). We have also been able to regenerate hairy roots from protoplasts and found a significant amount of somoclonal variation between the clones which were recovered (Furze et al. 1987). Some of the variants synthesized more alkaloid than the parent line. In addition, we have shown that disarmed binary vectors, derived from A. tumefaciens, are efficiently cotransferred with the T-DNA of wild type A. rhizogenes to produce kanamycin of hygromycin B-resistant hairy roots (Hamill et al. 1987). Therefore foreign genes, influencing secondary metabolism, can be efficiently introduced into hairy root genomes where their effects can be monitored in the absence of variation induced by growth in the undifferentiated state.

In conclusion it is clear that hairy roots, formed after infection of idcotyledonous plants by A. rhizogenes, represent a novel and useful tissue which can be used to study plant secondary metabolism. In addition, such tissue has marked advantages over cell suspensions with regard to its biotechnological exploitation.

Acknowledgements. Other members of our group, who have been involved in the work described in this report, are Lindsey Aird, Abbi Peerless, Martin Hilton, Judy Furze, Mohammed Iqbal, John Payne and Andrea Prescott (John Innes Institute).

REFERENCES

Aird ELH, Hamill JD and Rhodes MJC (1988) Plant Cell Tiss. Org. Cult. (in press)

Chilton MD, Tepfer D, Petit A, David D, Casse-Delbart F and Tempe J (1982) Nature 295:432-434

D'Amato F (1985) CRC Crit. Rev. Plant Sci. 3:73-112

Deus B and Zenk MH (1982) Biotech. Bioeng. 24:1965-1974

Deus-Neuman B and Zenk MH (1984) Planta Medica 50:427-431

De Paolis A, Mauro ML, Pomponi M, Cardarelli M, Spano L and Costantino P (1985) Plasmid 13:1-7

Furze JM, Hamill JD, Parr AJ, Robins RJ and Rhodes MJC (1987) J Plant Physiol 131:237-246

Gamborg OL, Miller RA and Ojima K (1968) Exp. Cell Res. 50: 151-158

Hamill JD, Parr AJ, Robins RH and Rhodes MJC (1986) Plant Cell Rep 5:111-114

Hamill JD, Prescott A and Martin C (1987) Plant Mol. Biol (In 9:573-584

Huffman GA, White FF, Gordon MP and Nester EW (1984) J. Bacteriol 157:269-276

Neuman D, Krauss G, Hieke M and Groger D (1983) Planta Medica 48:20-23

Parr AJ and Hamill JD (1987) Phytochemistry 26:3241-3245

Payne J, Hamill JD, Robins RJ and Rhodes MJC (1987) Planta Medica 53:474-478

Rhodes MJC, Robins RJ, Hamill JD and Parr AJ (1986) New Zeal. J. Technol. 2:59-70

Robins RJ, Hamill JD, Parr AJ, Smith K, Walton NJ and Rhodes MJC (1987) Plant Cell Reports 6:122-126

Yeoman MM and Street HE (1973) In Plant tissue and cell culture (Street He, ed) 137-176 Blackwell Scientific. Oxford

THE APPLICATION OF GERMPLASM STORAGE IN BIOTECHNOLOGY

Erica E. Benson and Lyndsey A. Withers
Department of Agriculture and Horticulture
University of Nottingham School of Agriculture
Sutton Bonington
Loughborough, Leics
LE12 5RD
UK

INTRODUCTION

Biotechnology has precipitated a vast change in the way in which plants are utilised. Plant tissue culture techniques and genetic engineering provide a new approach to plant breeding. Furthermore large scale *in vitro* culture is a medium through which plant metabolism can be exploited to provide products of potential benefit to industry. These developments have also affected the applications and needs of plant germplasm conservation, these can be categorised into several main areas:

1. The conservation of important genotypes (including cultures used in genetic engineering and industrial processes).
2. The commercial exploitation of cultures which need conserving for patenting purposes.
3. The control of time-related change (e.g. somaclonal and culture variation).
4. Reduction of handling risks (contamination hazards of long-term routine culture work).
5. Reduction of costs (maintenance costs of keeping plants in long-term culture at normal growth rates).

In economic terms, the implementation of storage techniques can reduce production costs and provide a secure basis for patenting. This is a necessity if the commercial exploitation of biotechnology is to become viable. This need is directly related to the labile nature of plants maintained *in vitro*. Scowcroft, (1984) has defined two types of variation; tissue

NATO ASI Series, Vol. H18
Plant Cell Biotechnology. Edited by M. S. S. Pais et al.

culture instability and somaclonal variation. Deus-Neumann and Zenk, (1984) suggest that the culture instability of high producing cell lines may be an obstacle to the technological utilisation of plant cells. Maintenance of these high yielding cell lines is best achieved by continuous clonal selection, a process which, in the long term, could be unrealistic and costly.

A number of factors are thought to promote variation in plant tissue cultures. Importantly, prolonged periods of culture result in increased variation (Bayiss, 1980) and loss of morphogenic potential. It is in the control of these time related changes that storage techniques may be the most useful. In the case of the secondary products industry, storage pools of high producing cell lines provide a long-term stable supply of inocula which can be reintroduced into the production system, when required. Similarly, in genetic engineering programmes the storage of cultures carrying important gene combinations will circumvent any time related variability induced by their maintenance in prolonged culture.

Successful storage depends on a number of factors:

1. The storage technique must not cause irreversible damage or change.
2. Selection pressures must be avoided.
3. A return to normal growth patterns and metabolic functions must occur.
4. Good quantitative recovery should be achieved.
5. The storage method must be reproducible.
6. Ease of utilisation without costly training programmes.
7. Good security and safety.
8. Cost effectiveness of the techniques employed.

To date, growth limitation and cryopreservation offer the most promising means of germplasm conservation in biotechnology.

GROWTH LIMITATION

The maintenance of plant tissue cultures under suboptimal conditions (i.e. physical and/or nutritional limitation) causes a reduction in growth rate. This property is exploited as a means of storage on the premise that: a reduction in cell division is a stabilising factor and thus reduces variability and, a lower frequency of subculturing promotes stability, reduces costs and minimises the risks of contamination.

The most frequently used limiting factor is low temperature; atmospheric modification, partial desiccation, nutrient limitation and the application of growth inhibitors provide other means of reducing growth. The success of the chosen method is based on the reduction in subculturing interval, the length of exposure to the limiting factor (before detrimental effects are observed) and the rapid recovery of normal functions after return to standard culture conditions. Details of plants and cultures stored using growth limitation are discussed in several reviews (Withers and Williams, 1985; Withers, 1983; Withers and Williams, 1982). The major methods employed can be described as follows:

Temperature

It is initially important to consider the range appropriate for normal growth and then choose a limiting temperature which will not prove damaging. For temperate species which in general have an optimal growth range of 20-25°C, storage at 5-10°C is common. Tropical species having a higher temperature requirement for normal growth will survive better at higher storage temperatures (e.g. 15-20°C). Plants established under normal growth temperatures are transferred directly to cold storage. However, there is evidence to suggest that timing of this transfer may be important. Hiraoka and Kodama (1984) observed that rapidly growing calli of tobacco could be stored at 4°C for six months but regrowth of calli stored in stationary phase was only possible after 4 months and not for longer periods. Interestingly, no difference in the recovery pattern of nicotine production by these calli was observed. In contrast no variation in either growth or secondary metabolite formation was found in cultures of different ages from Dioscorea tokoro stored at low temperatures. However, in general, variability between species is apparent. Calli from six different species showed differential recovery responses in both secondary product evolution and growth when stored at low temperatures (Hiraoko and Kodoma, 1984). Similarly, slow growth procedures applied to shoot cultures of a range of Solanum spp. also showed variability in recovery responses (Westcott, 1981a and b).

Oxygen

Oxygen is the second physical parameter used to limit growth. Availability of the gas can be reduced by:

1. a low pressure system which functions by decreasing the atmospheric pressure in the culture environment. This limits the partial pressure of all gases that are in contact with the plant.
2. a low oxygen system which reduces available oxygen by using an inert diluent gas (e.g. N_2) but still operates at atmospheric pressure, and
3. a mineral oil overlay which separates the cultures from the gaseous environment, thus reducing oxygen availability.

For discussion of these methods see Nitzche, (1983) and Bridgen and Staby (1983). The basis of slow growth in reduced O_2 environments is primarily related to a decline in respiratory activity. However, auto- and heterotrophic metabolism and the capacity of the tissues to follow C_3 or C_4 pathways of photosynthesis may be differentially affected by O_2 limitation and changes in $O_2 : CO_2$ ratios (Bridgen and Staby, 1983).

Investigations by the previous authors using differentiated and disorganised chrysanthemum and tobacco cultures have shown that oxygen partial pressures below 50 mm Hg. significantly reduce growth rates. No phenotypic difference was observed when plants were grown to maturity after storage.

Application of chemical growth retardants

A wide range of growth retarding additives can be incorporated into tissue culture media. Osmotically active compounds (e.g. mannitol) create a "water stress" environment which inhibits plant growth. Phenotypes develop which are characteristic of naturally drought tolerant species. Plant growth regulators such as succinic acid and 2,2 dimethyl hydrazine (B-9) also produce this effect. Studies, (Westcott, 1981a and b) using potato, have shown that the subculturing interval can be increased from several weeks to 1 year by using the growth retardants ABA and mannitol. In contrast, B-9 and phosphon-D (chlorophonium chloride) were less effective.

Overall, growth limitation offers a promising technique, particularly in the case of low temperature treatments. However, this method may not be suitable for long-term storage (ca > 2 years) particularly as selection pressures may arise. For longer periods, cryopreservation appears to be the most advantageous method.

CRYOPRESERVATION

It is usual to categorise cryopreservation storage methodology into several distinct phases:

Pregrowth and cryoprotection

The developmental characteristics of plant germplasm can greatly affect successful cryopreservation. In the case of organised structures (embryos, meristems) it is important to choose young meristematic tissues which are highly cytoplasmic and non-vacuolated (thus reducing available water for ice formation). In disorganised cultures the same rules apply and tissues used in freezing programmes are selected from the late lag/early log phase of growth.

When selecting secondary product evolving cultures for freezing, it is cautionary to note that the cryopreservation of cells already in the secondary metabolic state may have two detrimental consequences. (1) loss of compartmentalisation will disrupt enzyme function and, (2) the release of cytotoxic membrane-bound products of secondary metabolism will cause additional autoinjury. Cells of *Digitalis lanata* preloaded with purpurea glycoside A were shown to be less suited for cryopreservation than cells free of cardenolides (Diettrich *et al*., 1986). It may thus be advisable to select cell lines in the non-secondary metabolic state for storage.

Often, modification of the culture medium and/or pretreatment with protective additives during pregrowth can effectively enhance survival after freezing. Addition of osmotically active compounds (e.g. sorbitol and mannitol) reduces cellular water before freezing. Improved survival has also been achieved using pregrowth treatments of proline and/or ABA (Kartha et al., 1980; Withers, 1980a and b; Chen et al., 1985). In the case of shoot tips a recovery phase allowing the tissues to overcome dissection injury before freezing is also a routine practice (Henshaw et al., 1985).

Penetrating cryoprotectants have several beneficial modes of action, including the maintenance of cellular water in a liquid state at low temperatures. This colligative action prevents an excessive concentration of toxic solutes in the non-frozen cellular structures. Non-penetrating cryoprotectants act by promoting dehydration and reducing the amount of intracellular water available for ice formation. Other factors such as membrane stabilisation (Finkle et al., 1985) may be involved in cryopreservation.

The most commonly used cryoprotectants are dimethyl sulphoxide (DMSO), glycerol and sucrose. Other compounds are also used including: mannitol, sorbitol, proline, polyethylene glycol, polyvinylpyrrolidone and hydroxyethylstarch. Variations in cryoprotectant treatments for cell suspensions and organised structures are apparent. In most treatments for the latter, DMSO is used at concentrations of 5-15%. In the case of cell suspension cultures however, few species survive freezing using DMSO alone and 'cocktails', of DMSO, glycerol, sucrose or proline greatly enhance recovery.

Plant tissues are usually incubated for about 1 hour in the cryoprotectant before the onset of freezing. For cell suspensions, step-wise addition of cryoprotectants at chilling temperatures is advised. In contrast, successful cryopreservation of a range of shoot-tips from Solanum spp. has been achieved by one-step addition at room temperature (Towill, 1984).

When using cryoprotectants it is important to consider their possible cytotoxic effects. DMSO toxicity has been related in some part to the purity of the compound (Matthes and Heckensellner, 1981). And importantly, DMSO has been shown to generate genetic and epigenetic changes (Ashwood-Smith, 1985). Unfortunately, although its use is widespread in cryobiology, little is understood about the long-term effects of the compound.

Kartha, (1985) summarises the long-term effects of cryoprotectants on normal growth patterns in non-cryopreserved plant tissue cultures. As not all species demonstrated the same responses, it would be advisable to investigate the side effects of these components before initiating new freezing programmes.

Freezing

There are two approaches to freezing: rapid and slow. The former involves a direct immersion of the specimen in liquid nitrogen. Slow freezing rates of ca $< -1.0°C. min^{-1}$ are common and this can be maintained using a controlled freezing unit. One or several different freezing ramps can be incorporated to produce a step-wise freezing procedure. Usually in slow freezing, samples are cooled to an intermediate terminal temperature (e.g. -30 to -40°C) at which they can be maintained for a short period of time before plunging into liquid nitrogen or, alternatively, they can be transferred from the terminal temperature, directly into liquid nitrogen.

The attainment of an optimal cooling rate is largely dependent on dehydration injury and intracellular ice formation. In slow freezing, ice formation is initiated (nucleated) outside the cell and the cell wall acts as a barrier against internal ice growth. The external ice formation causes a vapour pressure deficit in relation to the cell and in order to equilibrate water leaves the plant tissues. This phenomenon can be cryoprotective as it reduces the amount of freezable water in the cell. However, if the dehydration effect is too great, cell damage can arise as a result of the reduced cellular water content. Cooling rate, nucleation and the holding time of terminal transfer temperatures can all affect protective dehydration. These parameters must thus be adjusted to suit the freezing characteristics of different culture types and species. In ultra-rapid freezing ice formation occurs both intra and extra cellularly before the cell has time to protectively dehydrate. Rapid freezing is probably best applied to tissues which are highly cytoplasmic and contain very little vacuolar water. In contrast, slow freezing may be more suitable for specimens which contain a relatively high water content (Withers, 1983). There are variations in the standard freezing protocols. The droplet freezing method (Kartha, 1985) was devised for cassava meristems. This is a slow freezing technique, but instead of the meristems being contained in an ampoule they are suspended in small droplets of cryoprotectant on aluminium foil. The success of this approach is thought to be due to the uniformity of freezing, resulting from the increased thermal conductivity of the metal surface.

Thawing and recovery

Rapid thawing in water at (+40°C) is advocated for cryopreserved plant tissues. By this approach the damaging recrystallisation of ice caused by slow thawing can be prevented. After thawing, the specimens can be either directly placed onto recovery media or washed before transfer. Washing is sometimes thought necessary to remove cryoprotectants which could have toxic effects after prolonged exposure. However, this practice may cause osmotic injury in previously dehydrated material and the damaged cell membranes are therefore more prone to leakage. It is suggested that the suspending solution containing cryoprotectants provides an important aid to recovery (Withers, 1980a). Recent studies (Benson and Withers, 1987) have shown that DMSO may have a protective role as a free radical scavenger in the post thaw stabilisation of cryopreserved *Daucus carota* tissues, thus providing yet another reason for not washing cells immediately after thawing.

The composition of the recovery media is a most important factor in promoting post-freeze recovery. In the case of cell suspensions it is advisable to transfer directly to semi-solid medium (Withers, 1980a and b). Once the cells are actively growing they can then be returned to liquid culture. The hormonal composition of the recovery medium is important, particularly for organised structures which often fail to recover in the absence of hormones. The use of auxin and cytokinin plant growth regulators in recovery medium improves survival considerably (Withers and Benson, 1987). These treatments must not, however, be used at the expense of regenerant quality as they can often promote destabilising phenomena such as callusing and adventitious shooting.

OTHER METHODS OF LOW TEMPERATURE GERMPLASM STORAGE

Preservation by undercooling has recently been applied to plant tissue cultures (Mathias et al., 1985). The objective of this approach is to maintain tissues at low temperatures (-10 to -20°C) but in the absence of ice crystallisation. The plant tissues are immersed in an immiscible oil and the emulsion thus formed can be undercooled to relatively low temperatures thereby circumventing ice formation, one of the most injurious consequences of low temperature storage. This method has been applied to Sainfoin cell suspensions and potato and pea meristems. Although good recovery has been reported this has only been achieved using a temperature of -10°C and for relatively short storage periods (6-48 hours). Douzou (1977) has indicated that enzyme activity can still take place at relatively low temperatures. As a consequence, total inhibition of metabolic activity may not occur in the undercooled state and deterioration in storage could be a problem. Although undercooling is a promising approach, it may be important to conduct stability evaluations before it can be applied routinely.

ASSESSMENT OF RECOVERY AND STABILITY AFTER CRYOPRESERVATION

One of the most important aspects of low temperature storage concerns the assessment of recovery and stability. Unfortunately many of the approaches used are under-developed, they often present practical difficulties and are limited in their capacity to provide a better understanding of the factors involved in promoting stability in storage. To date, a range of procedures have been used to assess cryostability.

Vital staining is one of the most common short-term methods of assessing early post-freeze recovery (Withers, 1980b). Unfortunately, this technique can be ambiguous. In the early stages of recovery, the post-thaw population contains a mixture of cells in a state of metabolic flux. Staining characteristics may therefore change as individual cells either recover fully or die. Vital staining has a useful application for cell suspensions. However, it is more difficult to assess larger organised structures. Microscopical analysis provides an important means of determining if recovery growth proceeds via destabilising adventitious development in organised structures. Similarly, electron microscopy can be used to assess damage at the cellular and subcellular level.

Non-destructive growth analysis can be used to evaluate the dynamic aspects of post-freeze recovery. For example, the duration of the lag phase may be an important factor to consider when a quick return to normal production capacities in cell lines is required.

Physiological approaches to the assessment of cryostability are limited. Studies on rice cells have shown that respiratory functions are impaired after storage in liquid nitrogen (Cella _et al_., 1982). This damage was not lethal and after a lag phase the cell returned to normal growth. Studies performed on animal tissues stored at low temperatures have shown that free radical damage may be important in cryoinjury. Benson and Withers (1987) suggest that lipid peroxidation occurs during the early post thaw recovery period of frozen _D. carota_ cells. In the same study, changes in the evolution of the stress hormone ethylene correlated with the onset of cellular recovery. Retention of secondary metabolic functions is a major factor determining the usefulness of cryopreservation in plant biotechnology. Studies have shown that biosynthetic functions remain unaltered after freezing. In _Digitalis lanata_ the production of glucosylated digitoxin was unchanged (Diettrich _et al_., 1985). Similarly steroid production by _Dioscorea deltoidea_ was stable after storage in liquid nitrogen (Butenko _et al_., 1984), as was biotransformation capacity in cryopreserved _D. lanata_ (Seitz _et al_., 1983).

Successful cryopreservation is usually evaluated numerically as total survival. However, this approach has one major drawback in that the quality of the regenerated material is ill defined. Qualitative analysis can relate to both phenotypic and genotypic characteristics. Studies using potato shoot-tips have shown that the morphological quality of the post freeze regenerants is dependent on the hormonal content of the recovery medium (Withers and Benson, 1987). The assessment of recovery in stored germplasm is usually approached in the short term as described above. However, the effects of freezing on long-term regenerant stability have not been determined in depth. One extensive study on strawberry plants regenerated from cryopreserved cultures was encouraging in that no abnormalities were observed in the mature regenerants (Kartha et al., 1980).

PROGRESS AND FUTURE AREAS OF DEVELOPMENT

Considerable progress has been made in the utilisation of conservation methods in plant biotechnology. If storage techniques are to become routinely available, safe and reproducible, several areas still require attention. Thus it will be important to consider the logistical aspects of conservation. Included in this area is the evaluation of regimes for accession numbers, the technical and production aspects of post storage recovery and the bulking up of conserved material to maximise production capacities.

The improvement of existing methods and the development of new approaches must also be achieved in conjunction with basic research studies. A greater understanding of cryoinjury, molecular/biochemical instability, growth retardant and cryoprotectant toxicity would, in the long term, aid the wider application of conservation in plant biotechnology.

REFERENCES

Ashwood-Smith MJ (1985) Genetic damage is not produced by normal cryopreservation procedures involving either glycerol or dimethyl sulphoxide : A cautionary note, however on possible effects of dimethyl sulphoxide. Cryobiology 22: 427-433

Bayliss MW (1980) Chromosomal variation in plant tissues in culture. Int Rev Cytol 11A: 173

Benson EE and Withers LA (1987) Gas chromatographic analysis of volatile hydrocarbon production by cryopreserved plant tissue cultures : A non-destructive method for assessing stability. Cryo - letters 8: 35-46

Bridgen MP and Staby GL (1983) Protocols of low-pressure

storage. In: Evans PA, Sharp WR, Ammirato PV and Yamada Y (eds). Macmillan Publishing NY, pp 816-827

Butenko RG, Popov AS, Volkova LA, Chemyak MD and Nosoy AM (1984) Recovery of cell cultures and their biosynthetic capacity after storage. Dioscorea deltoidea and Panax ginseng in liquid nitrogen. Plant Sci Lett 33: 285-292

Cella R, Colombo R, Galli MG, Nielson E, Rollo F and Sala F (1982) Freeze-preservation of rice cells : a physiological study of freeze-thawed cells. Physiol Plant 55: 279-284

Chen THH, Kartha KK and Gusta LV (1985) Cryopreservation of wheat suspension culture and regenerable callus. Plant Cell Tissue and Organ Culture 4: 101-109

Deus-Neumann B and Zenk MH (1984) Instability of indole alkaloid production in Catharanthus roseus cell suspension cultures. Planta Medica 50: 427-431

Diettrich B, Haack U and Luckner M (1985) Long term storage in liquid nitrogen of an embryonic strain of Digitalis lanata. Biochem Physiol Pflanzen 180: 33-43

Diettrich B, Haack U and Luckner M (1986) Cryopreservation of Digitalis lanata cells grown in vitro. Precultivation and recultivation. J Plant Physiol 126: 63-73

Douzou P (1977) Cryobiochemistry : An Introduction. Academic Press London

Finkle BJ, Zavala ME and Ulrich JM (1985) Cryoprotective compounds in the viable freezing of plant tissues. In: Kartha KK (ed). Cryopreservation of Plant Cells and Organs. CRC Press Inc Florida 75-114

Henshaw GG, O'Hara JF and Stamp JA (1985) Cryopreservation of potato meristems. In: Kartha KK (ed). Cryopreservation of Plant Cells and Organs. CRC Press Inc Florida 159-170

Hiraoka N and Kodama T (1984) Effect of non-frozen cold storage on the growth, organogenesis and secondary metabolism of callus cultures. Plant Cell Tissue and Organ Culture 3: 349-357

Kartha KK (1985) Meristem culture and germplasm preservation. In: Kartha KK (ed). Cryopreservation of Plant Cells and Organs. CRC Press Inc Florida 115-134

Kartha KK, Leung NL and Pahl K (1980) Cryopreservation of strawberry meristems and mass propagation of plantlets. J Am Soc Hort Sci 105: 481-484

Mathias SF, Franks F and Hatley RHM (1985) Preservation of viable cells in the undercooled state. Cryobiology 22: 537-546

Matthes G and Heckensellner KD (1981) Correlations between purity of dimethyl sulphoxide and survival after freezing and thawing. Cryo - letters 2: 389-392

Nitzche W (1983) Germplasm preservation. In: Evans PA, Sharp WR, Ammirato PV and Yamada Y (eds). Handbook of Plant Cell Culture. Macmillan NY 783-805

Scowcroft WR (1984) Genetic variability in tissue culture : impact on germplasm conservation and utilisation. IBPGR Report Rome

Seitz U, Alfermann AW and Reinhard E (1983) Stability of biotransformation capacity in Digitalis lanata cell cultures after cryogenic storage. Plant Cell Rep 2: 273-276

Towill LE (1984) Survival at ultra low temperatures of shoot tips from Solanum tuberosum groups : Andigena, Phureja,

Stenotomum tuberosum and other tuber bearing Solanum species. Cryo - letters 5: 319-326

Westcott RJ (1981a) Tissue culture storage of potato germplasm. 1. Minimal growth storage. Potato Res 24: 331-342

Westcott RJ (1981b) Tissue culture storage of potato germplasm. 2. Use of growth retardants. Potato Res 24: 343-352

Withers LA (1980a) The cryopreservation of higher plant tissue and cell cultures : an overview with some current observations and future thoughts. Cryo - letters 1: 239-250

Withers LA (1980b) Tissue Culture Storage for Genetic Conservation. IBPGR Publ 80/8 Rome

Withers LA (1983) Germplasm preservation through tissue culture : an overview. In: Proceedings of a workshop cosponsored by the Institute of Genetics, Academia Sinica and The International Rice Research Institute. Science Press Beijing China pp 313-342

Withers LA (1985) Cryopreservation of cultured plant cells and protoplasts. In: Kartha KK (ed). Cryopreservation of Plant Cells and Organs. CRC Press Inc Florida pp 243-267

Withers LA and Williams JT (1982) Crop Genetic Resources. The Conservation of Difficult Material. IUBS/IBPGR Pub SER B42 Rome

Withers LA and Williams JT (1985) Biotechnology in International Agricultural Research. Proc IARCS and Biotechnology IRRI Manila Phillipines

Withers LA and Benson EE (1987) In preparation

FUTURE TRENDS IN PLANT CELL BIOTECHNOLOGY

E. John Staba
Department of Medicinal Chemistry and Pharmacognosy,
University of Minnesota, Minneapolis, Minnesota 55455, U.S.A

INTRODUCTION

The plant tissue culture (ptc) technique has grown spectacularly since the 1930's when it was demonstrated that aseptic plant organs and explants could be subcultured (Gautheret 1983). A dramatic example of this growth is the large number of ptc research articles published in 1985, 4,200, which is seven times greater than that published in 1965 (Bhojwani et al 1986). Another even more significant growth indicator is the attendance of more than 1,500 scientists at the 1986 International Association of Plant Tissue Culture (IAPTC). The number attending exceeded by approximately one-third that attending the 1982 IAPTC meeting. It does appear that the ptc technique has evolved from an art to that of a science with principles that assure reasonable experimental reproducibility. However, the question of ptc as an art or a science is irrelevant as it has dramatically affected the practice of the plant sciences in academia, the government, and industry.

In the late 1970's recombinant DNA technology became not only of high interest to the scientist but to the commercial world. It is estimated that between 1979 and 1983 more than 250 biotechnology industries were established in the U.S. (Dibner 1986). In 1983, investments to commercialize biotechnology exceeded $1 billion (Anonymous 1984). The growth and development of biotechnology companies throughout the world has been slower and more financially risky than was first thought. Nevertheless, the potential corporate "winners" continue to complete for a part of the $12 billion biotechnology market predicted for 1990 and the $20 to $100 billion market predicted for the year 2000 (Anonymous 1984; Dibner 1986). No clear corporate winners are yet identified

NATO ASI Series, Vol. H18
Plant Cell Biotechnology. Edited by M. S. S. Pais et al.

for this market, and many now believe that biotechnology is to be more evolutionary than revolutionary.

The agri-biotech component of the biotechnology industry is very significant. It is estimated that this market will grow from almost nothing in 1985 to approximately $538 million in 1990 and $2.3 billion in 1995. The biotech seed market in 1995 is expected to be $650 million. The annual income of today's hybrid seed market is about $1 billion, today's retail flower market about $3 billion (Gebhart 1986), and the food coloring agents about $70 million of a $25 billion food market. For perspective, one should recognize that if all goes well biotechnology will represent less than 5% of the total agribusiness market in 1995 (Gebhart 1986)!

Some plants can be profitably replicated in a small space and in a short time by micro-propagation. A number of ornamentals and agronomic crops such as viral-free potatoes, coconut palm, and pyrethrum have been commercially micropropagated. It is expected that in the relatively near future ptc technology will have facilitated the development of ornamentals with unusual blue colors; the development of value-added plants such as high-solid tomatoes (DNA Plant Technology, Cinnaminson, NJ); value-added disease-resistant stem potato buds, celery seeds or plant embryoids encapsulated in a polymer gel containing fertilizers, pest-control agents, or beneficial nematodes (Plant Genetics, Inc., CA); micropropagated date palm trees (Twyford Plant Laboratory, England); the development of important agronomic and specialty crops such as corn, rice, and tobacco that are disease- and herbicide-resistant and possibly stress-tolerant; and improved nutrition, flavor, sweetness, or other chemical attributes. In addition, the commercialization of an in vitro process to grow ginseng biomass or to produce berberine, indole, or tropane alkaloids is possible and probable.

Most ptc problems can probably be found somewhere in one or more of the following statements: there is too much variability; there is too little variability; how can the system be synchronized; it costs too much; the cells and tissues don't grow and regenerate the way I want them to; how can I recognize and isolate specific cells and tissues. What we all wish is maximum control of the plant system. As observed by many and as so well stated by Professor E.C. Cocking at the VI-IAPTC meeting (Cocking 1986), many of the solutions to these problems will only come from fundamental plant research.

No one can predict the realities of the future as important biotechnical discoveries are made by accident, serendipity, hard work, and to some unknown degree by scientific design. One can only try to evaluate the available technology for today's needs and perhaps for the year ahead. As it is easier to ask questions than to give answers, I intend to raise questions to which many of us would like to know the answers. If we knew the answers to these questions we would in fact know how exciting the future may be!

FUNDAMENTAL PLANT PHYSIOLOGY

Plant tissue culture is an axenic model that enables us to study plant cells, tissues, organs, and if we wish, plants. These various models can be studied intracellularly or as tissues and organs.

Some intracellular materials are the cell's walls, plasmalemma, tonoplasts, vacuoles, nucleus, chloroplasts, and mitochondria. These biological components often interact with each other. As our ability increases to exchange such components at the cellular level we will better understand not only what regulates cell growth but also the processes that lead to cell death.

On the intracellular level, many of us would like to know better how lectins, phytoalexins and elicitors relate to the cell wall; what are the metabolic roles of adenosine 3,5'-cyclic monophosphate (cAMP), calmodulin, prolines, low M.Wt. peptides and the phosphorylated compounds; how to control the ionic gradient outside the cell, in the protoplast, and in the vacuoles; how is compartmentalization triggered; what is the physiological role of carbohydrate, protein, and lipid complexes and polymers; what is the dynamics and purpose of cytoplasmic streaming.

Significant similarities and differences exist between plant and animal systems. Concepts and theories from each will increasingly advance the other's understanding of biological processes. It is interesting to observe that today it is possible to use a plant potato disc assay to screen for human antitumor compounds (Ferrigni 1982).

Why do some cells within unorganized or organized tissues contain oil bodies (Takeda and Katoh 1981) while others differentiate into elongated fibers or embryoids? The answers to these questions might help us understand how terpenes are produced, how to produce quality cotton fibers (Klausner 1985), or how to make more 'normal' embryoids.

Why can't we more easily reorganize single plant cells or protoplasts into plants? Only recently have we been able to coax the milieu and environment about rice protoplasts to redifferentiate them into plants (Marx 1987). Perhaps the study of protoplasts isolated from specific plant sites such as the leaf guard cell or floral tissues, and how the protoplast is affected by the plant's pretreatment, genetics, or nurse tissues, will help us understand why protoplasts behave as they do.

On the more complex tissue and organ level, it is recognized but not adequately understood why callus maintained in culture is often refractive to reorganization, yet when one isolates approximately 6-7 layers of cells they sometimes dramatically form flowers and other organs (Tran Thanh Van 1981). Why is this observed phenomenon restricted to so few species? Also, why are monocots more refractive to culture than dicots, and why don't roots reform shoots as readily as shoots form roots?

Many physical, biochemical, biological and genetic factors that affect the plant's physiology and biochemistry are studied.

Some physical factors are diffusion/osmosis, electron density, gravity, cell position/restriction, radiation, and temperature. Diffusion and osmotic factors may affect how secondary plant products are produced, stored and released from cells (Morris 1985). Minute electric currents (1-2 microamps) stimulated shoot formation from tobacco callus cultures, a technique that may be useful to differentiate refractive monocot cultures and to "normalize" embryoids (Rathore and Goldsworthy 1985). Gravity affects the regeneration of tobacco protoplasts. In fact, protoplasts are suggested to be used to study space microgravity effects (Iverson 1985). Plant tissue culture technology may also complement the development of controlled environmental life support systems (CELSS) for lunar colonies and other long-term space environments once the plant systems needed are defined. Cell position and pattern are important for cell development, and it has been observed that externally applied pressure affected the development of tobacco pith and callus (Lintilhac and Vesecky 1984). Radiation and/or light is important to study photosynthesis (Edwards and Scott 1985) and to activate a number of metabolic processes, i.e., flavonoids (Fowler 1985). Temperature increases from 35-40°C will release heat shock proteins from plant cultures. The proper control of light and temperature is obviously critical for metabolic processes.

Protein topogenesis is the study of protein transportation and localization. Recent studies have demonstrated that specific proteins or enzymes can be genetically directed to be synthesized with an N-terminal extension that assures their transport across the chloroplast membrane (Netzer 1987). The polyamines, spermidine and spermine or their precursors can interact with nucleic acids, promote protein synthesis, maintain intracellular ionic balance and/or delay senescence (Galston and Smith 1985). The use of substituted amides and diamides may retard micropropagation abnormalities and increase the survival of cuttings (Anonymous 1986). Flower, fruit-set, or senescence induction and/or expression can be controlled to some extent by growth regulators such as gibberellic acid, ethephon, and benzyladenine. Nevertheless, the discovery, identity and specificity of florigens continues to be of high interest to some investigators -- and they should be.

Symbiotic and allelopathic interactive studies are done with plant cultures and the bacteria, fungi, viruses, and nematodes. These studies have established the identity of many biotic elicitors (DiCosmo and Misawa 1985), and that thiophenes, polyenes, and antibiotic biocides are being produced from plant cultures.

PLANT TISSUE CULTURE GENETICS

Unorganized plant cells and protoplasts are established from either haploid, diploid, or polyploid plant cells. Epigenetic variants (non-Mendelian traits) and somaclonal mutants are known to be present in tissue culture systems. Genetic variations originate in these cells from their nuclear, chloroplast and/or mitochondrial DNA pool. Such somoclonally derived variants are mutants, form spontaneously, and give rise to inheritable traits (Evans and Sharp 1983). Although most somaclonal variants are useless, some contain useful attributes that are discovered more rapidly than by conventional field genetics. Somaclonal-derived plants have been selected for herbicide resistance (atazine and chlorsulfuron), disease resistance, and to a limited degree for heat- and salt-tolerance, improved nutrition and flavor, and higher agronomic yields.

It has been observed that agronomic plants such as cassava, pineapple and garlic were improved when cycled as a tissue culture (Orr 1985). It also appears that variants can be selected from the cell cultures in a shorter time than by conventional breeding, particularly if the cultures were derived from pollen and subjected to selection pressures. However, the traits selected from tissue culture systems may be too specific and still require conventional cross-breeding to establish a quality plant. Regenerated plants should always be selfed for a number of generations to be certain that the desired trait is stable.

TRANSFORMATIONS

Potentially desirable gene elements for possible introduction into plant cells are available from underutilized food, legume, and tree crops (Vietmeyer 1986), and industrial and medicinal plants (Balandrin et al 1985). Such genetic transformations can be achieved biochemically, biologically or mechanically.

Synthetic protein bio-carriers may carry genetic and other materials into cells by endocytosis. Organelles such as mitochondria can also be introduced into plant cells by protoplast fusion and their segregations monitored (Flick et al 1985). It is known that mitochondria code for flower morphology and cytoplasmic male sterility (CMS). Dominant nuclear resistance marker systems have been established to select for somatic mitochondrial DNA hybrids (Kothari et al 1986). Companies are planning to use the CMS marker to protect plants derived from tissue culture from unauthorized exploitation (Orr 1985). Futher tissue culture organelle studies may help us understand why mitochondria DNA easily recombines and chloroplast DNA does not, and perhaps even how ribosomes control DNA expression.

The _Agrobacterium tumefaciens_ vector has been extensively used to introduce genetic materials into plants for herbicide resistance (glyphosate) to produce antibiotics, to confer insect resistance (_Bacillus thuringiensis_ toxins), and even to induce rubber production in tobacco plants (Anonymous 1987). Other _Agrobacterium_ species have been used to study both root and shoot differentiation processes. When appropriate, other vectors such as the cauliflower mosaic virus may be used (Crossway and Houck 1985).

Mechanical micromanipulation (Crossway et al 1986) and electroporation methods can be used with all plant species and is not restricted to use with the dicots as _A. tumefaciens_ transformations presently are. Approximately 25% of alfalfa protoplasts were transformed by intranuclear microinjection of Ti plasmids (Reich et al 1986), and the parallel electrode electroporation method enabled 3.0% of berberine and anthocyanin-containing protoplasts to fuse (Yamada and Morikawa 1985).

Two transformation systems of potential interest are a modified leaf-disk transformation-regeneration method in which cells of Arabidopsis were directly transformed by A. tumefaciens (Lloyd et al 1985) and the other is one in which DNA materials might be directly inserted into selectively prepared root-tip protoplasts on the plant (Cocking 1985).

SELECTION PROCESS

A significant problem is to be able to recognize, separate and stabilize variant cells from the tissue culture milieu. Procedures to address these concerns are often inefficient and time-consuming.

Where applicable, cell variants may be analyzed in real-time by potentiometric immunoassay (Monroe 1987). Direct analysis of cell-to-cell variability is sometimes possible microspectrophotometrically for some compounds, by nmr for phosphorylated compounds, and for some alkaloids by mass-analyzed ion kinetic energy (MIKE) (Cooks et al 1981).

Protoplasts which have fused can be recognized in as few as 1-50 cells with a radioactively labelled cauliflower mosaic virus (CaMV) probe (Crossway and Houck 1985). Cell sorters have also been used to separate somatic hybrid cells by fluorescence labeling as well as in cells that contain alkaloid fluorescence (Fujita and Tabata 1986). Cells have been genetically transformed to contain the firefly luciferase gene cells that can be directly recognized without light activation (Ow et al 1986).

Cryopreservation can conserve cell cultures and retard their variability. Dry freezing, period cold-hardening, storage at undercooled temperatures (-20°C), or pretreatment with abscisic acid, sodium butyrate or other compounds have been suggested as enhancements to conserve cells.

Commercial micropropagation systems have been used to

duplicate ornamental (orchids), agronomic (potato), and specialty crops (pyrethrum). Although micropropagate propagules appear uniform, one cannot be certain that they are until long-term field data is obtained. It is prudent to maintain more than one tissue culture clone in the field to be certain that possible field problems and variances can be coped with. Eucalyptus, Pinus radiata, date palm, and oil palm are some trees that have been micropropagated and selected for improved product yield, growth rate, insect resistance, timber quality, and tolerance to stress (dryness, heat, salinity). The demand for micropropagated date palm trees is said to now exceed 100,000 plants annually.

PLANT TISSUE CULTURE SYSTEMS

The reasons for establishing a ptc system to produce useful compounds is to be free of variables such as weather, plant disease, geography, and politics. A more important reason for many is to develop a profitable process. Recent estimates suggest that a profitable batch process requires a product valued at no less than $500/kg (Sahai and Knuth 1985). A fed-batch air-lift system may require a product valued at about $1,200/kg (Ten Hoopen et al 1985). Fixed capital investment, product market size, and process cycle time are variables that significantly affect the outcome of such projections. The value of shikonin, the first successful ptc system process, is more than $4,000/kg (Fujita and Tabata 1986; Sahai and Knuth 1985).

One needs to control a ptc process in order to assure its reproducibility. Unfortunately, we do not understand plant cell growth and metabolism, cell differentiation and organization, and plant genetics adequately to rigorously control the ptc system. However, some ptc systems are forgiving and even though we do not understand them, they are reasonably reproducible. In fact, the metabolism of the plant may be sufficiently different in the ptc system that we now look for new and novel compounds rather than conventional compounds. We may also use selected derepressed enzymes in cell systems to bioconvert digitoxin (Kreis and Reinhard 1986), anhydrovinblastine (Kutney 1985), and precursors to food flavors (Collin and Watts 1985). However, the future production of complex food flavors and perfumes will probably require organized differentiated tissues.

It is sometimes asked if ptc systems can be used to produce low-cost commodities such as starch, cis-polyisoprenes, or cotton fibers. One estimate is that an annual production of 100,000 to 1,000,000 kg of a commodity might result in a production cost of $20-$25/kg (Sahai and Knuth 1985).

Some interesting product targets are rosemarinic acid, catharanthus alkaloids, and ginseng. Although yields of rosemarinic acid may exceed that in the plant by more than twentyfold and be produced at greater than 0.5 g/liter/day output (Ulbrich et al 1985), it still remains a system in search of a market. The production of the dimeric indole alkaloids in _Catharanthus_ _roseus_ cultivars is believed by some (Kurz et al 1985) to require a differentiation process and the correct genotype. Ginseng appears closer to commercial success, as "informal permission" was received in 1986 in Japan to sell ptc-produced ginseng as a health food product. Approximately 300 kg of dried ginseng tissue is presently obtained each month from a 20-kl fermentor, and there are plans to build two additional fermentors sometime later this year (Ushiyama 1987).

The costs that relate to system design and processing have been summarized by Sahai and Knuth in 1985 and somewhat by Lee and An in 1986. An additional cost factor of significance is the preparation of data to obtain government approval for a biotechnology product to be used for animals, humans, or in the environment (Butts 1987).

POTENTIAL SYSTEM ENHANCERS

The ptc system may be enhanced by genetic improvements or by improved synchronization of cell and tissue growth. Not only are advances being made in the laboratory for ptc transformation, amplification, and somaclonal selection, but in software to assist in DNA sequencing, restriction studies, and map development (Dickson 1985). Plant synchronization has, to a limited extent, been achieved with ethylene gas and other stresses, but futher growth phase control is needed. Perhaps synchronized embryoids can better produce cardenolides, coco lipids (Wright et al 1982), or seedlings to produce morphinan alkaloids (Schuchmann and Wellmann 1983).

The immobilization of ptc in bio-gels and/or growth in hollow fibers, spin-filter systems, or in polyurethane (Navituna and Park 1986) has as its major advantage the uncoupling of ptc growth from product production. Its disadvantages are not unique to ptc, and they are possible variations in product production, relatively long process time, and a need to release the product extracellularly.

We should better understand how biotic and abiotic elicitors function. A major anticipated use of elicitors is to synthesize important products within a shortened fermentation time (Smith et al 1987). How do some elicitors turn on m-RNA synthesis in a matter of minutes? Can we establish a continuous rather than a short-duration elicitor response? Can elicitors significantly alter the major biosynthetic pathways in cells and tissues, i.e. from steroids to terpenes or from the acetate to the mevalonate pathway?

Another interesting experimental approach is to isolate enzymes for highly specific metabolic reactions from cell-free systems (Zito and Staba 1985; Banthorpe et al 1986) and then possibly immobilize them or use them for eventual genetic transformations.

Plant tissue culture systems will utilize more frequently computerized automation and robotics for micropropagation (De Bry 1986; Sluis and Walker 1985), for chemical analysis, and to control production systems. Innovative discoveries are the most important keys to unlocking the door to a more successful ptc system. Perhaps discoveries such as microbially induced hairy roots on ptc (Flores et al 1986), or organ culture that contain stems with profuse flowering and fruit set, or even perhaps systems that are protected and only partially aseptic are in the right direction.

Our applied industrial systems will become more successful when we know why and not just how the plant's processes operate!

REFERENCES

Anonymous (1984) Commercial Biotechnology: An International Analysis, Office of Technology Assessment, OTA-BA-218, Washington DC, p 3

Anonymous (1986) Improved micropropagation systems. Agricell Report 7(1): p 5

Anonymous (1987) Genetic Engineering News. University of California scientists investigate rubber production. January, p 41

Balandrin MF, Klocke JA, Wurtele ES, Bollinger WH (1985) Natural plant chemicals: Sources of industrial and medicinal materials. Science 228: 1154-1160

Banthorpe DV, Branch SA, Njar VCO, Osborne MG, Watson DG (1986) Ability of plant callus cultures to synthesize and accumulate lower terpenoids. Phytochemistry 25: 629-636

Bhojwani SS, Dhawan V, Cocking EC (1986) Plant Tissue Culture: A Classified Bibliography, Elsevier, New York

Butts ER (1987) A review of biotech regulations involving plants and microorganisms. Genetic Engineering News, February, p 24

Cocking EC (1985) Protoplasts from root hairs of crop plants. Bio/Technology 3: 1104-1106

Cocking EC (1986) Plant cell biology in the 21st century: The needs of plant cell and tissue culture. In: Somers DA, Gengenbach BG, Biesboer DD, Hackett WP, Green CE (eds) VI International Congress of Plant and Tissue and Cell Culture - Abstracts. University of Minnesota, Minneapolis p 2

Collin HA, Watts M (1985) Flavor production in culture. In: Evans DA, Sharp WR, Ammirato PV, Yamada Y (eds) Handbook of Plant Cell Culture - Vol I. Macmillan Pub Co, New York, p 729-747

Cooks RG, Kondrat RW, Youssefi M, McLaughlin JL (1981) Mass-analyzed ion inetic energy (MIKE) spectrometry and the direct analysis of coca. J Ethnopharmacology 3: 299-312

Crossway A, Hauptli H, Houck CM, Irvine JM, Oakes JV, Perani LA (1986) Micromanipulation techniques in plant biotechnology. BioTechniques 4(4): 320-334

Crossway A, Houck CM (1985) A microassay for detection of DNA and RNA in small numbers of plant cells. Plant Molecular Biology 5: 183-190

De Bry L (1986) Robots in plant tissue culture: An insight. International Association Plant Tissue Culture Newsletter 49: 2-22

Dibner MD (1986) Biotechnology in Europe. Science 232: 1367-1372
Dickson S (1985) Twyford uses tissue culture methods for breeding plants. Genetic Engineering News, October, p 17
DiCosmo F, Misawa M (1985) Eliciting secondary metabolism in plant cell cultures. Trends in Biotechnology 3(12): 318-322
Edwards GE, Scott R (1985) Photorespiratory metabolism in protoplasts. In: Pilet P-E (ed) op cit, p 267-276
Evans DA, Sharp WR (1983) Single gene mutations in tomato plants regenerated from tissue culture. Science 221: 949-951
Ferrigni NR, Putnam JE, Anderson B, Jacobsen LB, Nichols DE, Moore DS, McLaughlin JL, Powell RG, Smith CR Jr. (1982) Modification and evaluation of the potato disc assay and antitumor screening of Euphorbiaceae seeds. J Nat Prod 45(6): 679-686
Flick CE, Kut SA, Bravo JE, Gleba YY, Evans DA (1985) Segregation of organelle traits following protoplast fusion in Nicotiana. Bio/Technology 3: 555-560
Flores HE, Hoy MW, Pickard JJ (1986) Production of secondary metabolites by normal and transformed root cultures. In: Somers DA, Gengenbach BG, Biesboer DD, Hackett WP, Green CE (eds) op cit, University of Minnesota, Minneapolis, p 117
Fowler MW (1985) Plant cell culture - future perspectives. In: Neumann K-H, Barz W, Reinhard E (eds) Primary and Secondary Metabolism of Plant Cell Cultures. Springer-Verlag, New York, p 362
Fujita Y, Tabata M (1986) Secondary metabolites from plant cells: Pharmaceutical application and progress in commercial production. In: Somers DA, Gengenbach BG, Biesboer DD, Hackett WP, Green CE (eds) op cit, University of Minnesota, Minneapolis, p 2
Galston AW, Smith TA (eds) (1985) Polyamines in Plants, Nijhoff/Junk, Dordrecht. Reprint of Plant Growth Regulation 3: 1-422
Gautheret RJ (1983) Plant tissue culture: A history. Bot Mag Tokyo 96: 393-410
Gebhart F (1986) Genetic Engineering News, October 10, p 10
Iversen TH (1985) Protoplasts and gravireactivity. In: Pilet P-E (ed) op cit, p 236-249
Klausner A (1985) Researchers cotton to new fiber findings. Bio/Technology 3: 1049-1051
Kothari SL, Monte DC, Widholm JM (1986) Selection of Daucus carota somatic hybrids using drug resistance markers and characterization of their mitochondrial genomes. Theor Appl Genet 72: 494-502
Kreis W, Reinhard E (1986) Highly efficient 12B-hydroxylation of digitoxin in Digitalis lanata cell suspensions using a two-staged culture method. The Society for Medicinal Plant Research, 34 Congr, University Hamburg, September, p 4-5
Kurz WGW, Chatson KB, Constabel F (1985) Biosynthesis and accumulation of indole alkaloids in cell suspension cultures of Catharanthus roseus cultavars. In: Neumann K-H, Barz W, Reinhard E (eds) op cit, Springer-Verlag, New

York, p 143-153
Kutney JP (1985) Plant tissue cultures get around the search for natural drugs. Industrial Chemical News, November, p 20-21
Lee JM, An G (1986) Industrial application and genetic engineering of plant cell cultures. Enzyme Microb Technol 8: 260-265
Lintilhac PM, Vesecky TB (1984) Stress-induced alignment of division plant in plant tissues grown in vitro. Nature 307: 363-364
Lloyd AM, Barnason AR, Rogers SG, Byrne MC, Fraley RT, Horsh RB (1986) Transformation of Arabidopsis thaliana with Agrobacterium tumefaciens. Science 234: 464-466
Marx JL (1987) Rice plants regenerated from protoplasts. Science 235: 31-32
Monroe D (1987) Potentiometric immunoassay. American Clinical Products Review, March, p 31-39
Morris P (1985) Membrane transport in protoplasts. In: Pilet P-E (ed) The Physiological Properties of Plant Protoplasts. Springer-Verlag, New York, p 54-67
Navituna F, Park JM (1986) Improvements relating to biotransformation reactions. UK Pat Applic 2,168,721A, June 25
Netzer W (1987) Technologies and market forces shape the form of agribiotech products. Genetic Engineering News, February, p 16-17
Orr T (1985) Organelle transfer and mutagenesis in crop improvement. Genetic Engineering News, October, p 17
Ow DW, Wood KV, DeLuca M, de Wet JR, Helinski DR, Howell S (1986) Transient and stable expression of the firefly luciferase gene in plant cells and transgenic plants. Science 234: 856-859
Rathore KS, Goldsworthy A (1985) Electrical control of shoot regeneration in plant tissue cultures. Bio/Technology 3: 1107-1109
Reich TJ, Iyer VN, Miki BL (1986) Efficient transformation of alfalfa protoplasts by the intranuclear microinjection of Ti plasmids. Bio/Technology 4: 1001-1004
Sahai O, Knuth M (1985) Commercializing plant tissue culture processes: economics, problems, and prospects. Biotechnology Progress 1: 1-9
Schuchmann R, Wellmann E (1983) Somatic embryogenesis of tissue cultures of Papaver somniferum and Papaver orientale and its relationship to alkaloid and lipid metabolism. Plant Cell Reports 2: 88-91
Sluis CJ, Walker KA (1985) Commercialization of plant tissue culture propagation. International Association Plant Tissue Culture Newsletter 47: 2-12
Smith JI, Smart NG, Misawa M, Kurz WGW, Tallevi SG, DiCosmo F (1987) Increased accumulation of indole alkaloids by some cell lines of Catharanthus roseus in response to addition of vanadyl sulphate. Plant Cell Reports, in press
Takeda R, Katoh K (1981) Growth and sesquiterpenoid production by Calypogeia granulata Inoue cells in suspension culture. Planta 151: 525-530
Ten Hoopen HJG, de Jong MA, Stefess GC, Kossen NWF (1985)

Industrial production of secondary metabolites from plant cells in suspension culture: a feasibility study. Acta Agronomica 34(supplement): 11

Tran Thanh Van KM (1981) Control of morphogenesis in In Vitro cultures. Ann Rev Plant Physiol 32: 291-311

Ulbrich B, Wiesner W, Arens H (1985) Large-scale production of rosemarinic acid from plant cell cultures of Coleus blumei Benth. In: Neumann K-H, Barz W, Reinhard E (eds) op cit, Springer-Verlag, New York, p 293-303

Ushiyama K (1987) Nitto Electric Industrial Co, Ltd, Osaka, personal ltr dtd March 19

Vietmeyer ND (1986) Lesser-known plants of potential use in agriculture and forestry. Science 232: 1379-1384

Wright DC, Park WD, Leopold NR, Hasegawa PM, Janick J (1982) Accumulation of lipids, proteins, alkaloids and anthocyanins during embryo development in vivo of Theobroma cacao L. JAOCS 59(11): 475-479

Yamada Y, Morikawa H (1985) Protoplast fusion of secondary metabolite-producing cells. In: Neumann K-H, Barz W, Reinhard E (eds) op cit, Springer-Verlag, New York, p 255-271

Zito SW, Staba EJ (1985) Method for preparing a cell-free homogenate of Chrysanthemum cinerarieaefolium (Trev) Bocc. containing enzymes and methods of use. US Pat 4,525,455, June 25

ASPECTS OF SCREENING PLANT CELL CULTURES FOR NEW PHARMACOLOGICALLY ACTIVE COMPOUNDS

B. Ulbrich, H. Osthoff and W. Wiesner

Nattermann Research Laboratories, Biotechnology of Plant Cell Cultures, P.O. Box 35 01 20, D-Cologne 30, FRG

INTRODUCTION

Recently, the interest of international pharmaceutical industries has been directed more and more to high molecular compounds (peptides, proteins) with distinctive biological activities as forerunners of a new therapeutic era. On this bases some people are even dreaming of a new biosociety around the turn of this century (Howink 1985).

Meanwhile, at Nattermann Research Laboratories in Cologne we have been occupied with classical research using modern tools, i.e. plant culture technology. We believe that plant cell cultures as a source ob biologically active metabolites can play a role in such a biosociety.

Several groups throughout the world are trying to produce compounds for present markets. To a certain extent this concept is necessary for short-term success. But in the long run a real breakthrough has to be based on highly innovative compounds.

CALLUS COLLECTION

Six years ago, when we started to work with plant cells in order to find new metabolites of pharmacological relevance, we gradually learned to regard a stable plant cell population as an independent system, independent of the capabilities (secondary metabolism) of its original plant (donor plant).

Thus we decided to screen what we call "plant cell culture systems" and in order to screen a maximum diversity of gene pools we also decided to screen hundreds of species and not hundreds of strains or lines from several species.

NATO ASI Series, Vol. H18
Plant Cell Biotechnology. Edited by M. S. S. Pais et al.

To start the project the first step was to accumulate a sufficiently large callus collection as quickly as possible. At the beginning Prof. M.H. Zenk (Munich) helped us in providing about 250 plant species as established callus cultures. Meanwhile, we have sent back about 200 species of no pharmacological interest. Simultaneously with the activities involving Prof. Zenk's cultures, we established cultures of our own.

Table 1. Principles for selecting desired plant material for callus induction and subsequent screening

Check list of plant material	
Characteristics	Reasons
No medicinal plants !	Prescribed biological activity
Endemic Rare/seldom Protected Small-sized	Plant no easily available, little plant material to start, investigation, competitive situation to classical drug research
Chemotaxonomic aspects	Screening for further derivatives

We set up a check list of characteristics enabling us to select the desired plant material from all over the world. Our intention was not to work with medicinal plants or plants from folk medicine, because the biological activities of their metabolites are prescribed by the plant preparation and thereby can only be patented to a limited extend.

It is believed that about 250 000 angiospermal species exist on earth, whereas only about 21 000 species are identified as medicinal plants (< 10%). Obviously, there are enough other species throughout the plant kingdom which we can deal with. Consequently, we avoid the medicinal plants, thus reducing the risk of isolating a compound (either known or unknown) with an indirectly prescribed biological activity.

Characteristics such as "endemic" (geobotanical aspect), "rare/seldom" (population aspect), "protected" (legal aspect) and "small-sized" (morphological aspect) denote at least a limited availability to classical plant drug research (Table 1). Because we consider our work as an innovative section of this research, we work only with this limited plant material because in our cell technology we merely need a handful of seeds or some living explanted material. Experimental work with these plant cell cultures must to be performed by biotechnological or chemical methods, without danger to the natural stocks of plant species. Chemotaxonomic reasons play a minor role in our considerations (Table 2).

It is generally accepted as an axiom that cells from plant cell cultures are totipotent, i.e. they are able to regenerate an original plant with an original pattern of natural compounds. But the cells cultivated under artificial laboratory conditions produce different patterns of unpredictable combinations. Little is known about regulation phenomena, thus even today scientific work with plant cells mainly has an empiric character. Consequently, it is not recommended to select plant material on a chemotaxonomic basis.

Table 2. General comparison of the native plant system with a plant cell culture system

Comparison:	Cells of a native plant tissue / Cells of a plant cell culture
Genetic information (totipotency):	Equal
Environment of the cells:	Different
→Regulation of secondary metabolism:	Unknown
→Pattern of natural compounds:	Different
→Empiric screening:	Unpredictable

In summary, we are highly interested in plant species from all over the world which cannot be collected in large amounts and hence have not been investigated up to now. At present, we have more than 800 plant species under investigation, representing 85 plant families across the plant kingdom. In our callus collection about 400 plant species were cultivated as established callus cultures.

SCREENING

Generally, there are two concepts for screening plant cell cultures for new pharmacologically interesting compounds. On the one hand, major compounds of the secondary metabolite spectrum can be isolated from cells and can subsequently be tested for possible biological activities. Structures of the active compounds can be identified by modern analytical methods. This procedure is clearly substance-orientated and has the disadvantage of increasing the amount of compounds to be isolated to a large extent without any knowledge of their biological potency at the beginning.

On the other hand, crude extracts of callus cultures can be screened in special in vitro models and "active" cultures can be exclusively investigated. The active principle will be enhanced by stepwise, controlled purification in the same in vitro model.

This concept is primarily orientated towards biological activities and has the advantage of working with interesting cultures from the beginning. We decided to use this concept (Arens 1982) although we have found that it is a stony path.

Obviously, screening of crude callus extracts needs a powerful filter system. As we know from our own experience, in vitro models are exceptionally suitable when they possess special qualities.

Established callus cultures cultivated on growth medium normally produce low concentrations of secondary metabolites. A suitable test system (Table 3) must therefore have a high sensitivity. Assuming an average molecular weight of 400 we test concentrations of crude extract down to 1 μM.

Natural compound sources like plant cell cultures often contain a complex mixture of compounds consisting of major and trace metabolites. A high selectivity is needed to discover an active principle among the others. From experiments with internal standards we know that we can detect it at concentrations down to 0.1% crude extract.

When using specific in vitro models it is necessary to be aware of its relevance to the target indication. Classical models will give classical active principles, i.e. "me too" pre-

Table 3. General qualities of a suitable in vitro assay

Requirements for in vitro assays	
Quality	Reason Reason
High sensitivity	Low metabolite concentration
High selectivity	Complex mixtures of compounds
High relevance	Interpretation of the results
High throughput	Large numbers of cultures, rapidity
High reliability	Reproducible results

parations of low interest. Innovative models can give innovative principles of high interest to the market, but there is always a relatively strong risk because the relevance is often unclear.

In vitro models must have high throughput because successful screening means testing a large number of callus cultures in a short time. The limiting step should not be the screening.

Last, but not least, the systems offered should have a high reliability because repetitions are time-consuming, conficence-reducing and expensive.

Fortunately, several assays of the above mentioned quality are at our disposal and more in vitro systems will be created, since success also depends on a broad palette of screening possibilities.

CULTIVATION CONDITIONS

On the basis of our screening concept we exclusively generate suspensions from selected "active" callus cultures. They are treated in a regulated and standardized way: For logistic reasons we first try to adjust the growth cycle to a 7-day period (or 14-day period) by concentration or dilution of the biomass. Growth is indirectly controlled by measuring the sugar consumption. Good results were obtained by an off-line sterile measurement of the refraction indices of the culture medium. This method is easy, quick, reproducible and economical.

Below critical values the cell population is transferred to fresh culture broth. The transfer is easily performed by pouring 2 vol of fresh medium into 1 vol of aged culture, subsequent mixing by gentle shaking and distribution of equal volumes into three new empty shake flasks.

Under stringent and constant cultivation conditions a given cell population will gradually behave like an oscillator within defined limits. Any disturbance beyond critical values will destabilize such oscillators and the cell culture often becomes brown or it behaves strangely. Sometimes it is possible to restore the former state of the growth cycle, but often one has to begin anew.

Consequently, we establish what we call "basic suspensions" of a biologically active callus culture. Both callus and suspension are subcultured in a strictly constant manner. From these basic suspension cultures we branch off the biomass to perform all the necessary experiments and mass cultivation (Fig. 1).

All stock cultures are cultivated under identical physical conditions. Temperature is controlled at 25°C and the Gyratory (R) shakers for suspension cultures move at 150 rpm. Humidity is not important because there is already an atmosphere of saturated water vapour inside the culture containers. Light is also unimportant because we decided to give exclusive preference to heterotrophic cultures.

By handling hundreds of cultures, we attempt to diminish all the different media known to one basal medium (mineral salts and sugar). Other components like vitamins and hormones vary as the case arise. Once a primary callus is produced on such a medium the composition and identity of all components will remain unaltered during the whole lifetime of the stock cultures. Thus, major efforts in our plant cell culture laboratory are made to maintain all stock cultures under cultivation conditions which are as constant as possible.

We consider that the two-stage culture method plays a central role. It was successfully used in a real production process (shikonins) by the Fujita group of Mitsui Petrochemical Industries Ltd., Japan (Fujita 1981; Curtin 1983; Hara 1987). We also had some success with the pilot plant production of rosmarinic acid (Ulbrich 1985, 1986). Why is this concept prefered?

It is well known that plant cell cultures are poor producers of natural compounds (secondary metabolites of low molecular

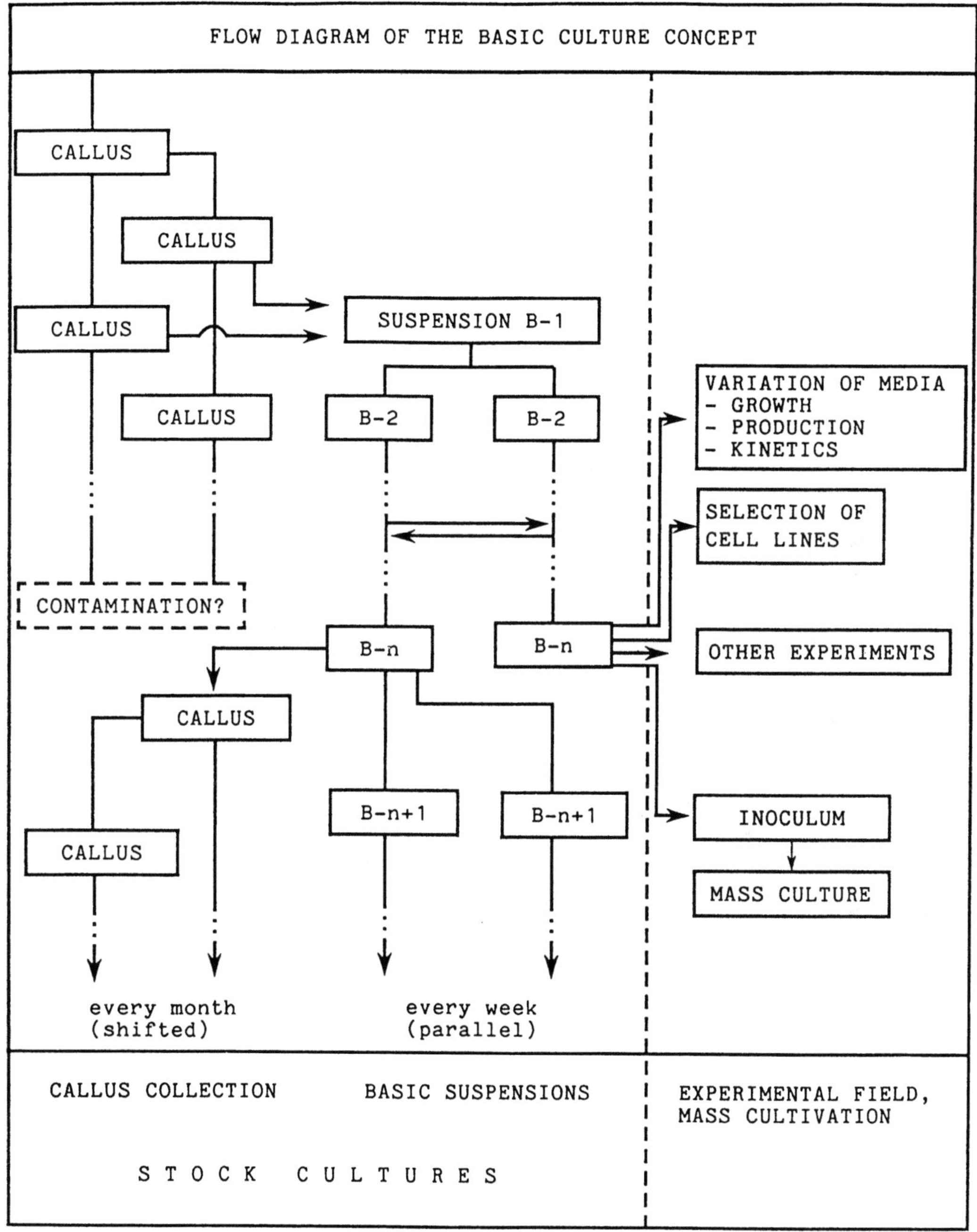

Fig. 1. The "basic suspension" concept for a certain callus line. B = basic suspension; n = number of transfers

weight) if they grow in an optimal environment with a highly specific growth rate (μ day^{-1}) (Berlin 1982). The nutritional compounds in the medium are channelled into the primary metabolism, causing cell enlargement and subsequent cell division. Under batch conditions several compounds in the medium are soon exhausted and environmental conditions become limited (stationary phase of the growth cycle). This can be the moment when secondary metabolite production is stimulated (Pirt 1975, Staba 1980). Consequently, it is logical to separate both phases and to optimize them separately. In this procedure an optimal growth medium shortens the time to cultivate the seed culture (high specific growth rate) and an optimal medium enhances the yield. Thus, to optimize the overall productivity (Wang 1979) of a given production process both aspects have to be considered.

Assuming that a given established callus culture consists of a cell population best adapted to the induction medium (= growth medium), we first try to elaborate a production medium by a method as quick and economical as possible.

It is a common fact that 2,4-dichlorophenoxyacetic acid (2,4-D) influences the secondary metabolite formation negatively (Staba 1980). Thus we avoid this phytohormone when screening for production media.

In a first step (variation) we transfer the biomass from one large inoculum, sufficient for the experiments, in equal quantities to a broad palette of heterogeneous media. The yield of secondary metabolites is measured relatively by comparing aliquots of the different extracts via HPLC analysis.

The best process passes on the next step (simplification). Here, we divide the selected production medium (first generation) into groups of macrosalts, microsalts, vitamins, phytohormones and sugars. Omitting one by one we try to single out ineffective groups or subgroups, i.e. those without influence on productivity. We want to cast the ballast overboard, thus gaining a production medium of the second generation.

In a last step (optimization) the triggering principle (physical, chemical parameter) is localized and finally optimized by testing a range of concentrations of a medium constituent or a range of physical parameters such as osmolarity, temperature, etc. The final production medium should be effective, simple and economical. By far the best production medium is water with one dissolved energy and carbon source (sucrose).

We met this aim in our rosmarinic acid program with *Coleus blumei*. We tested eight known media and registered an interesting effect. When plotting the productivity (product formation) of the cell population versus the total anion concentration of the macrosalts (nitrate, phosphate, sulphate and chloride) we found a relatively good correlation. The higher the anion concentration, the lower the productivity! We selected the HI medium (Heller 1953) as a production medium of the first generation (Ulbrich 1985).

In a second step we classified its components into four groups: macrosalts, microsalts, vitamins and phytohormones. Simplifying Heller's medium by omitting the last three groups (microsalts, vitamins and phytohormones) we could not observe any reduction in the productivity in comparison to the control. By omitting the chloride anions (KCl, $CaCl_2$) there was a reduction in productivity of about 30%. Finally, we used exclusively water and sucrose with almost the same effect and thus a very unusual production medium was created within a very short time (Ulbrich 1985).

FERMENTATION

Some time ago there was an idle debate whether an airlift fermentation system (AL) or a continuously stirred tank system (CSTR) was suitable for the mass culture of plant cells. In 1977 the airlift system based on investigations with *Morinda citrifolia* (Wagner 1977) was accepted and 4 years later this idea was rejected based on investigations with *Catharanthus roseus* (Vogelmann 1981).

As usual the truth is in the middle. With our investigation of the rosmarinic acid formation in cell cultures of *Coleus blumei* (Ulbrich 1985, 1986) we were able to demonstrate that both systems (AL and CSTR) are equivalent if reactor design and stirrer design are optimal (Table 4, Fig. 2).

We used a conventional reactor vessel with proportional dimensions of diameter to height equal to 1:3, known as DECHEMA-

Table 4. Three cultivation systems for Coleus blumei

	Working volume	Yield ($g\ l^{-1}$)	Productivity ($g\ l^{-1}\ day^{-1}$)	Overall productivity ($g\ l^{-1}\ day^{-1}$)
Shake flask[a]	70 ml	3.6	0.3	0.1
Airlift[b]	32 l	5.6	0.9	0.47
CSTR[c]	32 l	5.5	0.9	0.47

[a]300-ml Erlenmeyer flask with 70 ml; [b]As described in the German utility model G 83 14 233.9. [c]Biostat$_{(R)}$ 30D with helical blade impeller as described in the German utility model G 82 36 121.5. CSTR= continuously stirred reactor, overall productivity (see Fig. 2).

Norm. It is equipped with a special stirrer termed a helical blade impeller consisting of seven flexible modules (A. Nattermann 1982).

This stirrer system has the advantage of imitating the very sparing axial mixing characteristics of an airlift and additionally it allows a radial mixing of the cell broth, resulting in a short mixing time inside the bioreactor. This is important for maintaining a homogeneous medium for the plant cells at high concentrations of biomass. Even in solid substrate fermentation (SSF) the helical blade stirrer is a promising design (Tengerdy 1985).

From our experiments we are certain that plant cells can be successfully cultivated in large volumes in standard bioreactors with slow running agitators, as well as huge production bioreactors.

NEW COMPOUNDS

As often mentioned, plant cells in culture are a remarkable source of new pharmacologically interesting compounds (Arens 1982, 1985; Zenk 1978).

Recently we reported the isolation of three novel flavonoids, which we named Podoverine A, B and C, from a culture of Podophyllum versipelle (Arens 1986). These active principles were detected by testing methanolic callus extracts of Podophyllum in a specific in vitro cellular test system, the mouse macrophage

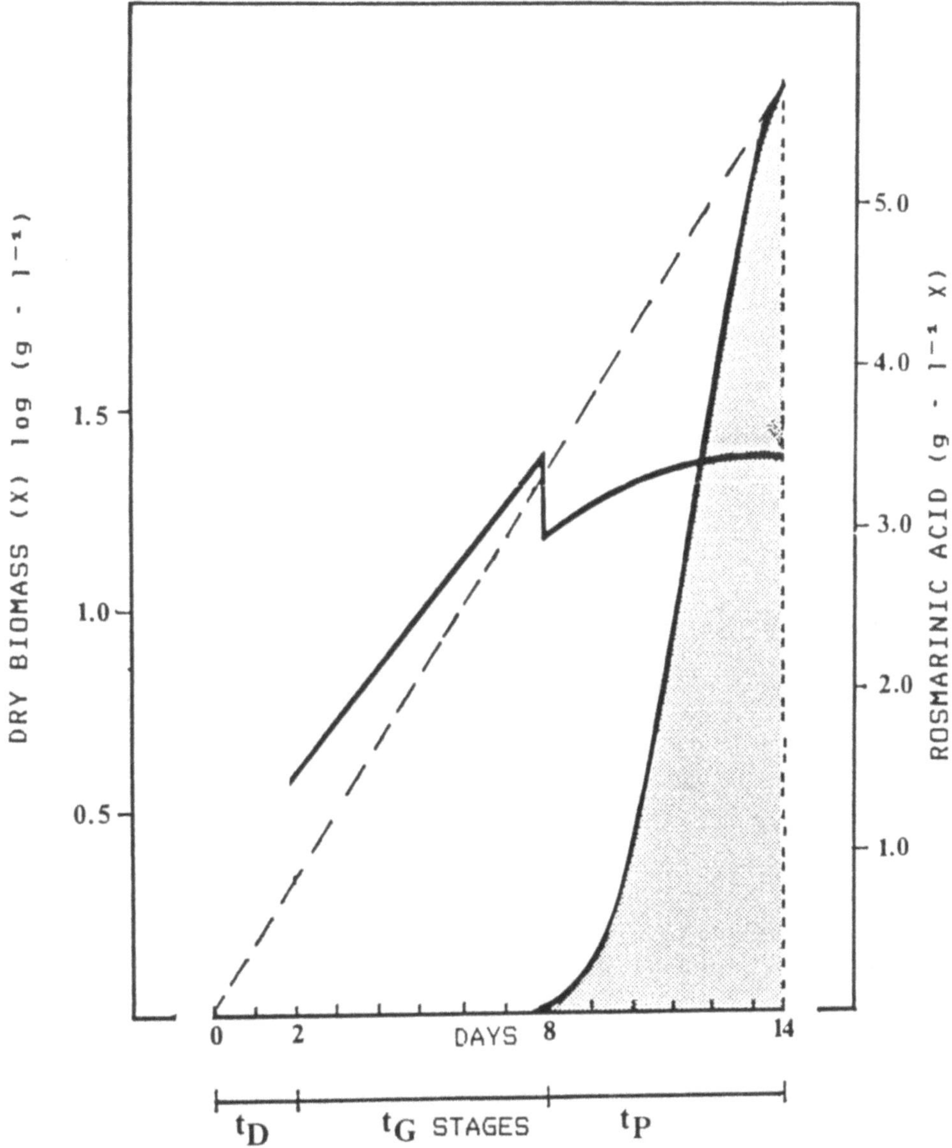

Fig. 2. Idealized diagrammatical presentation of the overall performance of the rosmarinic acid (RA) process. t_D Delay; t_G growth stage; t_P production stage; day 0 start of preparation of the equipment; day 2 start of inoculation; day 8 transfer of the biomass to production medium; day 14 harvest of the biomass, for more details see Ulbrich (1985, 1985); —— dry biomass concentraiton; ▒ rosmarinic acid concentration; - - - overall productivity

chemiluminescence assay (Parnham 1983). It permits the selection of anti-inflammatory compounds acting as anti-oxidants.

Applying the two-stage strategy, up to now, we have obtained an overall productivity of up to 150 mg podoverine l^{-1} day^{-1} (Table 5). Once more, we can demonstrate with our preliminary results that a continuously stirred fermentation (CSTR) is no worse than an airlift process.

Table 5. Preliminary results of the optimization of podoverine production

Fermenter type[a]	Working volume (l)	Content (mg g^{-1} dry weight)	Overall productivity (mg l^{-1} dry^{-1})
Airlift (AL)	32	47.4	68.4
Biostat (R) 30D (CSTR)	30	44.9	123.1
Biostat(R) 450D (CSTR)	390	50.7	150.8

[a]AL: see A. Nattermann (1983); 30D: see A. Nattermann (1982); 450D: continuously stirred tank reactor analogous to 30D but 450 litres total volume; 30D and 450D: equipped with a helical blade impeller, equal stirrer top speed in both cases

CONCLUDING REMARKS

Despite the gradually growing success of plant cell cultures as a source of interesting novel metabolites and as a tool for producing marketable natural compounds, these techniques still have a difficult position in industry, because other, more spectacular aspects of biotechnology, e.g. genetic engineering, dominate all other activities in the field of cell culture technology. In the last few years many groups have contributed the necessary knowledge for handling plant cells on a production scale. Thus, we already have the basis for a major breakthrough in this field.

Acknowledgements. The authors gratefully acknowledge the financial support of the Ministry of Research and Technology (BMFT) of the FRG (Bonn) and express their thanks to Miss G. Clermont, Mrs. G. Dvorszky, Miss A. Falderbaum, Mrs. B. Klein, Mrs. B. Kollejan, Miss M. Niggemeyer, Miss P. Reessing and Mrs. Ch.

Smeets for their technical assistance. Furthermore, the authors are grateful to Miss M. Leffin and Mrs. B. Meyer for typing the manuscript and to Dr.M.J. Parnham who kindly corrected the manuscript.

REFERENCES

Arens H, Borbe HO, Ulbrich B and Stockigt J (1982) Detection of pericine a new CNS-active indole alkaloid from picralima cell suspension culture by opiate receptor binding studies. Planta Medica 46:210-214

Arens H, Borbe H, Kesselring K and Ulbrich B (1982) Screening methods. Patent Offenlegungsschrift DE 32 30 527

Arens H, Fischer S, Leyck A, Romer A and Ulbrich B (1985) Anti-inflamatory compounds from Plagiorhegma dubium cell culture. Planta Medica 1:52-56

Arens H, Ulbrich B, Fischer H, Parnham MJ and Romer A (1986) Novel anti-inflammatory flavonoids from Podophyllum versipelle cell culture. Planta Medica 6:468-473

Berlin J and Bode J (1982) In: Prave P, Faust U, Sitting W and Sukatsch DA (eds). Handbuch der Biotechnologie. Akademische Verlagsgesellschaft 1982 Wiesbaden, p 105

Curtin ME (1983) Harvesting profitable products from plant tissue culture. Biotechnologie 1:649-657

Fujita Y, Hara Y, Suga C and Moripoto T (1981) Production of shikonin derivatives by cell suspension cultures of Lithospermum erythrorhizon II. A new medium for the production of shikonin derivatives. Plant Cell Reports 1:61-63

Hara Y, Morimoto T and Fujita Y (1987) Production of shikonin derivatives by cell suspension cultures of Lithospermum erythrorhizon V. Differences in the production between callus and suspension cultures. Plant Cell Reports 6:8-11

Heller R (1953) Recherches sur la nutrition minérale des tissus végétaux cultivés in vitro. These, Paris and Ann Sci Nat Bot Biol et Véget 14:1-223

Howink EH (1985) Biotechnology in 1984. 3rd Europ Congress on Biotechnol 1984 Proceedings 4:3-26

Nattermann A & Cie. GmbH, Cologne, FRG (1982) Wendelrührer für Bioreaktoren. German Utility Model G 86 36 121.5

Nattermann A & Cie GmbH, Cologne, FRG (1983) Bioreaktor mit vertikaler Flüssigkeitsumwälzung. German Utility Model G 83 14 233.9

Parnham MJ, Bitter CH and Winkelmann J (1983) Chemiluminescence from mouse resident macrophages: characterization and modulation by arachidonate metabolites. Immunopharmacology 5: 277-291

Pirt SJ (1975) Principles of microbe and cell cultivation. Blackwell Scientific Publications 1975 Oxford

Staba EJ (1980) Secondary metabolism and biotransformation. In Staba EJ (ed) Plant tissue culture as a source of biochemicals. CRC Press 1980 Boca Raton

Tengerda RP (1985) Solid substrate fermentation. Trends in biotechnology 34:96-99

Ulbrich B, Wiesner W and Arens H (1985) Large-scale production or rosmarinic acid from plant cell cultures of Coleus blumei Benth In: Neumann KH, Barz W and Reinhard E (eds) Primary and secondary metabolism of plant cell cultures. Springer 1985 Berlin, p 293

Ulbrich B (1986) Nutrition and environment of plant cells in bioreactors In: Kurhola M, Tuompo H and Kauppinen V (eds). Proceedings of Seventh Conference on Global Impact of Applied Biotechnology: Symposia on Alchohol Fermentation and Plant Cell Culture. Helsinki 1985. Foundation for Biotechnical and Industrial Fermentation Research 4 1986. Helsinki, pp 147-164

Vogelmann H (1981) Aspects on scale-up and mass cultivation of plant tissue culture. In: Moo-Young M, Robinson CW and Vezina C (eds). Advances in biotechnology. Pergamon Press 1981 Oxford New York 1:117-121

Wagner F and Vogelmann H (1977) Cultivation of plant tissue cultures in bioreactors and formation of secondary metabolites. In: Barz W, Reinhard E and Zenk MH (eds) Plant tissue and its biotechnological application. Springer 1977 Berlin pp 245-252

Wang DC, Cooney CL, Delain AL, Dunnill AE and Lilly MD (eds). Fermentation and enzyme technology. John Wiley & Sons 1979 New York, pp 81-82

Zenk MH (1978) The influence of plant tissue culture on industry. In: Thorpe TA (ed). Frontiers of plant tissue culture 1978. International Association for Plant Tissue Culture 1978 Calgary

ECONOMIC ASPECTS OF PLANT CELL BIOTECHNOLOGY

Toshihiro Yoshioka and Yasuhiro Fujita

Bioscience Research Center, Mitsui Petrochemical Industries LTD.
Waki, Kuga, Yamaguchi, 740 Japan

INTRODUCTION

Research on the production of useful secondary metabolites by plant cell culture (pcc) has been going on for the past 40 years, during which time technology has prgressed remarkably, resulting in the large-scale production of shikonin by cell cultures of Lithospermum erythrorhizon (Fujita et al. 1981a,b, 1982a,b, 1985 a,b). But, the fact that shikonin is still the only plant metabolite that is commercially produced, in spite of energetic studies by researchers throughout the world, indicates that the technology for industrial purposes is still at an immature stage.

In the research on the production of phytochemicals by pcc, few attempts have been made to find new compounds for use as new pharmaceuticals. The target of most of this research has been the production of existing compounds that are already widely used, however, it is necessary to produce the target compound at a lower cost than that for extraction from intact plants. Metabolites produced by pcc have the advantage of being products of uniform quality that can be stably produced without being affected by weather and other conditions.

If cells produced by pcc contain the same components as the intact plant, the processes for the extraction and purification of the intended compounds should be approximately the same as those used with intact plants. Consequently, cultured cells must be produced at a lower cost than that for harvested plants. It would be far more advantageous, from the view point of cost, if cultured cells contained much more of the target compound than the intact plant, or if the compound could be produced advantageously by a process possible only with cultured cells, for example, by the immobilized cell process.

NATO ASI Series, Vol. H18
Plant Cell Biotechnology. Edited by M. S. S. Pais et al.

Here, the factors that significantly influence the production costs of cultured cells and how that cost can be reduced will be discussed.

PRODUCTION COSTS OF CULTURED CELLS

The production cost for cells obtained by the culture method (ordinary method) widely used at present has been estimated. According to this method, culturing is done in a batch system with a known medium, e.g. Linsmaier-Skoog medium. Although cell yield varies, depending on the type of cell cultured, it is generally about 10-15 g l^{-1} of culture solution by dry weight, and the culture period is usually 2-3 weeks.

Assuming a cell yield of 15 g dry wt. l^{-1} and 20 batches of cultures per year, our calculations show that a culture tank with a volume of at least 40 m^3 is required to produce 10 t cells per year. Assuming that two operators are required, and that the depeciation period ist 7 years, production costs of the cultured cells is estimated to be \$70-80 US kg^{-1} at the plant gate. Figure 1 shows a graphic breakdown in percent.

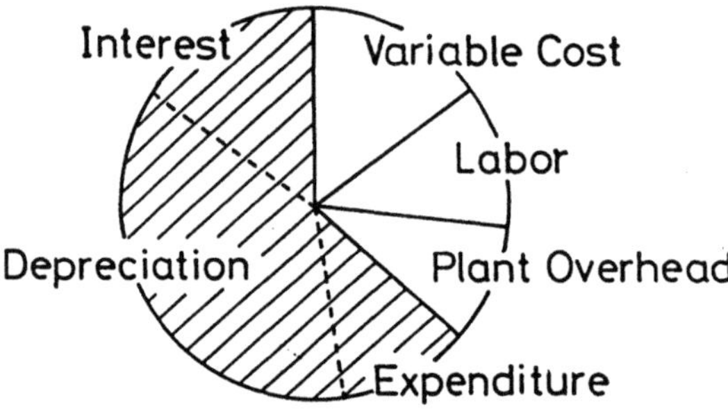

Fig. 1. Breakdown of production costs of cultured cells

Figure 1 also shows that the variable cost (raw materials and utilities) accounts for only 14%, whereas the fixed cost, particularly the cost attributable to the investment (interest, depreciation and expenditure; the shadowed protion in the figure), accounts for the greatest percent of the total cost of production.

Because this is only the cost for producing cultured cells, the cost of producing a desired compound will be much higher, although it will depend on the amount of the target compound present. The cost of the amount of cultured cells required to obtain 1 kg of an intended compound was estimated for different contents, on the assumption that 250 kg and 1 t of the compound are produced per year, and that 75% of the substance can be recovered from the cells. The results are shown in Fig. 2. Adding the recovery and purification costs, gives the actual production cost of the compound.

As the cost of a secondary metabolite produced by pcc is very high (Fig. 2) technological development is needed if commercial production is to be realized for many useful compounds.

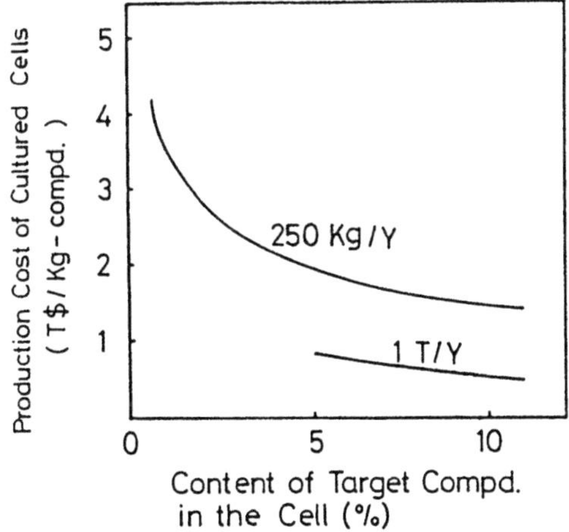

Fig. 2. Production costs of cultured cells required to produce 1 kg of the target compound

MEANS TO REDUCE THE PRODUCTION COSTS OF CULTURED CELLS

If we are to reduce the production costs of cultured cells, the cost of raw materials and labor must be lowered. The cost of raw materials can be reduced by using a cheaper carbon source, which is a major medium component, and by lowering the amounts of expensive hormones and vitamins used in the medium. Labor costs can be reduced by using highly automated equipment which would reduce the number of operators needed.

But, as stated at the beginning, the proportion of the cost attributable to the investment to the total cost is extremely high. This is because a very large culture tank is required for industrial production. Therefore, the most effective means of cost reduction would be to increase the product yield per unit volume of culture tank.

The basic ways to do this are (1) to enhance the productivity of the cell itself by selection or some other method and (2) to establish culture procedures and conditions that result in increased cell density. These production-enhancing methods are discussed on the basis of our experimental data.

ENHANCING THE PRODUCTIVITY OF THE CELL

There have been some reports (Matsumoto et al. 1981) of cell lines being subcultured for a long period in order to produce a stable production of secondary metabolites. There also have been many attempts that failed in spite of the best efforts of researchers. In some cases the desired compound is produced only when a cell differentiates into the roots or leaves; it is not produced by the cultured cell itself (Yamada and Endo 1984; Endo and Yamata 1985). Research on the production of secondary metabolites by cultured roots has been actively promoted since the recent report on the use of _Agorobacterium rhizogenes_ (Chilton et al. 1982), but the culture of differentiated organs presents problems such as a lower growth rate in comparison to that for the dedifferentiated cell, and there are difficulties in handling. Therefore, cell culture is the preferable method for the production of secondary metabolites.

Cell lines that do not produce the target secondary metabolites may be induced to produce them by such methods as the addition of an elicitor, the acquisition of a mutant cell line, or pin pointing the components in the medium which inhibit production of the desired compound. Only the enhancement of the productivity of the cell is described here.

Selection of a high-producing cell line from small cell aggregates separated from the parent line is effective when the parent line still has low productivity. The best method, however, is to select from cell lines derived from a single cell. We succeeded in obtaining a highly productive, stable cell line for shikonin derivative production from protoplast cultures of _Lithospermum erythrorhizon_ cells. In addition, we have research in progress on the use of a cell sorter to select high-producing cell lines from fused cells obtained by fusing two types of cells; cells with a high growth rate and cells with a high content of the needed compound.

CULTURE AT A HIGH DENSITY

With the usual method of cell culture the maximum yield is about 15 g l^{-1} when an established medium containing 3% sucrose is used. Assuming that 10% of the target compound is contained in the cells (such a high content is rare), the yield would be 1.5 g l^{-1}. Consequently, even with a huge tank that can be charged with 100 m^3 of medium, cells that produce only 150 kg of compound would be produced for 2-3 weeks. This amount is not economical except for the production of extremely limited and expensive compounds.

It is necessary to culture cells at a very high density in order to increase the production per unit volume of culture tank. The maximum cell density varies with the plant species and the culture conditions used because the water contents of plant cells differ.

For example, when our Coptis japonica cells were settled in the medium, the packed cell density was about 100 g dry wt. l^{-1}. But when the cell density was 70 g dry wt. l^{-1}, culture was physically possible in a commonly used aerated, agitation-type jar fermenter. Engineering problems such as how to agitate cells sufficiently without destroying them and how to supply sufficient oxygen must be solved in order to culture plant cells at such a high cell density. The critical element is the supply of nutrients because, as the number of cells increases under high cell density conditions, more nutrients are required; but, when the nutrient concentrations become too high, cell growth and the production of secondary metabolites are prevented.

An example of this is the C. japonica cells shown in Fig. 3. When the concentration of each component increased in the size of the inoculum, the cell yield also increased proportionally, but only up to 4 g dry wt. l^{-1} of the inoculum.

One solution to this problem is the "feed-batch culture". As this method does not raise the concentration of individual nutrients, inhibition of cell growth and secondary metabolite production is largely prevented. Also, in high density cultures secretions, such as superannuated cell substances, often inhibit the cell growth and production of secondary metabolites. In this case, it is necessary to renew a part of the medium continuously,

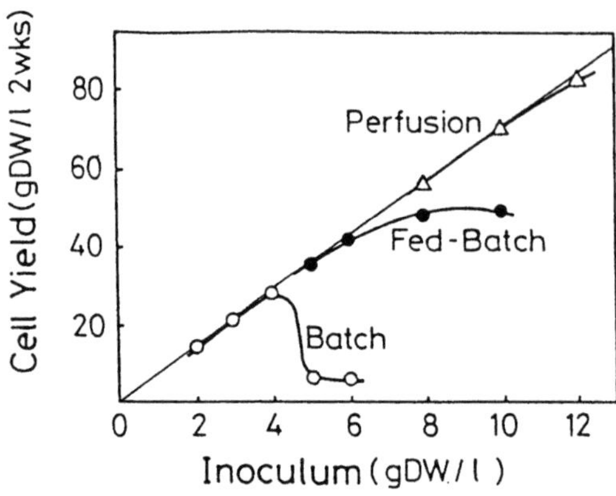

Fig. 3. Culture of C. japonica cells at a high cell density

or at intervals (perfusion culture). With this culture method, as shown in Fig. 3, we successfully cultured C. japonica cells, obtaining a 70 g l^{-1} cell yield (about five times the yield from ordinary density culture). This method can be used with many other plant cells.

The production costs of cells cultured at a density five times the ordinary density was estimated. The breakdown thereof is given in Fig. 4, which shows that the cost attributable to the investment cost was reduced from 64% (Fig. 1) to 42%.

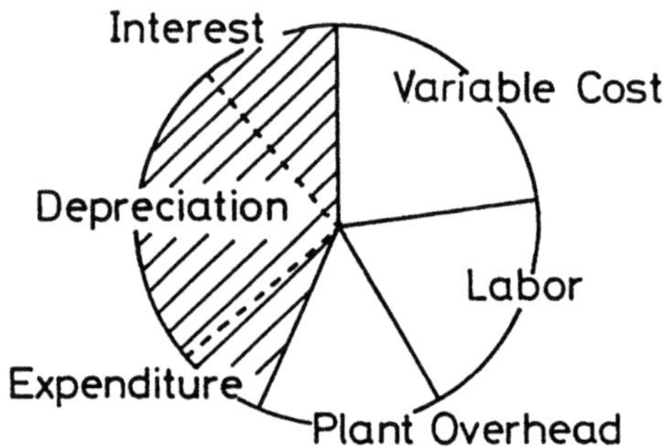

Fig. 4. Breakdown of the production costs of cultured cells (cell yield: 75 g dry wt. l^{-1})

CONCLUSIONS

A way to further reduce the costs of cultured cells is by continuous culture, in which the culture tank can be operated constantly with a full volume of cells, and the most suitable medium composition for the production of the desired compound is maintained. We cultured C. japonica cells continuously and succeeded in enhancing their productivity to more than two times that of culture at the high cell density described above.

Considerable research also has been done on culture with immobilized cells, and this method has a very high potential for reducing production costs. When immobilized cells are placed in a culture tank most of the tank is occupied by the material such as alginate immobilizing the cells. Therefore, the productivity per cell should be improved and, in fact, improvement has been reported (Brodelius et al. 1981). To use this process, cells are needed that produce the target compound stably over a long period and that release the product outside of the cell. In addition, means of immobilizing a large number of cells and of charging these cells into a large-scale tank must be developed. If such means can be developed, an excellent process would be established in which filtering and extraction of cells would be unnecessary for the recovery of the product, thus reducing costs.

Many engineering problems remain to be solved. But so far, we have been able to overcome engineering problems when we have faced industrial production. The problem for the future will be whether or not high-producing cell lines can be obtained.

At present, most pcc research being done on practical applications is concentrated on the production of expensive pharmaceuticals. The approval of the sale of these pharmaceuticals by national governments will greatly affect the practical application of biotechnology in this area. When relatively inexpensive materials that need not be regulated by the law, such as pigments and perfumes, can be produced economically, rapid technological development is to be expected.

Acknowledgements. I am grateful to Prof. Yasuyuki Yamada, Kyoto University, for his generous gifts of the *C. japonica* cell lines and expert advice, as well as to Prof. Mamoru Tabata, Kyoto University, for his generous gifts of the *L. erythrorhizon* cell lines and advice. In addition, I thank Prof. Arasuke Nishi, Toyama Medical and Pharmaceutical University, for his thoughtful suggestions and warm encouragement.

I also wish to thank the members of the Bioscience Research Center, Mitsui Petrochemical Industries LTD., for their invaluable discussions and help.

REFERENCES

Brodelius P, Deus B, Mosbach K, and Zenk MH (1981) Catalyst for production or transformation of natural products comprises biocatalysts of higher plant cell origin immobilized in polymer. European Patent Application 80850105.01

Chilton M-D, Tepfer DA, Petit A, David C, Casse-Delbart F, Tempe J (1982) Agrobacterium rhizogenes inserts T-DNA into the genomes of the host plant root cells. Nature 295:432-434

Endo T and Yamada Y (1985) Alkaloid production in cultured roots of three species of Duboisia. Phytochemistry 24:1233-1236

Fujita Y, Hara Y, Ogino T and Suga C (1981) Production of shikonin derivatives by cell suspension cultures of Lithosperumum erythrorhizon. I. Effects of nitrogene sourses on the production of shikonin derivatives. Plant Cell Reports 1:59-60

Fujita Y, Hara Y, Suga C and Morimoto T (1981) Production of shikonin derivatives by cell suspension cultures of Lithosperumum erythrorhizon. II. A new medium for the production of shikonin delivatives. Plant Cell Reports 1:61-63

Fujita Y, Tabata M, Nishi A and Yamada Y (1982) New medium and production of secondary compounds with the two-staged culture method. Proc. 5th Intl. Cong. Plant Tissue Culture, Maruzen, Tokyo, p399-400

Fujita Y, Takahashi S and Yamada Y (1985) Selection of cell lines with high productivity of shikonin derivatives by protoplast culture of Lithosperumum erythrorhizon cells. Agric. Biol. Chem. 49:1755-1759

Fujita Y and Hara Y (1985) The effective production by cultures with an increased cell population. Agric. Biol. Chem. 49: 2071-2075

Maeda Y, Fujita Y and Yamada Y (1983) Callus formation from protoplasts of cultured Lithosperumum erythrorhizon cells. Plant Cell Reports 2:179-182

Matsumoto T, Kanno N, Ikeda T, Obi T, Kisaki T, Noguchi M (1981) Selection of cultured tabacco cell strains producing high levels of ubiquinone 10 by a cell cloning technique. Agri. Biol. Chem. 45:1627-1633

Yamada Y and Endo T (1984) Tropane alkaloid production in cultured cells of Duboisia leichhardtii. Plant Cell Reports 3: 186-188

PLANT BIOTECHNOLOGY AND COMMUNITY DEVELOPMENT

E. Magnien[1]

Division of Biotechnology, Directorate of Biology, Commission of the European Communities (DG XII) Brussels, Belgium

The following thoughts should not be regarded as the authorized opinion of an expert in plant biotechnology: the author of this exercise is neither an expert in plant biotechnology nor a person authorized to express himself on behalf of the scientific community. He can only avail himself of a recent 5-year experience in trying, with others, to promote research and development in plant biotechnology by working on a scale which could hopefully circumvent some of the most penalizing academic and geographical barriers of the community. This contribution is therefore not as focused as the others on scientific demonstrations, but it will also, at the request of the organizers, address questions of a more general nature: questions which arise at the interface between the research domain of plant biotechnology, and some major development partners in biotechnology for the economic world. This is where the community dimension enters the scene, a concept which, for the purpose of this meeting, will be understood in its broad sense of "societal dimension"; which will be illustrated, however, against the background of European experience, and in the framework of the EEC in particular.

WHY SHOULD PLANT BIOTECHNOLOGY INTEREST THE COMMUNITY

Many observers believe that plant biotechnology is only at the verge of an ascending slope of growing importance. This is a result of the community facing correspondingly increasing problems: its agriculture does not appear able to afford the struc-

[1]The present article expresses only personal views.

NATO ASI Series, Vol. H18
Plant Cell Biotechnology. Edited by M.S.S. Pais et al.

tural evolutions required by unprecedented economic challenges, although it has traditionally supported the welfare of the western world. This does not imply that the scientific and technical solutions could simply replace structural changes, but that science and technology, more generally innovation, can facilitate some aspects of structural adaptation in the broad agricultural sector. For the community, the end of the 20th century is certainly a historical crossroad, where socioeconomic challenges of an exceptional magnitude are met by an explosion of biotechnology breakthroughs (Table 1).

Table 1. The pressure of growing economic problems against the multiplication of scientific and technical opportunities in agricultural production and transformation

Socioeconomic challenges	Scientific and technical opportunities
Food problems in parts of the world	Bioreactor technology
Surplus problems in parts of the world	Cell/organ/embryo culture in vitro
Environmental concerns (intensive agriculture)	Gene splicing
Energy cost	Host-vector systems
	Sequencing techniques
Renovation of traditional extraction and fermentation industries, etc.	Applications of monoclonal antibodies
	Computer science in biology, etc.

This coincidence of real problems and possible, even partial, solutions raises the expectation that agricultural production could be technically improved insofar as causing less pollution, conserving energy, meeting realistic market demands, etc. As with other technology-intensive sectors, methods of yielding and exploiting agricultural products will incorporate sooner or later an increased scientific and technical component. However, these expectations have not yet been met by corresponding achievements. It has not yet been fully realized how profound and far-reaching this present evolution of agricultural biotechnology could be. Of course, there has been too much over-selling, with the complicity of the media, and the confusion between nice laboratory products and real field varieties, or manufactured products, is still very diffuse. The distance to the market is regularly overlooked, leading to unfortunate misjudgements on the side of the community at large. Yet, one is dealing with a major quantum step, pulling an empiric technology, what plant biotechnology has always been, to the stage where it becomes a scientific technology. This certainly requires a time scale in the order of magnitude of a human generation or more. A vague misunderstanding between plant biotechnology and the community now exists. It has to be identified, explained, and possibly eliminated.

SCENARIOS FOR INTERRUPTED DEVELOPMENTS (IN APPEARANCE)

Many of the most promising plant biotechniques underwent the same cyclic scenarios (Fig. 1). Half a decade is often sufficient to bring these from sunrise to sunset, as perceived by an ordinary observer.

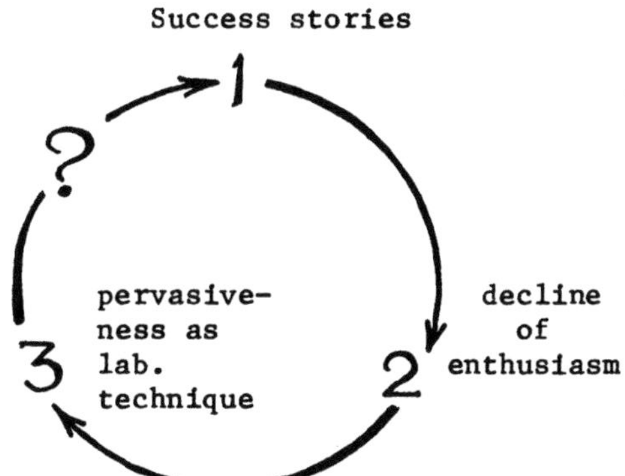

Fig. 1. Repetitive cycles of new biotechniques

Table 2 is an attempt at listing the most well-known biotechniques, with examples testifying to their cyclic occurrence.

What the ordinary observer will certainly miss in most cases is the third phase of these repetitive cycles: the pervasiveness of the techniques, as routine tools to improve the analytic performance of basic research in plant science. Yet, from the per-

Table 2. Cyclic occurrence of major biotechniques

Micropropagation	1. Success stories: floral species, orange tree, etc.
	2. Decline of enthusiasm: many ligneous species root with difficulty, problem of vitrification, unwanted rejuvenation of trees or vines, variations, etc.
	3. Pervasiveness as laboratory technique: genotype collections, sources of explants/cells for in vitro methods or biochemical extractions, for transformation, etc.
Mutagenesis	1. Success stories: dwarf wheat varieties, floral variations, etc.
	2. Decline of enthusiasm: blind approach, low frequencies, deleterious effects, most desired properties cannot be generated by mutation, etc.
	3. Pervasiveness as laboratory technique: collections of biochemical mutants for marker selections or for the investigation of biosynthetic pathways, etc.
Androgenesis	1. Success stories: first registered wheat variety which is a double haploid (Florin)
	2. Decline of enthusiasm: species/genotype limitations, low frequencies, deleterious effects and uncontrolled variations.
	3. Pervasiveness as laboratory technique: basic studies on variations derived from the commitment of the vegetative nucleus to embryogenic development, etc.
Mass production of plant secondary metabolites	1. Success stories: scaling-up of the production of shikonin in large volume fermenters, etc.
	2. Decline of enthusiasm: low concentrations compared to whole plant harvesting, cost of downstream processing, instability, etc.
	3. Pervasiveness as laboratory technique: for basic studies on continuous cell cultivation, cell cycle, photo-autotrophy, etc.
Somaclonal variations	1. Success stories: with potato (phenotypic characters), etc.

Table 2, continued :

	2. Decline of enthusiasm: species limitation, variability of an uncontrolled nature, instability of certain somaclones, etc. 3. Pervasiveness as laboratory technique: induction of instabilities in vitro for studies on genome structure and rearrangements, etc.
Multifactorial gene transfer (protoplast fusion, organelle/ chromosome transfer, etc.)	1. Success stories: transfer of male sterility from radish to oil seed rape 2. Decline of enthusiasm: chromosome/organelle elimination, gross instabilities, fertility problems, lack of marker selection systems with agricultural species, etc. 3. Pervasiveness as laboratory technique: study of the dynamics of organelle populations, chromosome mapping, etc.
Monofactorial gene transfer (Ti-plasmid-derived vectors, direct DNA transfer, micro- and macro-injection, etc.)	1. Success stories: were useful single gene characters are available (herbicide resistances, virus cross-protection,BT toxin, lectins, etc.) 2. Decline of enthusiasm: species limitations for natural micro-injectors, regeneration bottleneck, random integration with uncontrolled rearrangements and position effects, shortage of useful genes, etc. 3. Pervasiveness as laboratory techniques: homologous and heterologous expression systems coupled with deletion analysis.
REGENERATION	1. Success stories: few genotypes of tomato, one genotype of oil seed rape, one genotype of rice, etc. 2. Decline of enthusiasm: impossibility to generalize or transfer methods to useful species and genotypes, very little progress during the last 5 years, lack of rationale, black box approach. 3. Pervasiveness as laboratory technique: application to progeny analysis of mutants or transformants, basic studies on development, etc.

spective of the history of sciences, this third phase is more critical than any of the incidental applications or success stories mentioned in Table 2.

The quantum step from an empiric technology to a scientific technology will only be achievable through the pervasiveness of all newly developed biotechniques in the daily laboratory work. This in turn is the prerequisite for large-scale applications of plant biotechnology.

Under the perspective of the present community interest, however, it is obviously questionable whether this argument would be sufficient to mobilize policy-makers and professional groups, who will be generally more impressed by the immediate suspicion that many biotechniques are arrested in "phase 2" of the cycle: the decline phase.

Although this observation may not be entirely true for all biotechniques, it will be difficult to sustain the community's interest in the plant biotechnology area, at the same level as it has been placed in the early 1980s.

WHO IS STILL INTERESTED IN PLANT BIOTECHNOLOGY IN THE COMMUNITY?

Public debate on the prospects of plant biotechnology has been poor and misleading. It is certainly not sufficient to attribute the responsibility of the false image of plant biotechnology that the public has to the media alone; scientists themselves, science managers and policy-makers have their share of this responsibility. But, the situation today is not as black as one would imagine. On the contrary, some careful reconsiderations of the realistic opportunities arising from plant biotechnology are needed. Now that necessary critics are being heard, there is a new chance for the real interested partners to enter the scene and to base their relationships with the scientific world on sound conditions. These conditions will have to be established in clear and honest terms: who is doing what, for what objectives, for what time scale, for what sort of expected returns? This being said, a well-formulated demand for partnership in plant biotechnology can be identified from different socioeconomic components of the society: from the scientists themselves, from industrialists, and from public institutions.

As cooperation is still a major limiting factor in scientific progress, scientists themselves must continuously explore for cooperation partners in this fast-moving field. There may be several reasons, among which two are more directly relevant to the sector of plant biotechnology: the lack of a critical mass for work on plant functions and properties, as well as the requirement for broad multidisciplinary approaches involving wide combinations of skills.

The lack of critical mass is particularly obvious in the area of plant gene isolation and characterization. Figures of the 1985 Research Briefings for the Office of Science and Technology Policy (National Academy Press, Washington, D.C., 1985) mentioned 1100 mammalian genes being at least partially sequenced against only 86 plant genes investigated worldwide. The contribution of Europe to this work has been significant, considering that 20 different genes underwent partial characterization in the BEP Community program, during the period 1982-1986. The global figures may not be accurate, they are nevertheless indicative of the suboptimal effort which is placed in the plant kingdom, a kingdom of little glamour. No recent figure on the world statistics of plant genes could be given this year, but there are no reasons why the imbalance between agricultural and health sectors will not deteriorate further. All competent groups in Europe are now busy pursuing their efforts in characterizing the same genes and their regulatory mechanisms, therefore, most probably, the number of newly cloned plant genes will diminish: the research capabilities are becoming saturated. This imbalance is well known, and can be seen in other areas, such as protein chemistry, where the ratio between blood protein and all plant protein specialists is in the order of 100:1.

Another demanding reason for cooperation among plant biotechnologists stems from the diversified skills which the field requires. Many achievements of molecular genetics find little application under a reductionist approach. A whole-plant approach, integrating physiology, biochemistry, enzymology, etc., is often missing. Yet, this would be the only approach to shed some light on the functions or properties one hopes to manipulate. The classical bottleneck of molecular biology is after the isolation and structural characterization of a gene, when functions need to be predicted and described, in relation to the whole biology of the cell and the tissue to which it belongs. A whole panel of specialists, geneticists, molecular biologists, physiologists, biochemists, pathologists, etc., need to be attracted to work jointly on the same research topics.

The two subsequent programs, BEP (1982-1986) and BAP (1985-1989), have been attempts by the European Community to constitute a critical mass and to ascertain multidisciplinary cooperations in several areas of biotechnology. Research teams were brought together to constitute European Laboratories Without Walls (ELWWs) abiding by three major rules: constantly sharing foreground information, exchanging materials and/or staff, planning and evaluating experiments jointly.

The demand from the scientific community for this type of collaborative partnership was very high, considering the 1400 proposals submitted to the Commission of the European Communities in response to its call for proposals in 1985. The opportunities, in terms of the scientific challenge, are proportionately high. However, the ELWW's approach can only be considered as a model. The means allocated do not match in dimension the severe handicaps mentioned here.

A second type of partnership appears critical in plant biotechnology: i.e. one which should link academics with the many agro-industrial firms potentially interested in developing biotechnology breakthroughs. Here again, beyond all controversies, there is a real demand for technology transfer. This demand is difficult to evaluate in detail because of the usual secrecy in the industrial world. But there are clear indications that industries are expecting a lot from research, in particular from the type of mission-oriented research which is supported by the community program. Facts speak for themselves: 22 projects are in progress under community framework in the plant biotechnology area, 43 firms sent expressions of interest, 30 firms enrolled within an informal contact group to follow the work of the above 22 projects, 4 firms concluded transnational contract associations and others are still negotiating with the Commission's services. In addition, and parallel to their relationships with academic laboratories through the Community program, many of these firms have hired senior scientists directly from the same laboratories, to reinforce their own groups or to start new industrial research projects. Table 3 reveals some aspects of these limited statistics, which however do not provide a sufficient basis to derive conclusions, but confirm the reality of industrial expectation.

Despite the small size of the sample, these academic-industrial relationships have two obvious implications:

1. Firms are moving cautiously in the area of plant biotechnology, but they have their established strategies, they know what they want and they know what academic science can realistically offer. They will stress in particular that plant biotechnology does not replace plant breeding, but compliments plant breeding, thus raising new opportunities that will be channelled through field selection at any rate. Plant biotechnology unavoidably needs the filter of plant breeding, whereas the reverse is not necessarily true. The chance of developing plant biotechnology will be higher if firms are involved very early in a dialog with research scientists, a dialog allowing regular feedback evaluations from professionals.

2. Firms come from very heterogeneous professional groups (see Table 3) and do not speak in a single voice. The requirements as expressed by one firm in particular cannot account for the requirements of the whole community. Their attitudes with respect to the precompetitiveness of research and the role of public funding can vary to a large extent. Quite logically, industrial interests do not address themselves necessarily to societal needs. There would therefore be no reasons to overestimate the role of industries in defining priorities in plant biotechnology, unless their point of view is brought into harmony with a consistent policy approach. There is much work to be done in terms of communication for this harmony to be achieved.

COMMUNICATION ADJUSTMENTS BETWEEN COMMUNITY PARTNERS

All major partners mentioned above can easily meet on a very favourable premise: they have a common interest in the intelligent exploitation of an immense agricultural basis. But, they

Table 3. Reality of industrial interest towards biotechnological research; facts observed in implementing the community program BAP

	Number of firms enrolled in a contact group attached to the plant biotechnology area of BAP	Number of firms having directly hired senior scientists from contracting laboratories during the last two years[a]
R & D service companies	4	1
Seed companies	8	3
Agro-chemical firms	6	1
Agro-food companies	4	1
Oil companies	2	-
Chemical companies	2	2
Pharmaceutical companies	1	2
Agricultural cooperatives	2	-
Sugar manufactures	1	-

[a] An underestimate, as these events do not receive any publicity and are known only incidentally.

also have diverging opinions sometimes, with regards to the use they wish to make of this agricultural basis. They occasionally may refer to a time scale of a different order of magnitude, which is particularly striking in the case of developmental times, opposing academics to industrial scientists, but also opposing breeders to agro-chemists! Semantics may even add to the risk of confusion, e.g. it is now sufficiently obvious that a transgenic crop plant is not yet a transgenic variety. A favourable premise with unfavourable trends poses a problem of harmony. A communication link must be found. Table 4 shows the degree of divergence between two major community players in plant biotechnology.

A crucial question for plant biotechnology is whether it can contribute to alleviating the apparent antagonisms between the major options of public authorities, firms, and other recognized interlocutors (environmentalists, consumer associations, etc.). It is up to the scientist to look for appropriate answers.

Table 4. Some high-ranking objectives for the exploitation of the agricultural basis

As viewed from state administrations	As viewed from agricultural firms
Less surplus productions	High-yielding varieties
Farmers income	Intensive agriculture (high inputs)
Quality aspects of crops (food, feed, non-alimentary uses)	Profitable crops (maize, rice, wheat, etc., horticulture)
Alternative crops	Limited persistance of phytochemicals
Use of marginal land	
Environmentally safe agriculture	

Possibilities of linking several of these options are quite accessible.

Using items from Table 4, the following approaches could be suggested:

1. To develop alternative crops (=public authorities) that would be high yielding (=firms): e.g. by manipulating male sterility on Vicia faba and developing hybrid seeds for this candidate crop species;
2. To increase the nutritional quality of cereals (=public authorities) without affecting yield (=firms): e.g. by elucidating and modifying the regulation of seed protein synthesis in barley;
3. To improve the safety standards of agriculture by diminishing the load of phytochemicals (=public authorities) through rapid degradation (=firms): e.g. by engineering soil microorganisms that would metabolize the active compounds.

The above examples may be questioned by specialists, but the fact is that solutions of this type and others, elaborated through proper research approaches, can satisfy both policy objectives and commercial expectations: they render plant biotechnology highly desirable from a community point of view.

The community has much more to gain from plant biotechnology when it technically reconciles deverging options of its major partners. However, this position for plant biotechnology is also the least comfortable, as the scope left for exploratory research is gradually reduced by the accumulated pressures exerted by states, industries, and citizens. The route which is left open is a narrow one, but can a program be set up to support R & D along this trend?

CONDITIONS FOR AN R & D PROGRAM TO FIT COMMUNITY EXPECTATIONS

These conditions can best be described if they are set against the list of the above conclusions.

1. Plant biotechnology can be of community interest because many of the new biotechniques are seen as instrumental in the medium- to long-term elaboration of technical solutions to socioeconomic problems.
2. Plant biotechnology has not been able to generate rapid solutions in the short term. However, the decline of enthusiasm is the result of public misinformation, and a lack of measurable parameters to emphasize the qualitative change which is occurring in basic research.
3. Major partners are still aware of the medium- to long-term possibilities, and will be more cautions in formulating predictions. The condition for partnership will be more strict with regards to division of tasks, objectives, and the evaluation of progress.
4. Partnerships will be based on favourable premises (exploitation of the agricultural basis) but should also rest on reconciled policy options (states, industry, citizens).
5. The community and plant biotechnologists will be mutually enhanced if scientific progress can be directed towards economic targets that would alleviate the above policy tensions.

A community R & D activity will best fit in with the above recommendations if it gives the necessary attention to the following conditions:

1. The topics for priority research should link policy and economic options with scientists and technical opportunities;
2. Research work should invite multiple partnerships, favouring circulation of materials and staff across academic and geographical boundaries;
3. The correct division of tasks between academics, who generate knowledge, and professionals, who look for profitable deve-

lopments, should be established, on the basis of a mutual understanding of, and respect for, their respective missions: a program should not add to the confusion between their roles, but should rather be the platform where each partner can provide the right input to the other;

4. Multidisciplinary work needs further rationalization to avoid spreading over too many experimental systems: one should look for the few crop plants of strategic importance and build up on these the critical mass of multidisciplinary research.

For the community to move along this trend, there are a few questions to bear in mind. A new era for plant biotechnology lies where the answer to these questions can be made explicit, i.e. an era of welcome rationalization:

QUESTIONS

1. Why so few success stories?
2. Where is the relative weakness of plant biotechnology?
3. Is cooperation a necessary answer to the above weakness?
4. Is the field still in shortage of qualified staff?
5. How can we attract molecular biologists + genetic engineers + protein chemists in a field of no glamour?
6. Can we restore public esteem towards the scientific community?
7. What are measurable parameters for the evaluation of precompetitive research activities?
8. Are firms in a position to give impulses to plant biotechnology?
9. What sort of partnership with agriculture? A mining activity or an area for redistribution of work?
10. Can we afford to channel research money to a few strategic crop species (rice, barley, maize, soybean, sunflower, oil seed rape)?
11. Is quality improvement technically antagonistic to yield?

THE IBPGR IN VITRO CONSERVATION DATA BASES

Shelagh K. Wheelans and Lyndsey A. Withers

Department of Agriculture and Horticulture, University of Nottingham School of Agriculture, Sutton Bonington, Loughborough, LE12 5RD, UK

INTRODUCTION

At the beginning of the present decade, the International Board for Plant Genetic Resources (IBPGR) became aware of the considerable potential for the application of in vitro technology in plant genetic conservation. However, it was clear that this potential was under-exploited (Withers 1980; Withers and Williams 1982). There was a lack of published research on basic techniques and little liaison between tissue culturists and conservation workers. Before any future work could be directed appropriately, there was a need for a state of the art review of current research. To this end, IBPGR supported a survey by questionnaire of institutes working on tissue culture techniques with particular reference to genetic conservation. Information was sought on IBPGR priority crops such as staple roots and tubers, tropical and temperate fruits, and palms with respect to clonal propagation, research problems, characterization, storage and *in vitro* exchange. The survey showed that much valuable information, especially on storage techniques and biological problems encountered in research, was left unpublished. The results were published in report format (Withers 1982).

SURVEY DATA BASES

As a follow-up to this useful exercise, the IBPGR has, since 1983, sponsored a continuing information project at Nottingham University, UK, to carry out a biennial survey of in vitro conservation techniques. These surveys, and the original one, would be computerized to form data bases of current *in vitro* research to keep IBPGR and collaborators in its *In vitro* Con-

NATO ASI Series, Vol. H18
Plant Cell Biotechnology. Edited by M. S. S. Pais et al.

servation Programme (Withers and Williams 1986) informed of new developments applicable to conservation problems. The project would also use the survey information to provide an international IBPGR contact service and information bureau free of charge to the wider scientific community.

Data bases now available for consultation cover research carried out in 1980, 1983 and 1985. Over this period, the contacts' list has been revised and extended, the area of disease indexing has been included on the questionnaire, and a policy change now welcomes data on any species rather than just IBPGR priority crops. Consequently, the most recent data base contains 1341 records on 654 species, representing the work of 589 scientists in 67 countries.

The surveys show that much work has been carried out on woody and ornamental species as opposed to key crops from the point of view of conservation, for example, cassava and sweet potato. This continuing lack of appropriate research is likewise reflected in the low level of returns that recorded development of cryopreservation techniques. Biochemical characterization of regenerants, an area of importance in monitoring stability of cultures in storage and in germplasm evaluation, also showed a low frequency of use (Wheelans and Withers 1984).

DATA BASE MANAGEMENT

The data bases are generated and held in the ICL main frame computer at Nottingham University and managed using the FAMULUS data base management system. The information from each 1985 questionnaire return is entered in text form, with minimal interpretation, under the following 16 sections:

Label	Information content
NAME	Name of contributor
ADDR	Address of contributor
DATE	Date when questionnaire return was received
ACNO	Accession number assigned to each return which acts as a unique identifier
CROP	Species under investigation
APPL	Field of interest or application
EXPL	Explant material

PPGN	Clonal propagation procedures
PROB	Operational and biological problem
CHAR	Characterization of cultures and plants
DNDX	Disease indexing
STNG	Storage of cultures under normal growth conditions
STSG	Storage of cultures under slow growth conditions
STCR	Cryopreservation of cultures
EXCH	*In vitro* exchange or distribution of material
SUPP	Additional information or bibliographic references

SEARCH REQUESTS

A total of 320 searches have been performed on the survey data bases since their creation. Searches can be carried out at varying levels of complexity and the results printed in a variety of formats. For example, a list of all contacts in India with the crops under investigation can be supplied in tabular form. In another instance, all available information on a crop and/or technique, such as cryopreservation of *Solanum* spp., can be printed out under section headings. All searches are carried out free of charge. Unless otherwise requested, all searches will be processed against the most current data base.

IBPGR IN VITRO SURVEY 1987

The next phase of the information project is underway with the issue of the 1987 survey. Some 2,500 scientists are being sent a revised questionnaire that now includes a section on *in vitro* collecting techniques. All *in vitro* workers and other interested scientists who have not been contacted by the survey in the past are invited to write to the project for inclusion on the mailing list. In this way, it is hoped that the 1987 survey will provide a truly comprehensive data bases of international research which should be available for consultation early in 1988. The project will also be publicizing its services with the aim of encouraging scientists to make use of the survey and bibliographic information now available.

BIBLIOGRAPHIC DATA BASES

A further role of the project is to computerize literature of relevance to *in vitro* conservation, including the bibliographies of IBPGR-commissioned reports. This will generate a number of small data bases representing overviews of the literature on specific subjects. To date, specialist data bases have been constructed from bibliographies on isozyme analysis (see: Simpson and Withers 1986) and the *in vitro* conservation of temperate fruit species (see: Stushnoff and Fear 1985). This work will continue as further relevant reports are published. All such data bases are available for consultation upon request.

Acknowledgement. LAW gratefully acknowledges receipt of an SERC (UK) Advanced Fellowship.

REFERENCES

Simpson MJA and Withers LA (1986) Characterization of plant genetic resources using isozyme electrophoresis: a guide to the literature. International Board for Plant Genetic Resources, Rome

Stushnoff C and Fear CD (1985) The potential use of *in vitro* storage for temperate fruit germplasm: a status report. International Board for Plant Genetic Resources, Rome

Wheelans SK and Withers LA (1984) The IBPGR international data base on *in vitro* conservation. Plant Genetic Resources Newsletter 60:33-38

Withers LA (1980) Tissue culture storage for genetic conservation. International Board for Plant Genetic Resources, Rome

Withers LA (1982) Institutes working on tissue culture for genetic conservation, 2nd edn. International Board for Plant Genetic Resources, Rome

Withers LA and Williams JT (eds) (1982) Crop genetic resources - the conservation of difficult material. Proceedings of an international workshop held at the University of Reading, UK, 8-11 September 1980. IUBS Serie B42. IUBS/IBPGR, Paris

Withers LA and Williams JT (1986) IBPGR research highlights: *in vitro* conservation. International Board for Plant Genetic Resources, Rome

NATO ASI Series H

Vol. 1: **Biology and Molecular Biology of Plant-Pathogen Interactions.** Edited by J.A. Bailey. 415 pages. 1986.

Vol. 2: **Glial-Neuronal Communication in Development and Regeneration.** Edited by H.H. Althaus and W. Seifert. 865 pages. 1987.

Vol. 3: **Nicotinic Acetylcholine Receptor: Structure and Function.** Edited by A. Maelicke. 489 pages. 1986.

Vol. 4: **Recognition in Microbe-Plant Symbiotic and Pathogenic Interactions.** Edited by B. Lugtenberg. 449 pages. 1986.

Vol. 5: **Mesenchymal-Epithelial Interactions in Neural Development.** Edited by J.R. Wolff, J. Sievers, and M. Berry. 428 pages. 1987.

Vol. 6: **Molecular Mechanisms of Desensitization to Signal Molecules.** Edited by T.M. Konijn, P.J.M. Van Haastert, H. Van der Starre, H. Van der Wel, and M.D. Houslay. 336 pages. 1987.

Vol. 7: **Gangliosides and Modulation of Neuronal Functions.** Edited by H. Rahmann. 647 pages. 1987.

Vol. 8: **Molecular and Cellular Aspects of Erythropoietin and Erythropoiesis.** Edited by I.N. Rich. 460 pages. 1987.

Vol. 9: **Modification of Cell to Cell Signals During Normal and Pathological Aging.** Edited by S. Govoni and F. Battaini. 297 pages. 1987.

Vol. 10: **Plant Hormone Receptors.** Edited by D. Klämbt. 319 pages. 1987.

Vol. 11: **Host-Parasite Cellular and Molecular Interactions in Protozoal Infections.** Edited by K.-P. Chang and D. Snary. 425 pages. 1987.

Vol. 12: **The Cell Surface in Signal Transduction.** Edited by E. Wagner, H. Greppin, and B. Millet. 243 pages. 1987.

Vol. 13: **Toxicology of Pesticides: Experimental, Clinical and Regulatory Perspectives.** Edited by L.G. Costa, C.L. Galli, and S.D. Murphy. 320 pages. 1987.

Vol. 14: **Genetics of Translation. New Approaches.** Edited by M.F. Tuite, M. Picard, and M. Bolotin-Fukuhara. 524 pages. 1988.

Vol. 15: **Photosensitisation. Molecular, Cellular and Medical Aspects.** Edited by G. Moreno, R.H. Pottier, and T.G. Truscott. 521 pages. 1988.

Vol. 16: **Membrane Biogenesis.** Edited by J.A.F. Op den Kamp. 477 pages. 1988.

Vol. 17: **Cell to Cell Signals in Plant, Animal and Microbial Symbiosis.** Edited by S. Scannerini, D. Smith, P. Bonfante-Fasolo, and V. Gianinazzi-Pearson. 414 pages. 1988.

Vol. 18: **Plant Cell Biotechnology.** Edited by M.S.S. Pais, F. Mavituna, and J.M. Novais. 500 pages. 1988.

MIX
Papier aus verantwortungsvollen Quellen
Paper from responsible sources
FSC® C105338

If you have any concerns about our products,
you can contact us on
ProductSafety@springernature.com

In case Publisher is established outside the EU,
the EU authorized representative is:
Springer Nature Customer Service Center GmbH
Europaplatz 3, 69115 Heidelberg, Germany

Printed by Libri Plureos GmbH
in Hamburg, Germany